Springer Proceedings in Business and Economics

More information about this series at http://www.springer.com/series/11960

Madjid Tavana • Srikanta Patnaik
Editors

Recent Developments in Data Science and Business Analytics

Proceedings of the International Conference on Data Science and Business Analytics (ICDSBA- 2017)

Springer

Editors
Madjid Tavana
Department of Business Systems
and Analytics
La Salle University
Philadelphia, PA, USA

Srikanta Patnaik
Department of Computer Science
and Engineering
SOA University
Bhubaneswar, Odisha, India

ISSN 2198-7246 ISSN 2198-7254 (electronic)
Springer Proceedings in Business and Economics
ISBN 978-3-030-10259-3 ISBN 978-3-319-72745-5 (eBook)
https://doi.org/10.1007/978-3-319-72745-5

Printed on acid-free paper

This Springer imprint is published by the registered company Springer International Publishing AG part of Springer Nature.
The registered company address is: Gewerbestrasse 11, 6330 Cham, Switzerland

Preface

We live in a data-driven society. The search for innovative solutions and competitive advantage has led to digital revolution where a huge amount of data is produced every day. Today's organizations are facing new challenges in processing and analyzing this massive data, collected from different sources, containing inconsistent, incorrect, inadequate, and misleading information. These databases involve a great deal of complexities due to the variations in sources of data (such as mixture of media: text, graphics, audio, video) and velocity at which real-time data is being generated, collected, and processed. New data analytics patterns and trends are needed to analyze this enormous volume of structured and unstructured data. Big Data is a massive collection of data containing plentiful information, which is broadly defined by five dimensions: (i) volume: referring to the large size of the data; (ii) variety: representing unstructured data from heterogeneous sources such as traditional databases, transactional systems, Internet, and social media; (iii) velocity: the speed at which new data is generated and analyzed in real time; (iv) veracity: the authenticity or origin of the data; and (v) value: the value of data which depends on whether the data collected is old or recent.

One of the major problems encountered during the use of Big Data in industrial processes and applications is the hardware and software requirements to deal with such a huge amount of data from varying sources. Although the development of advanced technologies such as cloud computing, virtualization of processes and storages, new programming and data processing frameworks like MapReduce and Hadoop have been successful to some extent, still complex industrial applications need more improved technologies to deal with real-time requirements. In addition, the information about the authenticity and origin of the data collected is mandatory as inaccurate and erroneous data may result in inadequate and irrelevant information thus hampering prediction accuracy. These challenges have created a wide range of research opportunities including:

1. Algorithms for analyzing and optimizing Big Data
2. Alternative solutions for storing, accessing, and transmitting Big Data

3. Various communication techniques for fast processing and transmission of Big Data
4. Innovative solutions for controlling and monitoring machineries and processes
5. Data-driven solutions for supply chain management and logistic systems
6. Innovative software for remote distributed applications

Although a great deal of research is devoted to the realization of smart Industry, synergetic study of Big Data has not yet peaked up momentum. This conference provided a platform for discussion of various challenges and issues relating to Big Data and business analytics. This volume contains 55 research articles from diverse areas of data science and business analytics categorized into five sections:

(i) **Marketing and Supply Chain Analytics:** focuses on the integration of horizontal processes related to all value chain partners from various suppliers of raw materials to intermediate suppliers to final customers using tracking and tracing systems. There are eight research articles in this section.
(ii) **Logistics and Operations Analytics:** focuses on the digitalization, integration, and efficient management of various processes across the vertical value chain such as designing and planning of product, purchasing of raw materials, development of products, manufacturing of products, and logistics. There are 12 research articles in this section.
(iii) **Financial Analytics:** focuses on improving the financial health of the businesses and economy. There are 11 contributions in this section.
(iv) **Predictive Modeling and Data Analytics:** focuses on predictive modeling and digital solutions for integrated and data-driven platforms. These solutions are based upon optimization of customer access and interaction for generating additional revenue using predictive data analysis and modeling. There are 14 research articles in this section.
(v) **Communications and Information Systems Analytics:** focuses on smart factories and autonomous systems and intelligent equipment and machineries embedded with sensors and actuators and establishes the concept of smart manufacturing processes. High-level automation results in self-adaption and self-optimization of manufacturing processes from acquisition of raw materials to production of final products. Automatic data exchange and communication between machineries, infrastructures and human agents, and complete automation of machineries and processes are also a branch of this domain. There are 10 research articles in this section.

Philadelphia, PA, USA — Madjid Tavana
Bhubaneswar, Odisha, India — Srikanta Patnaik

Acknowledgments

The research articles included in this proceeding are the selected collection of the contributions made to the ICDSBA-2017. We are grateful to all contributors and the Editorial Director of the Springer Book series on "Springer Proceedings in Business and Economics," Nicholas Philipson for his support to bring out the conference proceedings of ICDSBA-2017. We are also thankful to Prof. Dr. Qin Xin, professor of Information and Communication Technology in the Department of Science and Technology at the University of the Faroe Islands, Faroe Islands (Denmark), for his keynote address.

The organizing committee is most grateful to Prof. Dr. Xilong Qu, the dean of School of Information Technology and Management, Hunan University of Finance and Economics, who is also the master degree student advisor for both Xiangtan University and Hunan Institute of Engineering. We owe our highest gratitude to:

- Hunan University of Finance and Economics
- Hunan Institute of Engineering
- Hunan Association for Science and Technology
- Financial Big Data Science and Technology Key Laboratory of Hunan Province
- Hunan Provincial 2011 Collaborative Innovation Center for Development and Utilization of Finance and Economics Big Data Property
- Hunan Province Higher Educational Institutions Key Laboratory "Information Technology and Information Security"

We look forward to the second International Conference on Data Science and Business Analytics (ICDSBA 2018) to be held in Changsha, China.

Contents

Part IV Predictive Modeling and Data Analytics

About the Editors

Madjid Tavana Dr. Tavana is Professor and Distinguished Chair of Business Analytics at La Salle University, where he serves as Chairman of the Business Systems and Analytics Department. He also holds an Honorary Professorship in Business Information Systems at the University of Paderborn in Germany. Dr. Tavana is Distinguished Research Fellow at the Kennedy Space Center, the Johnson Space Center, the Naval Research Laboratory at Stennis Space Center, and the Air Force Research Laboratory. He was recently honored with the prestigious Space Act Award by NASA. He holds an MBA, PMIS, and PhD in Management Information Systems and received his Post-Doctoral Diploma in Strategic Information Systems from the Wharton School at the University of Pennsylvania. He has published 12 books and over 250 research papers in international scholarly academic journals. He is the Editor-in-Chief of *International Journal of Applied Decision Sciences, International Journal of Management and Decision Making, International Journal of Communication Networks and Distributed Systems, International Journal of Knowledge Engineering and Data Mining, International Journal of Strategic Decision Sciences*, and *International Journal of Enterprise Information Systems.*

Srikanta Patnaik Dr. Patnaik is a Professor in the Department of Computer Science and Engineering, SOA University, Bhubaneswar, India. He received his Ph.D. in Engineering majoring in Computational Intelligence from Jadavpur University, India, and supervised 12 Ph.D. theses and more than 30 M.Tech theses in the areas of Machine Intelligence, Soft Computing Applications, and Re-Engineering. Dr. Patnaik has published more than 60 research papers in international journals and conference proceedings. He is the author of 2 textbooks and edited 12 books and a number of invited book chapters, published by leading international publishers like Springer-Verlag, Kluwer Academic, etc. Dr. Patnaik was the Principal Investigator of TAPTEC project "Building Cognition for Intelligent Robot" sponsored by the All India Council for Technical Education, New Delhi, and Major Research Project "Machine Learning and Perception using Cognition Methods" sponsored by the University Grants Commission. He is the Editor-in-Chief of *International Journal of Information and Communication Technology* and *International Journal of Computational Vision and Robotics* published by Inderscience Publishing and Editor-in-Chief of a Book Series on "Modeling and Optimization in Science and Technology" published by Springer. Dr. Patnaik has also edited many proceedings volumes, all published by Springer.

Part I
Marketing and Supply Chain Analytics

Chapter 1
Research on Differential Pricing and Coordination Mechanism of Second-Class Supply Chain of New Products and Remanufactured Products

Hubiao Li

Abstract Based on the price sensitivity of consumer products and the degree of substitutability of products, this paper analyzes the optimal pricing strategy of manufacturer, manufacturer and retailer. By comparing the total profit of decentralized decision and centralized decision supply chain, it is found that the decentralized decision can lead to the "marginal effect", which leads to the loss of supply chain efficiency. Therefore, the Shapley value method is used to study the coordination of supply chain. Finally, through the numerical simulation analysis of two kinds of consumer decision mode of unit product price sensitivity and the product can replace the influence on the sales price and the profit of the supply chain members profit changes.

Keywords Original equipment manufacturer (OEM) · Remanufacturer · Two echelon supply chain · Shapley value method

1.1 Literature

With the development of remanufacturing industry, more and more remanufactured products are made in the market. There are many vendors selling new products and remanufactured products at the same time. For example, the repair station of China National Heavy Duty Truck Group Co. LTD. sells their own new products as well as remanufactured engine produced by Jinan Fuqiang power co. LTD at the same time. Because there are differences between the component costs, technology and market acceptance. Remanufactured products and new products tend to have different market prices. So, how to differentiate these two types of products for vendors? What differences between Decentralized decision pricing strategy and the

H. Li (✉)
School of business, Zhengzhou University, Zhengzhou, China
e-mail: lihubiao123@126.com

M. Tavana, S. Patnaik (eds.), *Recent Developments in Data Science and Business Analytics*, Springer Proceedings in Business and Economics,
https://doi.org/10.1007/978-3-319-72745-5_1

centralized decision pricing strategy? And how to coordinate between supply chain members Under the optimal pricing strategy? This paper examines differential pricing strategy and coordination mechanism between manufacturers, remanufacturers and vendors, aiming to put forward the solution to these problems.

Currently, the pricing strategy and coordination mechanism between new products and remanufactured products ware main concentrated about two aspects: one is that the pricing strategy and coordination mechanism studying based on the different recovery function of the remanufacturing product, the other is that the pricing strategy and coordination mechanism studying based on closed-loop supply chain. Debo et al. [1] presents that manufacturers facing remanufacturing product pricing and technology selection problem. Ferrer and Swaminathan [2] established a new product and remanufacturing product cycle model, separately studied the two manufacturers of new products and remanufactured products pricing strategy. On this basis, the Ferguson and Toktay [3] introduced the recovery of function based on different consumer groups. The following study is about the differential pricing and new products and remanufactured products' pricing strategy under third parties participate in the remanufacturing proces. Ge and Huang [4] shows that who is responsible for the recovery of waste products based on the supply chain composed of a manufacturer and a retailer, and getting that profits under decentralized decision in the supply chain is lower than centralized decision. Based on the newsboy model and the bayesian model, Chen et al. [5] studied in the supply chain system with demand forecast update under asymmetric information, and coordinate upstream and downstream enterprises in supply chain. Liu and Chong [6] studied Coordination between two stage supply chain repurchase contract based on the newsboy model. Yan and Huang [7] focused on the second party's responsibility for recycling waste products, discussing the decentralized decision and supply chain optimization strategy where manufacturer, seller give priority to party for from the side. The comparison and analysis from the perspective of supply chain contract coordination is also studied. Bao et al. [8] constructed the closed-loop supply chain composed of a single manufacturer and single retailer, studied the differences between a single cycle of new products and remanufactured product pricing, and got the results as follows: decentralized decision makes the sales of new products and recycled waste products lossing 50%, relative value of the whole closed-loop supply chain efficiency loss is 25%, and giving supply chain coordination mechanism is presented. Zheng [9] assumed that the retailer was responsible to recycle based on game theory and studied closed-loop supply chain pricing strategy and contract coordination, under the circumstances that remanufactured product have price variance with new product and used product.

Most of papers on manufacturing/remanufacturing differential pricing assume that manufacturing enterprises and Remanufacturing enterprises are one. Meanwhile, these papers haven't studied the demand affecting parameters in detail. Based on the existing research, this paper assumes that manufacturing enterprises and Remanufacturing enterprises are not one and demand function is effected by the sensitivity of the consumer products to the unit price and Degree of alternative product. The optimal pricing between manufacturers, manufacturers and retailers strategy is analyzed. By numerical value simulating, it also analyze the effects of the sensitivity of the unit product price and Degree of alternative product on sales prices and profits under decentralized decision and centralized decision making.

1.2 Problem Description and Modeling

1.2.1 Problem Description and Assumptions

Considering a single manufacturer, a single manufacturer and a retailer, the two echelon supply chain. The original manufacturer makes the new product, the manufacturer makes the remanufactured product, and the new product and the remanufacturing product wholesale to the retailer. The goal of the original manufacturer, the manufacturer and the retailer is to maximize their profits. The order of decision is to determine the wholesale price of new products and remanufactured products separately from the original manufacturer and the manufacturer, and then the retailer decides the retail price of the two products.

Hypothesis 1.1 The original manufacturer is not the same as the manufacturer. However, the original manufacturer is in the leading position in the supply chain.

1.2.2 Parameters of the Model

w_m: The wholesale price of new products is the decision variable of the original manufacturer;
w_r: The wholesale price of remanufactured products is the decision variable of remanufacturing;
p_m: The retail price of new products is the decision variable;
p_r: The unit retail price of remanufactured products is the decision variable of remanufacturing;
c_m: Manufacturing cost per unit new products.
c_r: Manufacturing cost of unit remanufacturing products;
ϕ: Market capacity;
α: Express consumer sensitivity to price;
β: Represents the substitutability between new products and remanufactured products;
Q_m: Demand for new products in the market;
Q_r: Demand for remanufactured products on the market;

Considering the reality, we could find that $c_m \geq c_r$, $w_m \geq w_r$, $p_m \geq p_r$.

According to the literature [10] that we have, we could make sure the consumer demand for new and remanufactured products:

$$Q_m = \phi - \alpha p_m + \beta p_r, \quad Q_r = \phi - \alpha p_r + \beta p_m$$

1.2.3 Model Establishment

1.2.3.1 Decentralized Decision

$$\max \pi_M = (w_m - c_m)(\phi - \alpha p_m + \beta p_r) \tag{1.1}$$

$$\max \pi_R = (w_r - c_r)(\phi - \alpha p_r + \beta p_m) \tag{1.2}$$

$$\max \pi_T = (p_m - w_m)(\phi - \alpha p_m + \beta p_r) + (p_r - w_r)(\phi - \alpha p_r + \beta p_m) \tag{1.3}$$

1.2.3.2 Equation Solving

It easily knows that the (1.3) equation is a convex function about the variables p_m and p_r, so (1.3) gets the maximum value at the point which the partial derivative about p_m and p_r is equal to zero. Then we could get the optimal value through (1.3) as follows:

$$p_m{}^* = \frac{\phi + (\alpha - \beta)w_m}{2(\alpha - \beta)}, \quad p_r{}^* = \frac{\phi + (\alpha - \beta)w_r}{2(\alpha - \beta)}$$

Putting $p_m{}^*$ and $p_r{}^*$ into the equation of (1.1) and (1.2), Simulating to the (1.3), it could get the optimal value as follows:

$$w_m{}^* = \frac{(2\alpha + \beta)\phi + \alpha(2\alpha c_m + \beta c_r)}{4\alpha^2 - \beta^2}, \quad w_r{}^* = \frac{(2\alpha + \beta)\phi + \alpha(2\alpha c_r + \beta c_m)}{4\alpha^2 - \beta^2}$$

Then put the optimal value $\left(w_m^*, w_r^*\right)$ in to $\left(p_m^*, p_r^*\right)$, it could get the the optimal value as follows:

$$p_m{}^* = \frac{(2\alpha + \beta)(3\alpha - 2\beta)\phi + \alpha(\alpha - \beta)(2\alpha c_m + \beta c_r)}{2(\alpha - \beta)\left(4\alpha^2 - \beta^2\right)},$$
$$p_r{}^* = \frac{(2\alpha + \beta)(3\alpha - 2\beta)\phi + \alpha(\alpha - \beta)(2\alpha c_r + \beta c_m)}{2(\alpha - \beta)\left(4\alpha^2 - \beta^2\right)}.$$

The profit function of the OEM, the manufacturer and the retailer is as follows:

$$\pi_M{}^* = \frac{\alpha}{2}\left[\frac{(2\alpha+\beta)\phi + (\beta^2 - 2\alpha^2)c_m + \alpha\beta c_r}{4\alpha^2 - \beta^2}\right]^2$$

$$\pi_R{}^* = \frac{\alpha}{2}\left[\frac{(2\alpha+\beta)\phi + (\beta^2 - 2\alpha^2)c_r + \alpha\beta c_m}{4\alpha^2 - \beta^2}\right]^2$$

$$\pi_T{}^* = \alpha^2\frac{[(2\alpha+\beta)\phi - (\alpha-\beta)(2\alpha c_m + \beta c_r)]\{(2\alpha+\beta)\phi + [\beta\alpha c_r + (\beta^2 - 2\alpha^2)c_m]\}}{4(\alpha-\beta)(4\alpha^2-\beta^2)^2}$$
$$+\alpha^2\frac{(2\alpha+\beta)\phi - (\alpha-\beta)(2\alpha c_r + \beta c_m)\{(2\alpha+\beta)\phi + [\beta\alpha c_m + (\beta^2 - 2\alpha^2)c_r]\}}{4(\alpha-\beta)(4\alpha^2-\beta^2)^2}$$

The total profit is as follows:

$$\pi^* = \pi_M{}^* + \pi_R{}^* + \pi_T{}^*$$

1.2.3.3 Centralized Decision

$$\max_{p_m, p_r} \pi_C = (p_m - c_m)(\phi - \alpha p_m + \beta p_r) + (p_r - c_r)(\phi - \alpha p_r + \beta p_m) \tag{1.4}$$

Similar to the 1.2.3.2, it could get the optimal value about p_m and p_r as follows:

$$p_m{}^{**} = \frac{\phi + (\alpha-\beta)c_m}{2(\alpha-\beta)}, \quad p_r{}^{**} = \frac{\phi + (\alpha-\beta)c_r}{2(\alpha-\beta)}$$

Then it could get total profit as follows:

$$\pi_C^{**} = \frac{[\phi - (\alpha-\beta)c_m](\phi - \alpha c_m + \beta c_r)}{4(\alpha-\beta)} + \frac{[\phi - (\alpha-\beta)c_r](\phi - \alpha c_r + \beta c_m)}{4(\alpha-\beta)}$$

Conclusion 1.1 The profit of supply chain in centralized decision is greater than decentralized decision.

Proof of Conclusion 1.1 The second partial derivative of p_m, p_r about the total profit function (1.4):

$$\frac{\partial^2\pi}{\partial p_m{}^2} = -2\alpha, \quad \frac{\partial^2\pi}{\partial p_r{}^2} = -2\alpha, \quad \frac{\partial^2\pi}{\partial p_m\partial p_r} = \frac{\partial^2\pi}{\partial p_r\partial p_m} = 2\beta$$

Then it could get the Jacobi matrix as follows:

$$\begin{vmatrix} -2\alpha & 2\beta \\ 2\beta & -2\alpha \end{vmatrix} = 4(\alpha^2 - \beta^2) > 0, \ -2\alpha < 0.$$

Therefore, the function (1.4) reaches its maximum at.

$$(p_m^{**}, p_r^{**}) = \left(\frac{\varphi + (\alpha - \beta)c_m}{2(\alpha - \beta)}, \frac{\varphi + (\alpha - \beta)c_r}{2(\alpha - \beta)} \right).$$

The proof of Conclusion 1.1 being finished. It could be seen from the Conclusion 1.1 that decentralized decision reduces the supply chain benefit. So, it is meaningful and necessary to coordinate the supply chain. Then, the supply chain coordination strategy will be given in Sect. 1.3.

1.3 Supply Chain Coordination

1.3.1 Shapley Value Method

From the above analysis, we could see that decentralized decision leads to "marginal effect", which will reduce the profit of the whole supply chain. In order to avoid the loss of supply chain, the coordination of supply chain should be considered. In this paper, Shapley value method is used to coordinate supply chain. Shapley value method based on the size of the members of the alliance to achieve the overall interests of the alliance between the members of the fair and effective allocation. The basic principle of Shapley method is as follows:

Setting $<I, v>$ as union. $I = \{1, 2, \cdots, n\}$ represents the collection people in the game. Referring to the subset S, its profit function is compatible with those conditions:

$$v(\Phi) = 0$$
$$v(s_i \cup s_2) \geq v(s_i) + v(s_2) \text{ and } s_1 \cap s_2 = \Phi$$

In the collection of I, $v(s)$ can be seen as characteristic function, which represents the profits of S. The two equations reflect the basic system idea of "1 + 1 > 2", which means the profit peaks when the cooperation comes into being, and marks it as $v(I)$. We set $\phi_i(i = 1, 2 \cdots, n)$ as the deserved profit (the member i achieved from the collection I) . Therefore, the allocation of the maximum profit during the process of cooperation is $\phi(v) = (\phi_1(v), \phi_2(v) \cdots, \phi_n(v))$. At the same time, it should fulfill the following things:

$$\sum_{i=1}^{n} \phi_i(v) = v(I) \text{ and } \phi_i(v) > v(i), \quad i = 1, 2, \cdots, n$$

From the conditions above, the each member of the supply chain shapely value could calculate of.

$$\phi_i(v) = \sum_{i \in s(i)} w(|s|)[v(s) - v(s \backslash i)] \quad i = 1, 2, \cdots n$$

$$w(|s|) = \frac{(n - |s|)!(|s| - 1)!}{n!}$$

S(i) represents the subset of I (including company of i), $|s|$ means the amount of element in the subset S. N means the amount of element in the subset I. $w(|s|)$ is the weighting factor. In fact, it is the probability of contribution to the alliance. $v(s)$ is the effectiveness of S. $v(s\backslash i)$ is the available effectiveness of S(removing i).

1.3.2 *Part Alliance*

The original manufacturer and the manufacturer alliance, the alliance's revenue function is:

$$\max_{w_m, w_r} \pi_{MR} = (w_m - c_m)(\phi - \alpha p_m + \beta p_r) + (w_r - c_r)(\phi - \alpha p_r + \beta p_m)$$

$$\text{s.t.} \max_{p_m, p_r} \pi_T = (p_m - w_m)(\phi - \alpha p_m + \beta p_r) + (p_r - w_r)(\phi - \alpha p_r + \beta p_m)$$

$$w_m{}^* = \frac{\phi + (\alpha - \beta)c_m}{2(\alpha - \beta)}, \quad w_r{}^* = \frac{\phi + (\alpha - \beta)c_r}{2(\alpha - \beta)}$$

$$p_m{}^* = \frac{2\phi + (\alpha - \beta)c_m}{4(\alpha - \beta)}, \quad p_r{}^* = \frac{2\phi + (\alpha - \beta)c_r}{4(\alpha - \beta)}$$

$$\pi_{MR}{}^* = \frac{[\phi - (\alpha - \beta)c_m](2\phi - \alpha c_m + \beta c_r)}{8(\alpha - \beta)} + \frac{[\phi - (\alpha - \beta)c_r](2\phi - \alpha c_r + \beta c_m)}{8(\alpha - \beta)}$$

The original manufacturer and retailer alliance, the alliance's revenue function is:

$$\max_{p_m, p_r} \pi_{MT} = (p_m - c_m)(\phi - \alpha p_m + \beta p_r) + (p_r - w_r)(\phi - \alpha p_r + \beta p_m)$$

$$\text{s.t.} \quad \max_{w_r} \pi_R = (w_r - c_r)(\phi - \alpha p_r + \beta p_m)$$

Similar to the above, solution process can be obtained:

$$p_m{}^* = \frac{\phi + (\alpha - \beta)c_m}{2(\alpha - \beta)}, \quad p_r{}^* = \frac{(3\alpha - \beta)\phi + (\alpha - \beta)(\alpha c_r + \beta c_m)}{4\alpha(\alpha - \beta)}$$

$$w_r^* = \frac{\phi + \alpha c_r + \beta c_m}{2\alpha}$$

$$\pi_{MT}{}^* = \frac{[\phi - (\alpha - \beta)c_m]\left[(2\alpha + \beta)\phi - (2\alpha^2 - \beta^2)c_m + \alpha\beta c_r\right]}{8\alpha(\alpha - \beta)} + \frac{[(\alpha + \beta)\phi - (\alpha - \beta)(\alpha c_r + \beta c_m)](\phi - \alpha c_r + \beta c_m)}{16\alpha(\alpha - \beta)}$$

The alliance between the manufacturer and the retailer:

$$\max_{p_r, p_m} \pi_{RT} = (p_r - c_r)(\phi - \alpha p_r + \beta p_m) + (p_m - w_m)(\phi - \alpha p_m + \beta p_r)$$
$$\text{s.t.} \quad \max_{w_m} \pi_M = (w_m - c_m)(\phi - \alpha p_m + \beta p_r)$$

we could also reach the optimum solution.

$$p_m{}^* = \frac{(3\alpha - \beta)\phi + (\alpha - \beta)(\alpha c_m + \beta c_r)}{4\alpha(\alpha - \beta)}, \quad p_r{}^* = \frac{\phi + (\alpha - \beta)c_r}{2(\alpha - \beta)}$$

$$w_m^* = \frac{\phi + \alpha c_m + \beta c_r}{2\alpha}$$

$$\pi_{RT}{}^* = \frac{[\phi - (\alpha - \beta)c_r]\left[(2\alpha + \beta)\phi - (2\alpha^2 - \beta^2)c_r + \alpha\beta c_m\right]}{8\alpha(\alpha - \beta)} + \frac{[(\alpha + \beta)\phi - (\alpha - \beta)(\alpha c_m + \beta c_r)](\phi - \alpha c_m + \beta c_r)}{16\alpha(\alpha - \beta)}$$

According to the above, Shapley could get as Table 1.1:
The allocation of the Original manufacturer:

$$\phi_M{}^*(v) = \frac{1}{3}\pi_M{}^* + \frac{1}{6}[\pi_{MR}{}^* - \pi_R{}^*] + \frac{1}{6}[\pi_{MT}{}^* - \pi_T{}^*] + \frac{1}{3}[\pi_{MRT}{}^* - \pi_{RT}{}^*]$$

Similarly, the manufacturer and the retailer can gain a

$$\phi_R{}^*(v) = \frac{1}{3}\pi_R{}^* + \frac{1}{6}[\pi_{MR}{}^* - \pi_M{}^*] + \frac{1}{6}[\pi_{RT}{}^* - \pi_T{}^*] + \frac{1}{3}[\pi_{MRT}{}^* - \pi_{MT}{}^*]$$
$$\phi_T{}^*(v) = \frac{1}{3}\pi_T{}^* + \frac{1}{6}[\pi_{MT}{}^* - \pi_M{}^*] + \frac{1}{6}[\pi_{RT}{}^* - \pi_T{}^*] + \frac{1}{3}[\pi_{MRT}{}^* - \pi_{MR}{}^*]$$

Table. 1.1 The Shapley value of OEM

s	M	MR	MT	MRT
$v(s)$	$\pi_M{}^*$	$\pi_{MR}{}^*$	$\pi_{MT}{}^*$	$\pi_{\text{MRT}}{}^*$
$v(s\backslash M)$	0	$\pi_R{}^*$	$\pi_T{}^*$	$\pi_{RT}{}^*$
$v(s) - v(s\backslash M)$	$\pi_M{}^*$	$\pi_{MR}{}^* - \pi_R{}^*$	$\pi_{MT}{}^* - \pi_T{}^*$	$\pi_{\text{MRT}}{}^* - \pi_{RT}{}^*$
$\|s\|$	1	2	2	3
$w(\|s\|)$	1/3	1/6	1/6	1/3
$w(\|s\|)[v(s) - v(s\backslash M)]$	$\pi_M{}^*/3$	$(\pi_{MR}{}^* - \pi_R{}^*)/6$	$(\pi_{MT}{}^* - \pi_T{}^*)/6$	$(\pi^* - \pi_{RT}{}^*)/3$

Conclusion 1.2 If $(p_m, p_r, w_m, w_r) = ({p_m}^{**}, {p_r}^{**}, c_m, c_r)$, it makes Supply chain coordination.

The Proof of Conclusion 1.2 According to ${p_m}^{**} = \frac{\phi+(\alpha-\beta)c_m}{2(\alpha-\beta)}$, ${p_m}^{*} = \frac{\phi+(\alpha-\beta)w_m}{2(\alpha-\beta)}$ ${p_r}^{**} = \frac{\phi+(\alpha-\beta)c_r}{2(\alpha-\beta)}$ and ${p_r}^{*} = \frac{\phi+(\alpha-\beta)w_r}{2(\alpha-\beta)}$, it could get $(w_m, w_r) = (c_m, c_r)$.

The proof of Conclusion 1.2 being finished.

1.4 Numerical Analysis

According to literature [11], it could choose $\phi = 2000$, $c_m = 200$, $c_r = 150$, and the α and β could choose values are as follows:

According to Tables 1.2, 1.3, and 1.4, the result could get as follows:

1. According to the numerical example, it is necessary to reduce α, and increases β, i.e., the different between α or β should keep small. When α or β is given and the difference between α and β is smaller: when it is decentralized decision the wholesale and retail prices of the both products is bigger, and the profit of the OEM, the remanufacturer and the retailer becomes larger; when it is centralized decision, the profit of supply chain becomes larger as the retail price increases.
2. When the different between α or β is fixed, the profit of the OEM, the remanufacturer and the retailer becomes larger when α and β becoming smaller and the retail price increasing. The wholesale price of new products and remanufactured products is less than the wholesale price of decentralized decision. The contribution of the retailer is the largest, the second is remanufacturer, and the OEM is the least accounting to the profit of supply chain for centralized decision.
3. When the supply chain is coordinated, the coordination contract requires that the wholesale price is equal to the production cost, then the sales profit will be zero. However, the profit of the supply chain is greater than that of decentralized decision. The main reasons is that one is the centralized decision-making unit, the retail price of the product is much smaller than decentralized decision-making, will increase the products, and the centralized decision avoids the marginal effect of decentralized decision, the entire supply chain has been reached optimal.

Table. 1.2 The profit changes after supply chain coordination

α	β		π_M^*	π_R^*	π_T^*	π^*
9	5	Decentralized decision	32346.4	59737.7	101919.9	194004.0
		Centralized decision	32840.9	65494.9	117289.2	215625.0
	6	Decentralized decision	51801.8	86372.1	204624.0	342798.0
		Centralized decision	55182.87	95514.87	216594.0	367291.7

Table. 1.3 The optimal decision changes when α and β changing

		Decentralized decision								Centralized decision		
α	β	p_m^*	p_r^*	w_m^*	w_r^*	π_M^*	π_R^*	π_T^*	π^*	p_m^{**}	p_r^{**}	π^{**}
9	4	332.8	322.6	265.6	245.1	19355.9	40723.61	52983.8	113063.3	300	275	130,625
	5	392.4	382.6	284.8	265.2	32346.4	59737.7	101919.9	194004.0	350	325	215,625
	6	487.0	477.6	307.3	288.5	51801.8	86372.1	204,624	342798.0	433.3	408.3	367291.7
6	3	507.8	497.8	348.9	328.9	66503.7	96003.7	161607.4	324114.8	433.3	408.3	365416.7
8		336.0	325.5	272.1	251.0	20773.3	40813.8	48538.4	110125.6	300	275	130,000
10		258.6	247.7	231.5	209.7	4948.0	17831.6	15612.9	38392.4	242.9	217.9	46964.3
8	5	492.6	483.1	318.6	299.6	56277.8	89481.3	192380.2	338139.3	433.3	408.3	366666.7
7	4	499.4	489.7	332.2	312.8	61189.5	92738.1	178195.9	332123.5	433.3	408.3	366041.7
5	2	1244.0	1235.1	488.1	470.2	207497.2	256381.1	1,156,826	1,620,704	1100	1075	1,668,125

Table. 1.4 The optimal decision changes when it is different alliance strategies

		MR					MT				RT			
α	β	p_m^*	p_r^*	w_m^*	w_r^*	π_{MR}^*	p_m^*	p_r^*	w_r^*	π_{MT}^*	p_m^*	p_r^*	w_m^*	π_{RT}^*
9	4	350	337.5	300	275	56,250	300	315.3	230.6	86822.9	322.2	275	244.4	117291.7
	5	425	412.5	350	325	97,500	350	370.8	241.7	158906.3	376	325	252.8	196822.9
	6	550	537.5	433.3	408.3	172083.3	433.3	459.7	252.8	295989.6	463.9	408.3	261.1	342083.3
6	3	550	537.5	433.3	408.3	172083.3	433.3	479.2	291.7	275104.2	485.4	408.3	304.2	316588.5
8		350	337.5	300	275	56,250	300	318.8	237.5	84062.5	326.6	275	253.1	113066.4
10		264.3	251.8	242.9	217.9	16607.14	242.9	245.4	205	24276.8	254.1	217.9	222.5	43167.4

1.5 Conclusions

Based on the analysis of the sensitivity of consumers to the product price and the degree of substitution of the product, this paper analyzes the optimal pricing strategy of the OEM, the remanufacturer and the retailer under the two decision modes. Shapley value method is used to coordinate the supply chain, and the coordination contract is given to optimize the supply chain. Finally, the results are verified by numerical simulation. In this paper, we can further study the recycling of waste products, so that the supply chain becomes a closed loop supply chain.

References

1. Debo, L. G., Toktay, L. B., & Van Wassenhove, L. N. (2005). Market segmentation and product technology selection for remanufacturable products. *Management Science, 51*(8), 1193–1205.
2. Ferrer, G., & Swaminathan, J. M. (2006). Managing new and remanufactured products. *Management Science, 52*(1), 15–26.
3. Ferguson, M., & Toktay, L. B. (2006). The effect of competition on recovery strategies. *Production and Operations Management, 15*(3), 351–368.
4. Ge, J. Y., & Huang, P. Q. (2008). Price decision analysis for closed-loop supply chain based on game theory. *Journal of Systems Engineering, 23*(1), 111–115.
5. Chen, J. L., Song, H., & Yu, X. U. (2010). Supply chain contract coordination model in demand forecast update conditions with asymmetry information. *Chinese Journal of Management Science, 18*(1), 83–89.
6. Liu, J. G., & Chong, W. U. (2010). Study of a tow-level supply chain returns policy model based on the newsboy model. *Chinese Journal of Management Science, 18*(4), 73–78.
7. Yan, N. N., & Huang, X. Y. (2008). Models of closed-loop supply chain with third-party reverse logistics and their applications. *Journal of Management Sciences in China, 11*(4), 83–93.
8. Bao, X. Y., Tang, Z. Y., & Tang, X. W. (2010). Coordination and differential price strategy of closed-loop supply chain with product remanufacturing. *Journal of Systems & Management, 19*(5), 546–552.
9. Zheng, K. J. (2012). Study on pricing decision and contract consideration of closed-loop supply chain with differential price. *Operations Research & Management Science, 21*(1), 118–123.
10. Yao, Z., Leung, S. C. H., & Lai, K. K. (2008). Manufacturers revenue-sharing contract and retailer competition. *European Journal of Operational Research, 186*, 637–651.
11. Aumann, R. J. (1990). The Shapley value. In *Game theory and applications (Columbus, OH 1987). Econom. theory econometrics math econom* (pp. 158–165). San Diego: Academic Press.

Chapter 2
A Study on Cooperation Strategies of Dual Channel Supply Chain Based on Service Level

Guojun Ji and Xi Liu

Abstract Dual channel operation under the competition condition not only requires to improve service level, but needs to achieve the collaborative service among channels, which amplifies the conflict of the manufacturer and the retailer and cooperation of the whole supply chain. Based on the service level under the dual channel supply chain, considering that cooperative strategy between network direct-channels and retail channels, to realize supply chain coordination, this paper builds providing the same services and differentiation service model based on the centralized decision and decentralized decision, through the theoretical model and numerical computation to explore the different service levels and network channels that effect on demands, prices, profits. Our conclusions show that to provide differentiated services to enable the system to achieve the optimal profit, manufacturers and retailers to adopt decentralized decision or centralized decision are affected by the acceptance degree of direct marketing network.

Keywords Service level · Dual Channel supply chain · Cooperative strategy

2.1 Introduction

With the popularity of the Internet, e-commerce, mobile commerce and other emerging e-marketing channels rapid development, companies have successfully achieved from the traditional channels to the network channels and physical retail channels dual channel transformation. A variety of channels of supply chain system will not only help to improve the efficiency of enterprise supply chain management, but also help to meet customer differentiation preferences, enhance customer loyalty, improve profitability and market coverage(2004).GOME Electrical Appliances in early 2014 announced a strategic transformation, the full implementation of the

G. Ji (✉) · X. Liu
School of Management, Xiamen University, Xiamen, Fujian, China
e-mail: jiking@xmu.edu.cn

M. Tavana, S. Patnaik (eds.), *Recent Developments in Data Science and Business Analytics*, Springer Proceedings in Business and Economics,
https://doi.org/10.1007/978-3-319-72745-5_2

"O2M Omni-channel retailer strategy mode, namely "combined offline store with online e-commerce and mobile terminal", return to the commercial essence that is customer demand-oriented. According to Wikipedia statistics (2014), Dell's 30%, Apple's 43% of the net income comes from the sale of services, IBM through continuous service innovation to obtain customer's consistent affirmation, become one of the world's successful transformation of service enterprises. Therefore, it is one of the most important issues for dual channels to study the service level and the cooperation between the supply chain participants in order to coordinate the conflicts between channels.

The existing research, which is directly related to this article, focuses on channel pricing and service. (1) Channel pricing. Tsay and Agrawal [1] have studied consumers' sensitivity to price and sales support, and found that manufacturers' increased online channels is not necessarily harmful to retailers. Cai Guangshu [2] discusses the impact of discount pricing strategies under dual channels. Liu Hanjin et al. [3] discusses the effect of different pricing strategies on the channel when the manufacturer introduces the online channel on the basis of taking into account the different channel distribution efficiency and the online channel acceptance of the product in view of the channel structure of the retailers are price leaders. Chen Guopeng et al. [4] analyze the pricing situation of dual channel retailers, establish the price demand model, and test the results of the example according to the price situation between channels. (2) Channel service. For example, Dumrongsiri believes that consumers choose which channels to shop mainly depends on the price and service two factors. Kaya's [5] study shows that the level of service provided by manufacturers and retailers in a dual-channel supply chain is higher than that of a single channel. Dan [6] study the optimal price and service decision of retailers in the dual-channel supply chain, the results show that the retailer's services have a huge impact on manufacturers and retailers. Ding Feng [7] thinks that the improvement of service level will lead to manufacturers wholesale prices and the channel sales and channel demand increase, the impact on channel competition exists spillover effect, and the extent of the price and service depends on the impact of demand. Most of these documents are considered that the manufacturer not to provide services but the retailers to provide services, or retailers and manufacturers to cooperate by the retailers to provide services and the manufacturers provide services based on retailers to give some compensation, but rarely consider the manufacturer provide individual service. In fact, manufacturers also provide services, including electronic payment, after-sales support, regular updates the commodity types and many other services.

Based on the consideration of manufacturers and retailers both provide services, this paper takes into account the impact of service level on commodity prices, demand and profit, and focuses on the influence of different decision-making methods and acceptance of the network channels.

2.2 Model Description and Optimal Decision Making

2.2.1 Model Hypothesis

Suppose that there is a dual supply chain composed of a manufacturer and a supplier (as shown in Fig. 2.1), where manufacturers sell the same product through direct online channels with retailers. Manufacturers and retailers can provide the same or differentiated service through the direct online channels. The consumer's channel preference is concerned with the quality of the service other than the price. For example, due to the growing competition between online sales and retail channels, network direct sales channels can improve the quality of service by providing similar retailer returns, providing independent packaging, customized services, or products that are conducive to increase the consumer's experience.

Let i = r, d, represent the decision variables of traditional channels and internet direct marketing channels respectively. Suppose the wholesale price of the product is w, the consumer's value for the product is v, v obeys the uniform distribution on interval [0,1]. Assume that the cost of production is zero [8]. According to the literature [10], the utility obtained by consumers buying through two channels is: $U_i = V - P_i + \theta S_i\ (i = \text{r, d})$. According to the literature [9], $\mu, \theta \in [0, 1]$. According to the literature [1], Service cost is: $C(S_i) = \eta S_i^2/2$, where η represents a unit of service cost ($\eta > 0$). Assume that service level S∈[0,1], 0 means that the manufacturer or retailer does not provide service, 1 means that the manufacturer or retailer offers the perfect service. The meaning of variables is shown in Table 2.1.

When $U_r = U_d$, there is no difference that the consumer choose which channel to buy products; when $U_r > U_d$, consumers choose the retail channels; when $U_r < U_d$, consumers choose the network direct selling channel. Considering the two channel demand situation, it shall be satisfied that $U_r < U_d$ and $U_d > 0$. When $U_r = U_d$, there is $V = \frac{P_r - P_d + \theta(S_d - S_r)}{1-\mu}$. When $U_d = 0$, we have $V = \frac{P_d - \theta S_d}{\mu}$. Therefore,

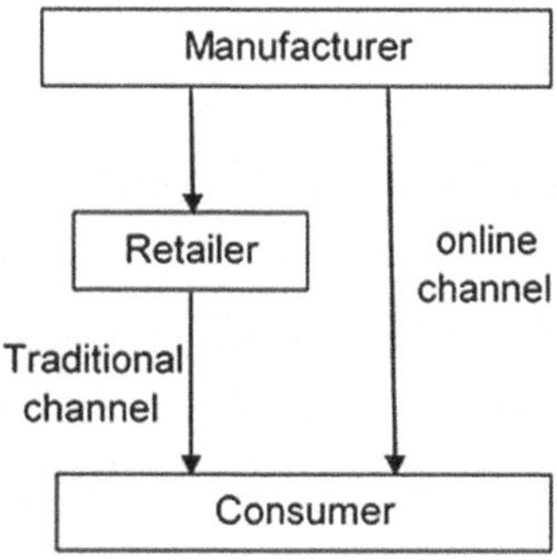

Fig. 2.1 Dual channel manufacturers and retailers' channel structure

Table 2.1 Main variables of dual channel supply chain model

Variable	D_i	P_i	S_i	U_i	θ	μ	η
Implication	Demand for products in channel i	Price of products in channel i	Level of service in channel i	The utility of a product purchased by a customer from a channel i	Service sensitivity coefficient	Acceptance of network direct selling channel	Channel service cost factor

$$D_d = \int_{\frac{P_d - \theta S_d}{\mu}}^{\frac{P_r - P_d + \theta(S_d - S_r)}{1-\mu}} f(v)dv = \frac{P_r - P_d + \theta(S_d - S_r)}{1-\mu} - \frac{\mathrm{P_d} - \theta S_d}{\mu} \tag{2.1}$$

$$D_r = \int_{\frac{P_r - P_d + \theta(S_d - S_r)}{1-\mu}}^{1} f(v)dv = 1 - \frac{P_r - P_d + \theta(S_d - S_r)}{1-\mu} \tag{2.2}$$

2.2.2 *Optimal Decision Under Decentralized Decision*

Consider that only one product is sold in the dual channel supply chain, and the manufacturer is the sole producer of the product, assuming the manufacturer is the leader in the Stackberg game. Therefore, the reverse solution is adopted: the first stage assumes that the manufacturer direct price P_d and wholesale price w are known, and the retailer determines the retail price P_r. The second stage manufacturer maximizes its revenue and identifies the direct selling price P_d, and wholesale price w.

1. Manufacturers provide the same services as retailers

When the manufacturer The manufacturer's e-channel and retail channel provide same services (Sd = Sr = S), so $C(S_d) = C(S_r) = \eta s^2/2$. Let C0 = ηS2/2, we can get the optimal decision variables are obtained by using the reverse solution (with superscript *) which is shown in Table 2.2:

The total profits of the manufacturer, retailer and supply chain system are:

$$\begin{cases} \Pi_{d1}^* = \dfrac{(A+2\mu)^2}{16\mu} + \dfrac{1-\mu}{8} \\ \Pi_{r1}^* = \dfrac{1}{16}(1-\mu) \\ \Pi_{s1}^* = \dfrac{(A+2\mu)^2}{16} + \dfrac{3}{16}(1-\mu) \end{cases}, \text{where } \mathrm{A} = 2\theta \mathrm{S} - \eta S^2. \tag{2.3}$$

2. Manufacturers and retailers offer differentiated services

The manufacturer's e-channel and retail channel offer different services ($S_d \neq S_r$). Similarly, the value of each optimal decision variable (expressed by superscript *) is shown in Table 2.3:

Table 2.2 Optimal decision variables with the same service under decentralized decision making

Variable	P_{d1}^*	W_1^*	P_{R1}^*	D_{d1}^*	D_{r1}^*
Optimal value	$\frac{\theta S + C_0 + \mu}{2}$	$\frac{\theta S - C_0 + 1}{2}$	$\frac{3 - \mu + 2\theta S + 2C_0}{4}$	$\frac{1}{4}\left(1 - \frac{2C_0 - 2\theta S}{\mu}\right)$	$\frac{1}{4}$

Table 2.3 Optimal decisions for differentiated services under decentralized decision making

Variable	P'^*_{d1}	W'^*_1	P'^*_{r1}	D'^*_{d1}	D'^*_{r1}
Optimal value	$\frac{\theta S_d + C_d + \mu}{2}$	$\frac{\theta S_r - C_r + 1}{2}$	$\frac{3 - \mu + C_d + C_r + \theta(3S_r - S_d)}{4}$	$\frac{1}{4} + \frac{B_1 - B_2}{8(1-\mu)} + \frac{B_1}{4\mu}$	$\frac{1}{4} - \frac{B_1 - B_2}{8(1-\mu)}$

$$\begin{cases} \Pi'^{*}_{d1} = \dfrac{1+\mu+B_1+B_2}{8} + \dfrac{(B_1+B_2)^2}{32(1-\mu)} - \dfrac{B_1^2}{16\mu} \\ \Pi'^{*}_{r1} = \dfrac{[2(1-\mu)-(B_1+B_2)]^2}{64(1-\mu)} \\ \Pi^{*}_{s1} = \dfrac{1+\mu+B_1+B_2}{8} + \dfrac{(B_1+B_2)^2}{32(1-\mu)} - \dfrac{B_1^2}{16\mu} + \dfrac{[2(1-\mu)-(B_1+B_2)]^2}{64(1-\mu)} \end{cases} \tag{2.4}$$

where $B_1 = 2\theta S_d - 2C_{\mathrm{d}}$, $B_2 = 2\theta S_r - 2C_{\mathrm{r}}$.

2.2.3 Optimal Decision Under Centralized Decision Making

Manufacturers and retailers form strategic alliances under centralized decision-making. They work together to set the online price and retail price to maximize the profit of the whole system.

1. Manufacturers provide the same services as retailers

Assume that the manufacturer and the retailer provide the same service S. The total profit of the system is that: $\Pi_{s2} = (P_{d2} - C_0)D_{d2} + (P_{r2} - C_0)D_{r2}$. So the optimal decision variables are obtained (with superscript *), as shown in Table 2.4:

$$\Pi^{*}_{s2} = \frac{1-\mu}{4} + \frac{(\mathrm{E}+2\mu)^2}{16\mu}, \text{ where } \mathrm{E} = 2\theta\mathrm{S} - \eta S^2.$$

2. Manufacturers and retailers offer differentiated services

Manufacturers and retailers offer differentiated services (S_r, S_d and $S_r \neq S_d$) under centralized decision making. The optimal decision variables are(with superscript *) shown in Table 2.5:

Here $\Pi'^{*}_{s2} = \dfrac{(F_2+2)^2}{16} + \dfrac{(F_1-\mu F_2)^2}{16\mu(1-\mu)}$ where $F_1 = 2\theta S_d - 2C_{\mathrm{d}}$, $F_2 = 2\theta S_r - 2C_{\mathrm{r}}$.

Table 2.4 Provide same service optimal decisions under Centralized Decision Making

Variable	P^{*}_{d2}	P^{*}_{r2}	D^{*}_{d2}	D^{*}_{r2}
Optimal value	$\frac{\theta S+C_0+\mu}{2}$	$\frac{\theta S+C_0+1}{2}$	$\frac{E}{4\mu}$	$\frac{1}{2}$

Table 2.5 Provide differentiated service optimal decisions under Centralized Decision Making

Variable	P'^{*}_{d2}	P'^{*}_{r2}	D'^{*}_{d2}	D'^{*}_{r2}
Optimal value	$\frac{\theta S_d+C_d+\mu}{2}$	$\frac{\theta S_r+C_r+1}{2}$	$\frac{F_1-\mu F_2}{4\mu(1-\mu)}$	$\frac{1}{2}-\frac{F_1-F_2}{4(1-\mu)}$

2.3 Model Conclusion and Numerical Analysis

2.3.1 Model Results Analysis

By analyzing the service level S, the acceptance of network direct selling channel μ, and the influence of the service sensitivity coefficient θ in the system, we can get the following conclusions:

Proposition 2.1 *The price of network direct selling channels and retail channels is positively related to the service level, whether in decentralized decision-making or centralized decision-making.*

Proof As $\frac{\partial p_{d1}}{\partial s_d} = \frac{\partial P'_{d1}}{\partial s_d} = \frac{\partial p_{d2}}{\partial s_d} = \frac{\partial P'_{d2}}{\partial s_d} = \frac{\eta S_d + \theta}{2} > 0.$

$$\frac{\partial p_{r1}}{\partial s_r} = \frac{\partial p_{r2}}{\partial s_r} = \frac{\partial P'_{r2}}{\partial s_r} = \frac{\eta S_r + \theta}{2}, \frac{\partial P'_{r1}}{\partial s_r} = \frac{\eta S_r + 3\theta}{4} > 0.$$

Proposition 2.2 *When* $S_d \in \left(0, \frac{\theta}{\eta}\right)$*, In any case, there is a positive correlation between the demand for direct marketing channels and the level of service; when* $S_d \in \left(\frac{\theta}{\eta}, 1\right)$*, there is a negative correlation between the demand of the direct marketing channel and the service level.*

Proof As $\frac{\partial D_{d1}}{\partial s_d} = \frac{\theta - \eta S_d}{2\mu}, \frac{\partial D'_{d1}}{\partial s_d} = (\theta - \eta S_d)\left[\frac{1}{4(1-\mu)} + \frac{1}{2\mu}\right] < 0;$

$$\frac{\partial D_{d2}}{\partial s_d} = \frac{\theta - \eta S_d}{2\mu}, \frac{\partial D'_{d2}}{\partial s_d} = (\theta - \eta S_d)\left[\frac{1}{2(1-\mu)} + \frac{1}{2\mu}\right] < 0.$$

Proposition 2.3 *In the case of the same service provided by the manufacturer and retailer, the retailer's demand is independent of the service level, whether it is decentralized decision making or centralized decision making. When retailers and manufacturers provide differentiated services,* $S_r \in \left(0, \frac{\theta}{\eta}\right)$*, there is a positive correlation between the demand for retail channels and the level of service; when* $S_r \in \left(\frac{\theta}{\eta}, 1\right)$*, there is a negative correlation between the demand for retail channels and the level of service.*

Proof As $\frac{\partial D'_{r1}}{\partial s_r} = \frac{\theta - \eta S_r}{8(1-\mu)}, \frac{\partial D'_{r2}}{\partial s_r} = \frac{\theta - \eta S_r}{2(1-\mu)}.$

Proposition 2.4 *In any case, the price of network direct selling channels is positively related to the acceptance of direct marketing channels. In the case of the same* *service, whether it is decentralized decision-making or centralized decision-making, when* $S_d \in \left(0, \frac{2\theta}{\eta}\right)$, *there is a negative correlation between the demand of network direct channel and the acceptance of network direct channel. When* $S_d \in \left(\frac{2\theta}{\eta}, 1\right)$, *there is a positive correlation between the demand for direct channel channels and the degree of direct channel acceptance (this is because of the higher degree of acceptance of the network direct channel, the higher the pricing of the firm).*

Proof As $\frac{\partial p_{d1}}{\partial \mu} = \frac{\partial P'_{d1}}{\partial \mu} = \frac{\partial p_{d2}}{\partial \mu} = \frac{\partial P'_{d2}}{\partial \mu} = \frac{1}{2}; \frac{\partial D_{d1}}{\partial \mu} = \frac{\partial D_{d2}}{\partial \mu} = \frac{S_d(\eta S_d - 2\theta)}{4\mu^2}$.

Proposition 2.5 *In the decentralized decision-making premise, the degree of network direct channel acceptance and retailer prices were negatively correlated. In the case of centralized decision-making, the degree of network direct channel acceptance has no effect on retail prices. In the case of the same service, the retailer 's demand is not related to the degree of network direct channel acceptance.*

Proof As $\frac{\partial p_{r1}}{\partial \mu} = \frac{\partial P'_{r1}}{\partial \mu} = -\frac{1}{4}$.

Proposition 2.6 *In any case, the manufacturer 's pricing is positively related to the service sensitivity coefficient. In the case of the same service, the network direct channel demand is positively related to the service sensitivity coefficient; the retailer 's demand is independent of the service sensitivity coefficient.*

Proof As $\frac{\partial p_{d1}}{\partial \theta} = \frac{\partial P'_{d1}}{\partial \theta} = \frac{\partial p_{d2}}{\partial \theta} = \frac{\partial P'_{d2}}{\partial \theta} = \frac{S_d}{2}; \frac{\partial D_{d1}}{\partial \theta} = \frac{\partial D_{d2}}{\partial \theta} = \frac{S_d}{2\mu}$.

2.3.2 *Numerical Analysis*

The following is a numerical analysis of the relationship between service levels and profits, and compares the impact of service levels on the level of acceptance of direct marketing channels across different networks. Here is the assumption that $\theta = 0.1, \eta = 2$.

2.3.2.1 When the Acceptance of Network Direct Marketing Channel Is Higher ($\mu = 0.8$)

1. Comparing Figs. 2.2, 2.4, 2.6 and 2.8, regardless of the acceptance of the network is high and low, manufacturers provide the same service under decentralized decision making profits is always greater than the retailer's profit, and the system benefits from centralized decision making at the same level of service. In the case

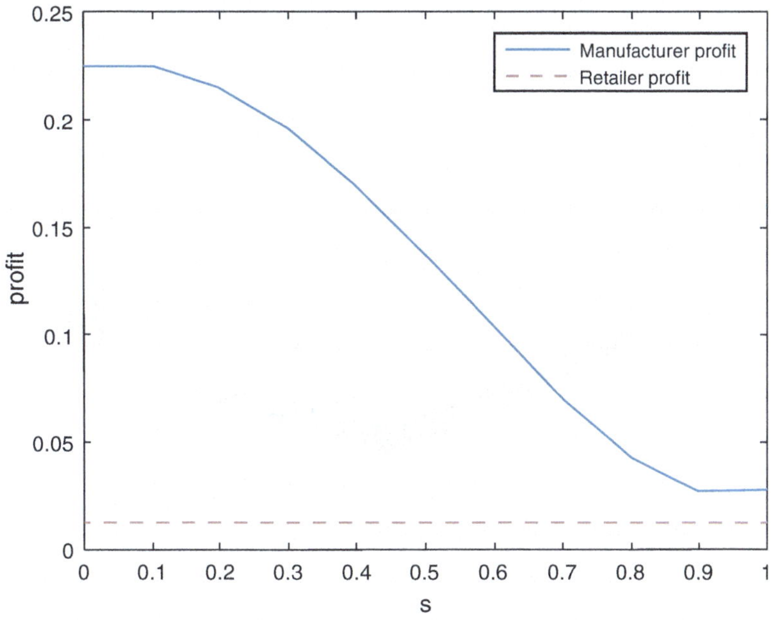

Fig. 2.2 When the acceptance of network direct marketing channel is high (μ = 0.8), the profit of the two parties under decentralized decision-making and same service

of a higher degree of the acceptance of the internet, the profit of the manufacturer decreases with the increase of the service level. On the contrary, the profit of the manufacturer increases with the increase of the service level when the degree of direct channel acceptance is low.

2. As shown in Fig. 2.3, the profitability of manufacturers shows a trend of decreasing first and then increasing with the level of their service level and retail channel service. The retailer's profit increases with the increase of the service level of the electronic channel, and decreases with the increase of the service level.
3. As shown in Fig. 2.5, the total profit of decentralized decision-making system increases with the increase of network direct channel and retail channel service level when the acceptance of the network direct channel is higher. The total profit of the system under the centralized decision-making system increases with the increase of the network direct sales channel service level and the retail channel service level, and decreases with the increase of the network direct sales channel service level and the retail channel service level.

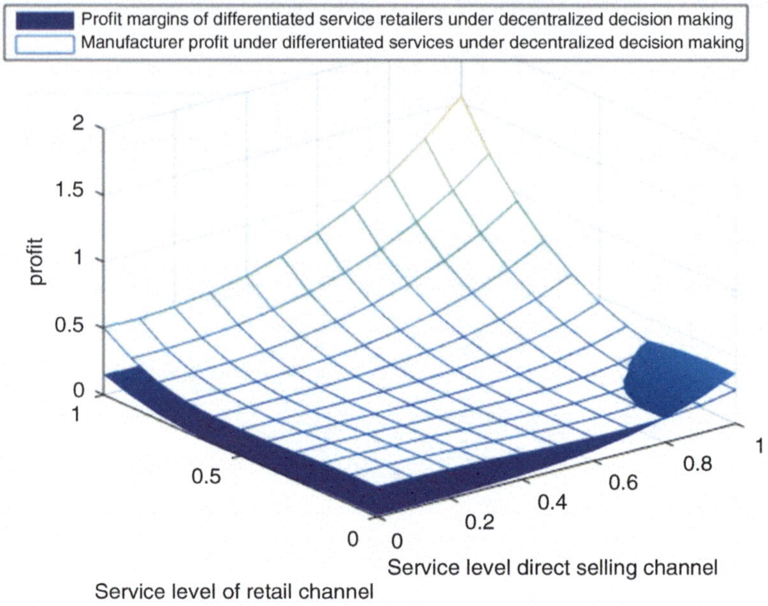

Fig. 2.3 When the acceptance of network direct marketing channel is high (μ = 0.8), the profit of the two parties under decentralized decision-making and differentiated service

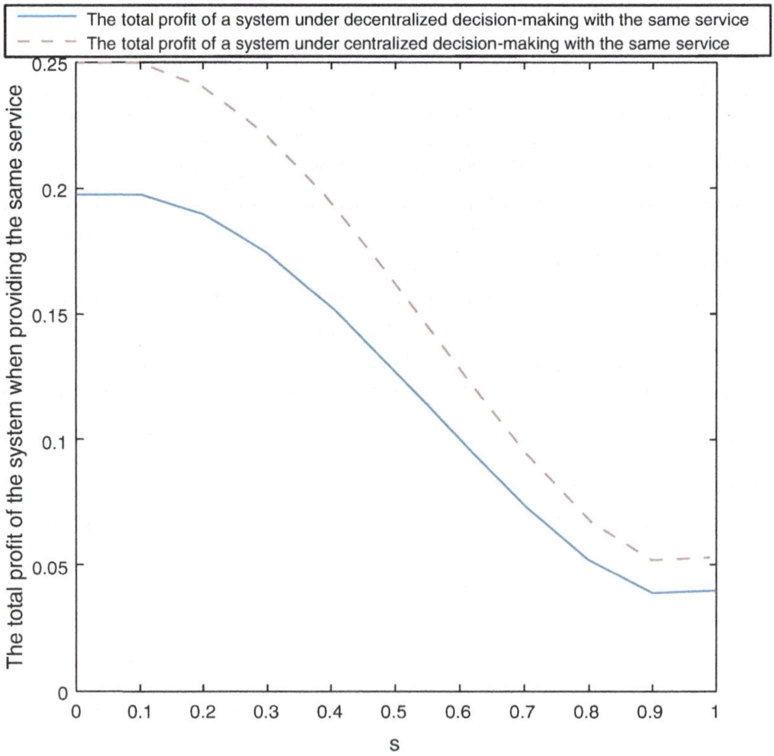

Fig. 2.4 When the acceptance of network direct marketing channel is high (μ = 0.8), the profit of the system under decentralized decision-making and same service

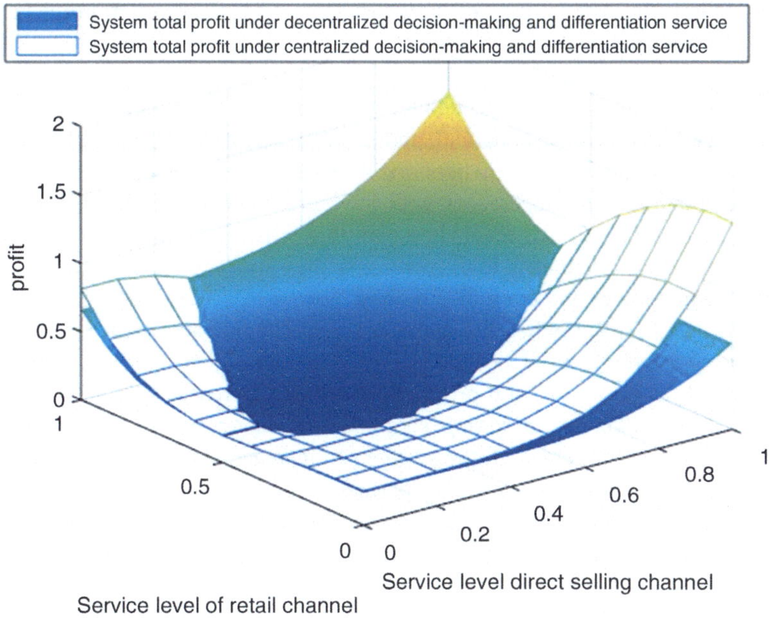

Fig. 2.5 When the acceptance of network direct marketing channel is high (μ = 0.8), the system profit under differentiated service

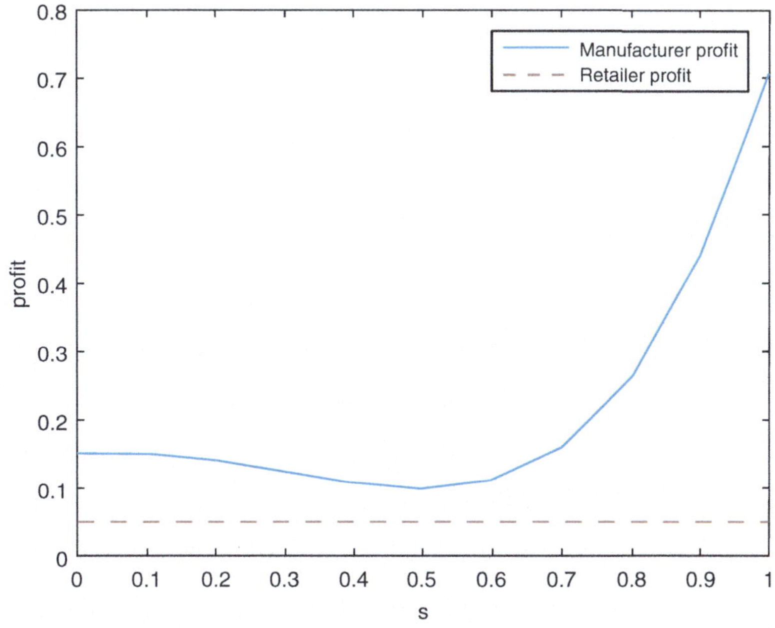

Fig. 2.6 The profits of both parties are provided under the decentralized decision and same service when the acceptance of network direct marketing channel is lower (μ = 0.2)

2.3.2.2 When the Acceptance of Network Direct Marketing Channel Is Lower ($\mu = 0.2$)

1. From Fig. 2.7, the profit of the retailer increases with the increase of the service level of the network direct channel, and decreases with the increase of the service level when the acceptance of network direct marketing channel is lower. The profit of manufacturers decreases with the increase of direct marketing channel and retail channel service level. It shows that when the acceptance of network direct marketing channel is low, increasing the service will make the manufacturers' profits lower and even lose money.
2. As shown in Fig. 2.9, when the network direct channel acceptance is lower, the total profit of the system decreases with the increase of service level, under centralized decision, the total profit of the system increases with the increase of service level of direct marketing channel, and decreases with the increase of service level of retail channel, the system profits reached the maximum when the direct marketing channels provide perfect service and retail channels do not provide service. As a result, when the acceptance of network direct marketing channel is lower, the system profit is higher under the centralized decision.

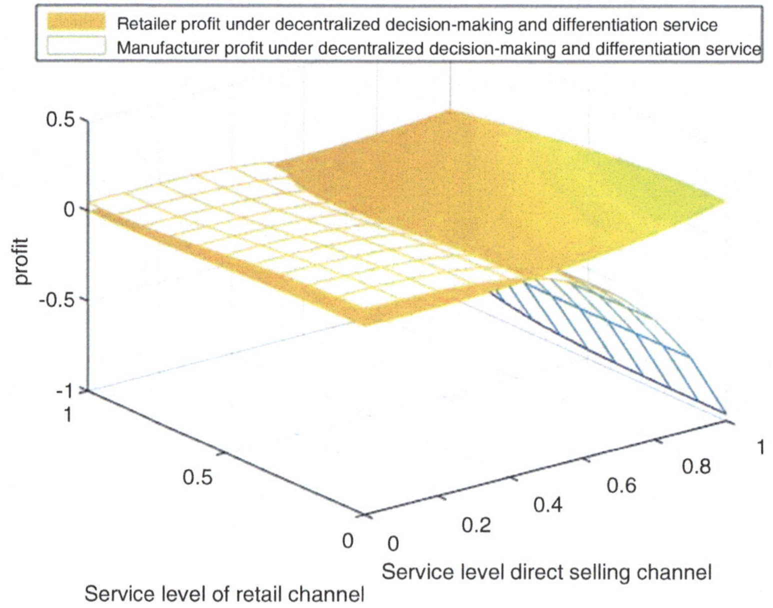

Fig. 2.7 The profits of both parties are provided under the decentralized decision and differentiated service when the acceptance of network direct marketing channel is lower ($\mu = 0.2$)

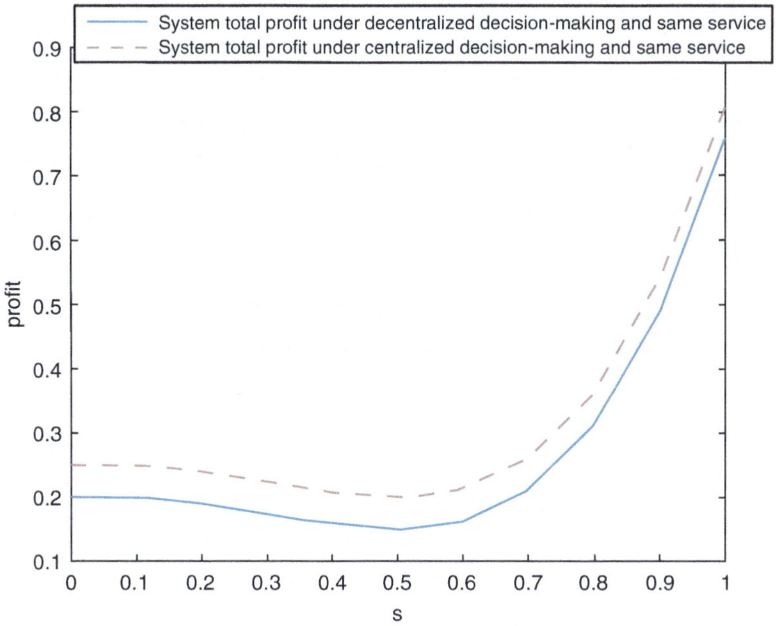

Fig. 2.8 The system profit with the same service when the acceptance of network direct marketing channel is lower (μ = 0.2)

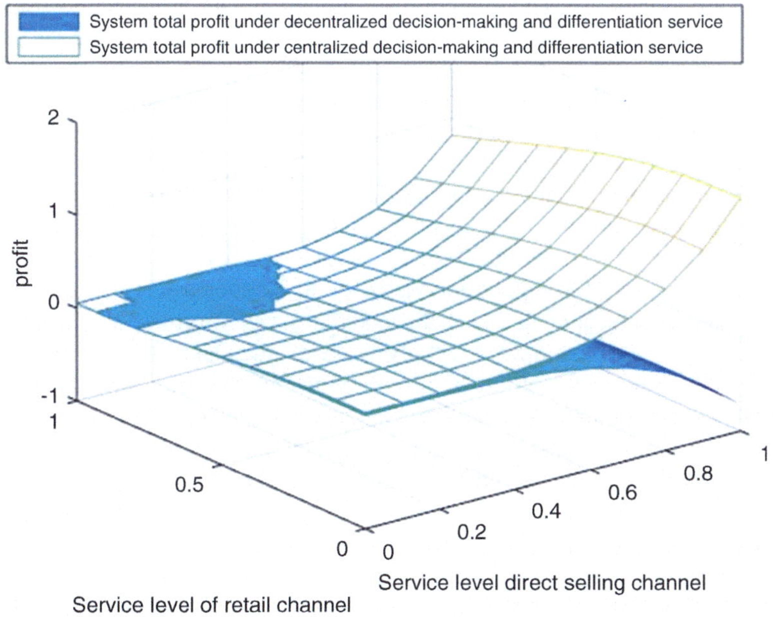

Fig. 2.9 The system profit with the differentiated service when the acceptance of network direct marketing channel is lower (μ = 0.2)

2.4 Conclusion

This paper studies the impact of service level on price, demand and profitability of supply chain under dual channel conditions, and analyzes the impact on the profitability of the supply chain when the acceptance of the online channel is higher or lower. Through theoretical analysis and numerical calculation, it is concluded that both the decentralized decision-making and the centralized decision-making network direct channel price are positively correlated with the service level and the acceptance of the direct channel. The relationship between the demand and the service level of the network direct marketing shows a "first increase, then decrease" relationship. There is always a positive correlation between the price of the retail channel and the service level, in decentralized decision-making, the price of retail channel is negatively related to the acceptance of direct marketing channel, in the case of centralized decision-making, the retail price has nothing to do with the acceptance of the direct marketing channel. When the manufacturer and retailer provide the same service, the retailer's demand has nothing to do with the level of service, when manufacturers and retailers offer differentiated services, the relationship between retailer demand and service level of retailers increases first and then decreases. For the whole supply chain, no matter the level of the acceptance of the network direct selling channel, the differential service provided by the manufacture and the retail can make the profit of the system reach the optimum.

The conclusions can provide the following management implications: for manufacturers, when the acceptance of the network direct channel is higher, they can raise the price of products to obtain higher profits, the service level of the network direct selling channel should subdivide the product category according to the customer's sensitivity to the service, and reasonably set up the different service level and its proportion. For retailers, product prices are bound to increase as service levels improve, in decentralized decision-making, the products with higher acceptance of direct marketing channels should increase their profits through appropriate price reductions. When manufacturers and retailers offer differentiated services, retailers and manufacturers need to increase sales by setting reasonable levels of services and their share. From the whole supply chain, manufacturers and retailers should provide differentiated services through the provision of differentiated services, alleviate the competition between channels and channel conflicts, cater to consumers' demand for personalized services, optimize the supply and demand balance of the entire system, and ultimately improve the profits of the entire system. As for the strategy should be decentralized or centralized decision-making, it need to be based on the level of the acceptance of the online direct channel. For consumers, when choosing a product, the service level and price should be integrated and balanced.

Acknowledgments This research was supported by the National Natural Science Foundation of China (NSFC)(Project Nos. 71571151 and 71371159).

References

1. Tsay, A., & Agrawal, N. (2004). Channel conflict and coordination in the E-commerce age. *Production and Operations Management, 13*(1), 93–110.
2. Guangshu, C., Zhang, Z. G., & Zhang, M. (2009). Game theoretical perspective on dual-channel supply chain completion with price discounts and pricing schemes. *International Journal of Production Economics, 117*(1), 80–96.
3. Hanjin, L., Xiaojun, F., & Hongmin, C. (2015). Study on dual channel pricing strategy under retailer price leadership structure. *Chinese Management Science*, (19), 91–98.
4. Guopeng, C., Congwu, X. (2016). Research on pricing strategy of dual channel retailers in "internet +" environment[J]. *Price Theory and Practice*, (5), 152–155.
5. Kaya, M. (2006). Essays in supply chain contracting: dual channel management with service competition and quality risk in outsourcing. California: Stanford University.
6. Dan, B., Xu, G. Y., & Liu, C. (2012). Pricing policies in a dual-channel supply chain with retail service. *International Journal of Production Economics, 139*(1), 312–320.
7. Feng, D., & Jiazhen, H. (2006). Research on the influence of service level on coordination strategy of dual channel supply chain. *Chinese Management Science, 11*(22), 485–490.
8. Yuangao, C., & Nan, L. (2011). Research on coordination of dual channel supply chain with differentiated products. *Journal of Management Engineering, 25*(2), 239–244.
9. Chiang, W. K., Chhajed, D., & Hess, J. D. (2003). Direct marketing, indirect profits: A strategic analysis of dual channel supply-chain design. *Management Science, 49*(1), 1–20.
10. Ruiliang, Y., & Zhi, P. (2009). Retail services and firm profit in a dual-channel market. *Journal of Retailing and Consumer Services, 16*(4), 306–314.

Chapter 3
The Quality Management of Food Supply Chain in Perspective of Food Safety

Gang Liu and Li Guo

Abstract Food safety is become more and more valued by the public and the government. Due to a large number of stakeholders of food safety, it is very necessary for business managers and researchers to improve the quality of food from the perspective of food supply chain. Consequently, the study introduces the concept of supply chain quality management (SCQM) into the food safety management, taking SCQM as an important practice tool for quality control in order to achieve food safety. The paper develops the supply chain quality management (SCQM) models for food safety, and takes Yili as a case to verify the model. We propose a Quality-SCM framework that including supplier selection, supplier participation and QM practice in order to implement the quality control for the whole food supply chain.

Keywords Food safety · Quality management · Supply chain management · Yili

3.1 Introduction

Food safety is now under public scrutiny. As the consumers become more conscious about health and more informed, consumers, regulators and other stakeholders are now paying more attention to food safety issues and food scandals, especially after the scandals of Melamine, and clenbuterol. Food safety is becoming an important issue for all stakeholders in food production, including farmers, processors, retailers et al. A typical food supply chain network is fairly complicated, from farmland to plate, achieving the food safety in the whole supply chain is not easy. Any risk or moral hazard in food supply chain will lead to the food safety issues. Application of supply chain quality management (SCQM) can make a powerful contribution towards achieving food safety.

G. Liu (✉) · L. Guo
School of Economy and Management, Tianjin Agricultural University, Tianjin, China
e-mail: liugang_tianjin@163.com

M. Tavana, S. Patnaik (eds.), *Recent Developments in Data Science and Business Analytics*, Springer Proceedings in Business and Economics,
https://doi.org/10.1007/978-3-319-72745-5_3

3.2 Literature Review

As any other supply chain, the food supply chain is also the network of cooperation between different orgnizations in order to provide qualified food to consumers and meet the demand of consumers [2]. As food safety control systems in processing phase, all input elements must be well understanded by the manufacturing process, and the quality of ingredients and raw materials must meet the requirement of further processing [8]. Kaynak and Hartley [3] stated that supplier quality management and customer focus are two important factors which influence quality-related performance. Their study identified eight QM practices in supply chain including quality management for supplier, quality training, customer focus, the design of product or service, quality data and reporting, process management, the emphasis on quality by top management, and employee relations. Roth et al. [6] developed a conceptual frame work to identify the key elements which are critical to ensure food safety in supply chain including transparency, trust, training, traceability, testability and time. The main reason for food scandals maybe the poor strategy of vertical control which is not available to manage the quality risk of food supply chain [1]. Manning et al. [4] developed a quality assurance model for food supply chain, the QA model including pre-requisite programme and benchmarking protocol. This study intends to analyze how to achieve food safety through supply chain quality management.

3.3 SCQM Models for Food Safety

The food supply chain involves many stakeholders who have different types, scale and target. It is very necessary for us to design a scientific method for food safety in supply chain. As the asymmetry information about the quality of food, whether farmers or processors have no motivation to control the quality of food. From the perspective of food safety, supply chain quality management is an important tool for integrating independent activities of various suppliers in order to assure the quality of materials in each stage of food supply chain. This paper gives an supply chain quality management model based on food safety (Fig 3.1), Fig. 3.1 presents the factors used to quality control in food supply chain, including supplier selection, supplier participation, and quality management practice.

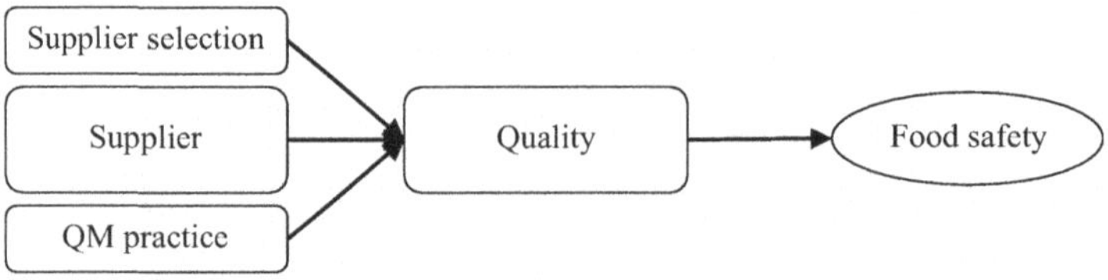

Fig. 3.1 Supply chain quality management for food safety

3.3.1 Supplier Selection

For core enterprise in supply chain, there have two different orientations to select cooperative partners, suppliers can be selected based on quality or cost. What differentiates food supply chain from other supply chains is that the quality and safety is the primary factor must be considered by all stakeholders in supply chain [7]. Many food safety scandals directly stem from supplier's moral hazard, such as melamine incident. Selecting suppliers based on food quality is very important for food supply chain. Suppliers should be selected based on their capacity to meet the safety standards of the core enterprise. It is very necessary for core enterprise to establish the system of evaluation and selection for suppliers. To select prospective suppliers, scientific criteria must be well established. Supplier criteria should be designed to measure some aspects of the supplier's business: production environment, technical ability, management approach, financial strength, especially quality control systems.

3.3.2 Supplier Participation

More and more food sectors have recognized advantages in mutual cooperation among partners to achieve the benefits of food safety and efficient food supply business [8]. Supplier participation is the process of bringing together different orgnizations, business and fuctions in food supply chain by management or information technology, to work jointly for the purpose of food safety. It is very important for core enterprise in food supply chain to strengthen communication with suppliers, strengthen specific investments for suppliers in the area of food quality improvement. Some management activities, such as proactive support for food processing or agricultural products cultivation, supplier participation in the food quality improvement teams, and supplier involvement in the product design are representative indicators of supplier participation in food supply chain management activities.

3.3.3 Quality Management Practice

Quality management (QM) practice include series of procedures and methods for food quality management. We focus on several QM practices, including hazard analysis critical control point (HACCP), traceability system, and good manufacturing practice (GMP). HACCP is a very important method for food quality control, which is based on the prevention of problems [5]. Traceability system is to trace and record the history of food by identification code. At practice level, traceability system involves what must be traced and how the trace-back should be carried out [9]. GMP is concerned with a series of mandatory standards which are related to raw materials, personnel, facilities, equipment, production process, packaging, transportation and quality control in order to produce safe food [10].

3.4 Case Study of Yili

In order to illustrate our SCQM for food safety, we employ a case study of the firm named as Yili. This firm, located in Inner Mongolia in China, is China's largest dairy producer which has the most complete productlines. Green industrial chain plan is the fair indicator of food safety and environmental sustainability for Yili, which requires cleaner production processes in the whole food supply chain. The paper discusses the measures that Yili has taken, describes how Yili control the quality of whole supply chain. Yili has made great efforts to improve the quality of food, especially launched the strategy of quality leadership which extended the quality control to every stakeholders.

3.4.1 Suppliers Management

The suppliers are required to pay high attention to the quality of raw milk, it is the basis for long-term and stable cooperation between Yili and suppliers. Yili carried out series of backward vertical integration in order to better control for the quality of raw milk. Yili have more than 2400 pastures, including self-constructive pastures, under-constructive pastures, and co-operative pastures, all of them are large-scale and concentrated-breeding pastures.

3.4.2 International Quality Control Standard

Yili and SGS, LRQA, Intertek have reached strategic cooperation in order to continuously updated the global quality and safety management systems. To further enhance the ability of quality risk control, most products of Yili including liquid milk, powdered milk, Yogurt et al. have passed the FSSC22000 food safety system certification. Yili is the first dairy enterprise in china that all products have passed this global food safety management standard system certification.

3.4.3 Process Quality Control System

In the process of quality control, Yili adds a lot of quality control indexes, and sets strict limits on each index. HACCP and Traceability system are critical components in Yili's process quality control system, which promotes the inspection for every link of food supply chain. In Yili's global quality management system, food supply chain can be traced back to every stage, from raw milk production and transportation to process, by whole course traceability system for quality. Early warning system for food safety has been established by Yili, which can analyze the risk from the whole industrial chain, and identify the key risk points and risk levels.

3.5 Discussion and Conclusions

The paper has discussed supply chain quality management for food safety, and developed the theoretical model. The SCQM model for food safety including supplier selection, supplier participation and QM practice in order to implement the quality control for the whole supply chain. The model deal with establishing effective guidelines for managing suppliers and implementing quality control. The key lesson that food supply chain seeking to develop the system of quality assurance can learn from this study is the need for a good strategy to support the supplier management and supply chain quality coordination.

Acknowledgements This research was financially supported by the quality improvement project for postgraduate education of Tianjin agricultural university in 2017(Project No: 2017YJG007), and also supported by the innovation and entrepreneurship training program for college students in Tianjin (Project No: 201510061029), China.

References

1. Chen, C., Zhang, J., & Delaurentis, T. (2014). Quality control in food supply chain management: An analytical model and case study of the adulterated milk incident in China. *International Journal of Production Economics, 152*(6), 188–199.
2. Christopher, M. (2005). *Logistics and supply chain management*. London: Prentice Hall.
3. Kaynak, H., & Hartley, J. L. (2008). A replication and extension of quality management into the supply chain. *Journal of Operations Management, 26*(4), 468–489.
4. Manning, L., Baines, R. N., & Chadd, S. A. (2006). Quality assurance models in the food supply chain. *British Food Journal, 108*(2), 91–104.
5. Mortimore, S. E. (2001). How to make HACCP really work in practice. *Food Control, 12*(4), 209–215.
6. Roth, A. V., Tsay, M. E., Pullman, M. E., & Gray, J. V. (2008). Unraveling the food supply chain: Strategic insights from China and the 2007 recalls. *Journal of Supply Chain Management, 44*(1), 22–39.
7. Salin, V. (1998). Information technology in agri-food supply chains. *International Food and Agribusiness Management Review, 1*(3), 329–334.
8. Stringer, M. F., & Hall, M. N. (2007). A generic model of the integrated food supply chain to aid the investigation of food safety breakdowns. *Food Control, 18*(7), 755–765.

9. Hobbs, J. E. (2004). Information asymmetry and the role of traceability systems. *Agribusiness, 20*(4), 397–415.
10. Konecka-Matyjek, E., Turlejska, H., Pelzner, U., & Szponar, L. (2005). Actual situation in the area of implementing quality assurance systems GMP, GHP and HACCP in Polish food production and processing plants. *Food Control, 16*(1), 1–9.

Chapter 4
Strategic Customer Behavior with Risk Preference for a Supply Chain Management Based on Double Channel

Wanjin Hou and Quansheng Lei

Abstract This paper consider a supply with a manufacturer, a retailer and an independent strategic customer. Strategic customer with risk preference can buy products through double channels which are physical channel of offline and internet channel of online. We use rational expectation equilibrium related knowledge, analyze the retailer best order quantity to reach profit maximization, analyze the best quantity of products delivered in the internet channel to reach profit maximize without risk preference. We find the retailer's maximum profit is decreasing in risk preference in the physical channel as well as manufacturer's in the internet channel. In order to ease the profit reduction, manufacturer and retailer should reduce price and quantity in their respective channel at the normal period. This is not consistent with the fact that price is raised in the normal period.

Keywords Double channel · Strategic customer · Risk preference · Supply Chain

4.1 Introduction

In recent years, the development of e-commerce has gave birth to double channel marking system with offline physical channel and online internet channel. More manufacturers begin to choose double channel to distribute products. On the one hand, manufacturer sells products though retailer to customer in physical channel. On the other hand, manufacturer sells products to customer directly in the internet channel. However, consumer can choose period of sells: normal sell period and discount sell period. In other words, consumers have strategic behavior. The consumer involved in our paper is strategic customer. There are some examples in reality, online double 11 activity of Taobao, discounting activities of physical store at the end of season. Based on this background, how does the manufacturer operate

This work was supported by the National Science Foundation of China Grant No. 71171022.

W. Hou (✉) · Q. Lei
Beijing University of Posts and Telecommunications, Beijing, P. R. China
e-mail: 18811603606@163.com

M. Tavana, S. Patnaik (eds.), *Recent Developments in Data Science and Business Analytics*, Springer Proceedings in Business and Economics,
https://doi.org/10.1007/978-3-319-72745-5_4

its own channel strategy and how retailer make price and quantity strategy according to strategic customer. When does customer choose to purchase and what's channel to choose to maximize utility. What's more, different strategic customers have different attitude towards risk.

Researchers do not combine double channel and strategic customer for research currently. Our model builds on the classic newsvendor model. We introduce strategic customers into double channel. Next, we discuss the impact of risk preference on strategic customer. We divide strategic customers into risk-seeker, risk-neutral and risk-averse. Determine the optimal order quantity for the retailer, the optimal quantity to put on the internet channel for manufacturer.

4.2 Literature Review

Alba J, et al. examine the implications of electronic shopping for consumers, retailers, and manufacturers n [1]. Ryan, et al. analyze impact of competition in an online marketplace system [2]. Lu H and Chen Y investigate the interaction between the capabilities of introducing the internet channel, the pricing strategies, and the channel structure [3]. Chen Y G, Liu N studies that the manufacturer sells the products through the traditional online retail channel and the online direct channel in order to alleviate the conflict [4]. Hahn P, et al. studies an effort toward modeling competition in the multiple channel environment from a strategic view point [5]. Chiang W Y K, et al. analyze strategic of double channel supply chain design [6]. Our paper attempts to fill the gap between double channel and strategic customer. Though our studies, we find that retailer's inventory and profit has unique maximum in physical channel. Manufacturer's delivery quantity in the internet channel has unique maximum, as well as profit. In addition, we find the retailer's best profit is decreasing in risk preference in the physical channel as well as manufacturer's in the internet channel.

4.3 Model of Strategic Customer

4.3.1 Model Setup and Rational Expectations Equilibrium

Our paper consider a supply with a manufacturer, a retailer and an independent strategic customer. Strategic customer can buy products through double channels. The manufacturer produces products though internet channel. The retailer operates physical channel which is offline. The manufacturer provides products to the retailer, the retailer sells the products to the strategic customer through the physical channel. At the same time, the strategic customer can choose the purchase periods: the normal sales period and discount sales period.

In our model, we use t demote period of sales, the normal sales period ($t = 1$) and discount sales period ($t = 2$). Each unit product cost is c and wholesale price of per product is w. However, each unit of product is valued by strategic customer are v in the physical channel and v_i in the manufactuer-owned channel. We use r and r_i denote strategic customer reservation price in physical channel and in the manufacturer-owned internet channel, which is their willingness to pay in the normal sales period. We assume the mass of market demand is $X(X \geq 0)$. We use p and Q denote the retail price and quality in the physical channel of the normal sales period. We use p^d denote discount price in the physical channel and assume $p^d < p < v$. Similarly, w_i and Q_i^M denote the retail price and quality in the manufacturer-owned internet channel, w_i^d denote discount price in the manufacturer-owned internet channel, $w_i^d < w_i < v_i$. This paper use $X(X \geq 0)$ denote random aggregate demand of market. We use $F(f)$ denote distribution function (destiny function) of demand ($\bar{F} = 1 - F$). We use Π^R to denote the retailer's payoffs in the physical channel and Π_i^M to denote the manufacturer's payoffs, which contains profit supplied to retailer and payoffs in the manufacturer-owned internet channel.

The Sequence of events is adopted as following. Frist, Manufacturers and retailers guessing strategic customer's reservation price is r and r_i in the normal period, retailer and manufacturer to determine the price of products for p and w_i. According to the principle of profit maximization, retailer determines its own order quantity Q, and manufacturer determines the sells quantity Q_i^M in the manufactuer-owned channel. The possibility of not selling at the end of normal period is $\boldsymbol{F}(Q)$, $\boldsymbol{F}(Q_i^M)$. Second, strategic customers guessing the possibility of obtaining products is φ, φ_i at the discount period ($t = 2$), and then to determine the reservation price of products r, r_i. Strategic customer can observe the retail price of products, but cannot observe the number of products. Finally, all remaining products are sold at the price p^d in the physical channel and w_i^d in the manufacturer-owned internet channel. The strategic customer's expected surplus in the physical channel, as in (4.1). The strategic customer's expected surplus in the manufacturer-owned internet channel, as in (4.2) (Table 4.1).

$$\max\{v - p, (v - p^d)\varphi\} \tag{4.1}$$

$$\max\{v_i - Q_i^M, (v_i - w_i^d)\varphi_i\} \tag{4.2}$$

The strategic customer choose buying or waiting in the normal period and will buy if and only if $(v - p) \geq (v - p^d)\varphi$ or $(v_i - Q_i^M) \geq (v_i - w_i^d)\varphi_i$ in the normal period. That is to say strategic customer's reservation price is r and r_i on physical channel and internet channel, respectively. We assume the proportion of strategic customer for the α in the Internet channel, then the physical channel on the demand

Table 4.1 Expected utility of strategic customer

Period	Physical channel	Internet channel
$t = 1$	$(v - p)$	$(v_i - w_i)$
$t = 2$	$(v - p^d)\varphi$	$(v_i - w_i^d)\varphi_i$

for $(1-\alpha)X$, the Internet channel demand for αX. Q^* is the optimal order quantity for the retailer.

$$r = v - (v - p^d)\varphi \tag{4.3}$$

$$r_i = v_i - (v_i - w_i^d)\varphi_i \tag{4.4}$$

$$\Pi^R = (p - p^d)E(\min\{(1-\alpha)X, Q\}) - (w - p^d)Q \tag{4.5}$$

$$\Pi_i^M = (w_i - w_i^d)E(\min\{\alpha X, Q_i^M\}) - (c - w_i^d)Q_i^M + (w - c)Q^* \tag{4.6}$$

Rational expectations equilibrium satisfied the following conditions:

(i) $r = v - \varphi(v - p^d), r_i = v_i - \varphi_i(v_i - w_i^d)$
(ii) $p^d = \xi_r, w_i^d = \xi_{r_i}$
(iii) $Q = \text{argmax}\ \Pi(Q, p), Q_i^M = \text{argmax}\ \Pi_i^M(Q_i^M, w_i)$
(iv) $p = r,\ w_i = r_i$
(v) $\varphi = \boldsymbol{F}(Q), \varphi_i = \boldsymbol{F}(Q_i^M)$

Among them (i–iii) represent retailer, manufacturer and strategic customer are choosing their own utility to maximize (iv, v) represent the expectation must be consistent with the results, and we believe that the predicted value is the same as the actual value. (iv) Note that retailer and manufacturer must correctly predict the strategic customer's reservation price in the physical channel and the internet channel. (v) Note that $\varphi(\varphi_i)$ must be consistent with the actual probability of obtaining products though offline (online).

4.3.2 Retailer's and Manufacturer's Strategies

We assume the strategic customer is risk-neutral. In the physical channel, strategic customer's utility in normal period is $(v - p)$. The probability of obtaining the product at the price p^d is φ in the discount period where $\varphi = \boldsymbol{F}(Q)$. Then, we consider the decision of the retailer, the retailer determines the quantity Q and the normal selling price $p$and the order quantity $Q(p) = \text{arg max}\ (\Pi^R)$. The profit of the retailer is Π^R. In the internet channel, strategic customer's utility in normal period is $(v_i - w_i)$. The probability of obtaining the product at the price w_i^d is φ_i in the discount period where $\varphi_i = \boldsymbol{F}(Q_i^M)$. Then, we consider the decision of manufacturer, the manufacturer determines the quantity Q_i^M and the normal selling price w_i and the order quantity $Q_i^M(w_i) = \text{argmax}(\Pi_i^M)$. The profit of the manufacturer is Π_i^M.

Proposition 4.1 *In the RE equilibrium. When the strategic customer has no risk preference characteristics, the retailer's order quantity Q and the profit Π^R have the only optimal value in the physical channel.*

Proof of proposition 4.1:

$$\Pi^R = (p - p^d)E(\min\{(1-\alpha)X, Q\}) - (w - p^d)Q$$

$$\Pi^R = (p - p^d)\left\{(1-\alpha)\left[\int_0^{\frac{Q}{1-\alpha}} xf(x)dx + \frac{Q}{1-\alpha}\int_{\frac{Q}{1-\alpha}}^{\infty} f(x)dx\right]\right\} - (w - p^d)Q$$

$$\because p = v - (v - p^d)F(Q) \ \therefore p - p^d = (v - p^d)\bar{F}(Q)$$

$$\Pi^R = (v - p^d)\bar{F}(Q)\left[Q - (1-\alpha)\int_0^{\frac{Q}{1-\alpha}} F(x)dx\right] - (w - p^d)Q$$

The derivative of Π^R at Q : $\dfrac{\partial \Pi^R}{\partial Q} = (v - p^d)\bar{F}(Q)\left(1 - F\left(\dfrac{Q}{1-\alpha}\right)\right) - (w - p^d)$

The second derivative of Π^R at Q : $\dfrac{\partial^2 \Pi^R}{\partial Q^2} = -(v - p^d)\bar{F}(Q)f\left(\dfrac{Q}{1-\alpha}\right)\dfrac{1}{1-\alpha} < 0$

So, there exist a unique optimal value Q^*.

Q^* satisfy $(v - p^d)\bar{F}(Q^*)\bar{F}\left(\dfrac{Q^*}{1-\alpha}\right) - (w - p^d) = 0$

Proposition 4.2 *In the RE equilibrium, when the strategic customer has not risk preference characteristics, manufacturer's quantity Q_i^M and the profit Π_i^M have the only optimal value in the physical channel.*

Proof of proposition 4.2 is similar to the proposition 4.1

4.4 Model of Strategic Customer with Risk Preference

In this section, we introduce risk preference. The risk-seeker choose the higher risk of the purchasing discount period when the two periods' expected utility are same. Risk-neutral keep a neutral attitude towards risk. The only criterion for risk-neutral is expected utility regardless of risk situation, because two periods with the same expected returns will give them the same effect. Risk-averse, when two periods' expected utility are same, tend to choose the normal period of low-risk, namely the customer hates the loss of losing opportunity to buy products. We use η note the strategic customer's risk preference. $\eta > 1$ is risk-seeker, $\eta = 1$ is risk-neutral, $0 < \eta < 1$ is risk-averse. In this situation, the customer's utility is shown in the following Table 4.2.

Table 4.2 Expected utility of strategic customer

Period	Physical channel	Internet channel
$t = 1$	$(v - p)^\eta$	$(v_i - w_i)^\eta$
$t = 2$	$(v - p^d)^\eta \varphi$	$(v_i - w_i^d)^\eta \varphi_i$

$$p = v - (v - p^d)(F(Q))^{\frac{1}{\eta}}, \quad w_i = v_i - (v_i - w_i^d)(F(Q_i^M))^{\frac{1}{\eta}}$$

Proposition 4.3 *If the strategic customer has the risk preference* η*, F(Q), Q, p,* Π^R*is decreasing in* η*, with* η *is increasing, the profit in the physical channel is decreasing.*

Proof of proposition 4.3: $\frac{\partial \Pi^R}{\partial Q} = (p - p^d)\bar{F}\left(\frac{Q}{1-\alpha}\right) - (w - p^d) = 0$

$$p = \frac{w - p^d}{1 - F\left(\frac{Q}{1-\alpha}\right)} + p^d, \quad p = v - (v - p^d)(F(Q))^{\frac{1}{\eta}}$$

Simultaneous above two formulas: $w - p^d = (v - p^d)\left[1 - (F(Q))^{\frac{1}{\eta}}\right]\left[1 - F\left(\frac{Q}{1-\alpha}\right)\right]$

$$\text{Let}: G(\eta, F(Q)) = (v - p^d)\left[1 - (F(Q))^{\frac{1}{\eta}}\right]\left[1 - F\left(\frac{Q}{1-\alpha}\right)\right] - (w - p^d)$$

$$\frac{\partial G}{\partial \eta} = (v - p^d)\left[1 - F\left(\frac{Q}{1-\alpha}\right)\right](F(Q))^{\frac{1}{\eta}}\frac{1}{\eta^2} lnF(Q)$$

$$\frac{\partial G}{\partial F(Q)} = (v - p^d)\left[-\frac{1}{\eta}\left[1 - F\left(\frac{Q}{1-\alpha}\right)\right](F(Q))^{\left(\frac{1}{\eta}-1\right)} - \left[1 - (F(Q))^{\frac{1}{\eta}}\right]\frac{dF\left(\frac{Q}{1-\alpha}\right)}{dF(Q)}\right]$$

$$\therefore \frac{\partial G}{\partial \eta} < 0, \ \frac{\partial G}{\partial F(Q)} < 0, \therefore \frac{\partial F(Q)}{\partial \eta} = -\frac{\left(\frac{\partial G}{\partial \eta}\right)}{\left[\frac{\partial G}{\partial F(Q)}\right]} < 0$$

That is $F(Q)$ is decreasing in η.

Hence there exist a positive correlation between $F(Q)$ and Q, Q is decreasing in η. $p = \frac{w-p^d}{1-F\left(\frac{Q}{1-\alpha}\right)} + p^d$, p is decreasing in η.

Next, we consider the relationship Π^R with η. Let $\eta_1 > \eta_2$, then $p_1 < p_2, Q_1 < Q_2$

$$\Pi_1^R = (p_1 - p^d)E(\min\{(1-\alpha)X, Q_1\}) - (w - p^d)Q_1$$

$$\Pi_1^R > (p_1 - \mathrm{p}^\mathrm{d})E(\min\{(1-\alpha)X, Q_2\}) - (w - p^d)Q_2$$

$$\Pi_1^R > (p_2 - \mathrm{p}^\mathrm{d})E(\min\{(1-\alpha)X, Q_2\}) - (w - p^d)Q_2$$

$(p_2 - p^d)E(\min\{(1-\alpha)X, Q_2\}) - (w - p^d)Q_2 = \Pi_2^R$. So, Π^R is decreasing in η.

Proposition 4.4 *If the strategic customer has the risk preference* $\eta, F(Q_i^M), Q_i^M, w_i$*,* Π_i^M *is decreasing in* η*, with* η *is increasing, the profit in the manufacturer-owned internet channel is decreasing.*

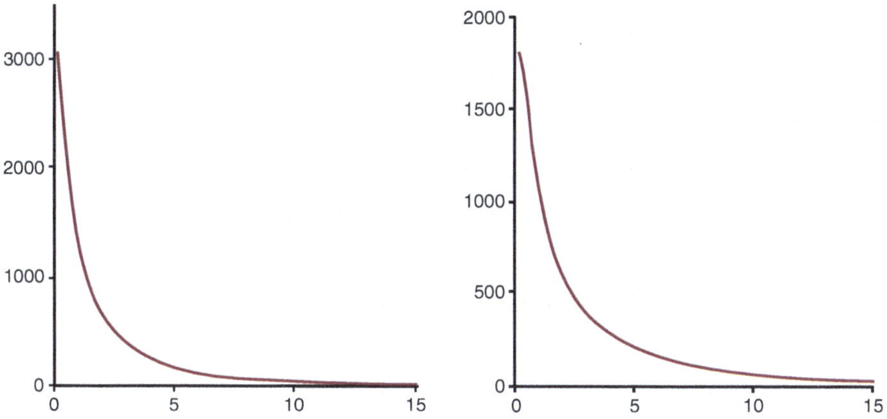

Fig. 4.1 Retailer 's profit function Π^R and Manufacturer's price function $\Pi_i^M - (w - c)Q^*$

Proof of proposition 4.4 is similar to the proposition 4.3. This result of propositions 4.3 and 4.4 state, both physical channel and internet channel, that reducing quantity of products which make product scarce can guide the strategic customer to purchase in the normal sales period ($t = 1$). What's more, reducing price of products can guide the strategic customer to purchase in the normal sales period ($t = 1$) and reducing the loss of profit Π^R in the physical channel. We use numerical examples to illustrate.

Let $c = 65,\ \ w = 70,\ \ v = 130,\ \ v_i = 115,\ \ p^d = 60,\ \ w_i = 95,\ \ w_i^d = 60,\ \ \alpha = 0.4$ and market demand X obey Uniform Distribution U[0,200]. Profit of physical channel (Fig. 4.1-left) are decreasing in risk preference in the physical channel, profit of internet channel (Fig. 4.1-right) are decreasing in risk preference in the physical channel.

Acknowledgments This paper attempts to fill the gap between double channel and strategic customer. We discuss the impact of strategic customer behavior under the double channel on retailer's and manufacturer's profit. The strategic customer's risk characteristic impacts inventory, price and profit under the double channel. Though our studies, we find that retailer's inventory and profit has unique maximum in physical channel. Manufacturer's delivery quantity in the internet channel has unique maximum, as well as profit. In addition, we find the retailer's maximum profit is decreasing in risk preference in the physical channel as well as manufacturer's in the internet channel.

In the future research, not only manufacturer can introduce internet channel, retailer can introduce retailer-owned internet channel. Follow research can consider the introduction of risk preference for manufacturer and retailer.

References

1. Alba, J., Lynch, J., Weitz, B., et al. (1997). Interactive home shopping: Consumer, retailer, and manufacturer incentives to participate in electronic marketplaces. *Journal of Marketing, 61*(3), 38–53.
2. Ryan, J. K., Sun, D., & Zhao, X. (2012). Competition and coordination in online marketplaces. *Production and Operations Management, 21*(6), 997–1014.
3. Lu, H., & Chen, Y. (2014). Strategic motive for introducing internet channels in a supply chain. *Production and Operations Management, 23*(1), 36–47.
4. Chen, Y. G., & Liu, N. (2011). Coordination of dual-channel supply chain under product differentiation. *Journal of Industrial Engineering & Engineering Management, 2*, 239–244.
5. Hahn, P., Grant, T., Constantino, M., et al. (1998). Mail versus mall: A strategic analysis of competition between direct marketers and conventional retailers. *Marketing Science, 17*(3), 181–195.
6. Chiang, W. Y. K., Chhajed, D., & Hess, J. D. (2003). Direct marketing, indirect profits: A strategic analysis of dual-channel supply-chain design. *Management Science, 49*(1), 1–20.

Chapter 5
Competition and Coordination in Single-Supplier Multiple-Retailer Supply Chain

Jiahang Du and Quansheng Lei

Abstract We analyze a dual-channel supply chain consisting of a single supplier and multiple retailers. The retailers make sales effort to attract the potential costumers. The retail channels compete in price at the end market. We find out the equilibrium solution in centralized case and decentralized case. Then we analyze the effect of total supply chain profit with respect to effort level. We show that the supply chain efficiency decreases and then weakly increases as the number of retailer increases. We also coordinate the supply chain with two-part tariff mechanism and find out the corresponding parameters. Finally, a numerical example is given to illustrate our model.

Keywords Dual-channel · Effort level · Multiple retailers · Two-part tariff mechanism

5.1 Introduction

Recent years, the development of electronic commerce has encouraged a growing number of suppliers to operate a direct channel. For example, Apple has both online shop and physical stores where the end customers can buy the products. In fact, a lot of suppliers choose to operate a direct channel as well as sell through traditional retail channel. For instance, one can buy Sony's product from both the official website and the retailers like Carrefour and Wal-Mart. This kind of supply chain is called dual-channel supply chain.

The relationship between supplier and retailer can be more complex in a dual-channel supply chain. The development of direct channel can decrease the market demand of retail channel. Therefore, many retailers make sales effort (advertising, store decoration, training of the sellers, special offers, etc.) to improve their quantity of sale.

This work was supported by the National Science Foundation of China Grant No. 71171022.

J. Du (✉) · Q. Lei
Beijing University of Posts and Telecommunications, Beijing, P. R. China
e-mail: 18310811979@163.com

M. Tavana, S. Patnaik (eds.), *Recent Developments in Data Science and Business Analytics*, Springer Proceedings in Business and Economics,
https://doi.org/10.1007/978-3-319-72745-5_5

The retailer's sales efforts will attract more prospective customers. What's more, the potential customer may get the information about this product from the retailers' advertisements and then buy online. Under this circumstance, when the retailers make the sales effort, the demand of direct channel increases.

In this paper, we consider a supplier as an exclusive source of a single product. The supplier sells his products through his own direct channel and multiple retailers. Retailers compete in price at the terminal market and make sales effort to increase the demands.

5.2 Literature Review

Coordination in dual-channel supply chains has been researched for a long time. Huang and Swaminathan [1] consider a supply chain consisting of a traditional channel and an Internet channel and find the optimal decision. Chen Zhang and Sun [2] analyze the supplier's price strategy in dual-channel supply chain and find that wholesale price contract can coordinate dual-channel supply chain. David and Adida [3] study competition and coordination in a supply chain in which a single supplier both operates a direct channel and sells its product through multiple differentiated retailers.

A related steam of literature study in the supply chain with multiple retailers. Most of the suppliers sell through more than one retail channel in real life. Sinha and Santanu [4] optimize the single-vender multi-buyer supply chain with discount pricing policy. Mateen and Chatterjee develop analytical models for various approaches through which a single supplier–multiple retailers system may be coordinated through vendor managed inventory.

We also interested in recent literatures focused on sales effort. Yao and Liu [5] show that there exists an optimal wholesale price under a different market structure that could be used to encourage the retailer to accommodate the additional e-tail channel. Li and Liu [6] show that two-part tariff contract can effectively coordinate the supply chain with price and sales effort dependent demand.

Our work differs from all the above mentioned literatures in that we assume the demand to be endogenous and consider a multi-retailer supply chain affected by the retailers' sales effort.

5.3 Model

5.3.1 Basic Setup

We consider a dual-channel supply chain that is composed of a single supplier [he] and N retailers [she]. The supplier can sell his products through both the direct channel and the retail channels. The retailers purchase the product from supplier at wholesale price w. The market demand is affected by sales effort and price. It is distributed in both retail channels and direct channel. Retailers compete in price with each other.

We assume that the demand is endogenous. The demand of direct channel is dependent on the direct sales prices, while the retailer's demand depends on its own price and other retailers' prices. And the retailers' sales effort can benefit both retail channel and direct channel. Therefore, we can get the linear function of retail channel's demand, D_i, and direct channel's demand, D_0:

$$D_i = \theta_i a + \lambda_r e_i - \beta p_i + \gamma \sum_{\substack{j=1 \\ j \neq i}}^{N} p_j \tag{5.1}$$

$$D_0 = \theta_0 a + \lambda_s \sum_{i=1}^{N} e_i - \beta p_0 \tag{5.2}$$

where a is basic demand of the market, θ_0 is the direct channel's share of basic demand, θ_i represents the *ith* retailer's share of basic demand, these coefficients should satisfy $\theta_0 + \sum_{i=1}^{N} \theta_i = 1$. We define e_i to describe the effort level of retailer i. The coefficient $\lambda_r \geq 0$ represents the effort sensitivity in each retail channel, and $\lambda_s \geq 0$ represents the effort cross-sensitivity in direct channel. The demands are affected by sale price of direct channel, p_0, and retail price, p_i, of retailer i. The coefficient $\beta \geq 0$ represents the price sensitivity of demand at each channel, and $\gamma \geq 0$ represents the price cross-sensitivity of demand at retail channel.

The retailers have to pay for her own sale effort. We define the cost of *ith* retailer's effort as $\frac{\eta e_i^2}{2}$, which should be strictly convex in e_i. Similarly, the supplier has a unit production cost of c. Therefore, we can get the supplier's profit, π_0, the profit of retailer i, π_i, and total supply chain profit, π_T.

$$\pi_0 = D_0(p_0 - c) + \sum_{i=1}^{N} D_i(w - c) \tag{5.3}$$

$$\pi_i = D_i(p_i - w) - \frac{\eta e_i^2}{2} \tag{5.4}$$

$$\pi_T = \pi_0 + \sum_{i=1}^{N} \pi_i = D_0(p_0 - c) + \sum_{i=1}^{N} \left[D_i(p_i - c) - \frac{\eta e_i^2}{2} \right] \tag{5.5}$$

We make the following assumptions to develop our model.

1. As a change of competitor's price affects the retailer's demand less than the change of its own sales price, we restrict that $\beta \geq \gamma$.
2. In reality, the demand in each channel should be positive, otherwise the channel's existence will become meaningless. Thus, we define $D_i \geq 0$, $D_0 \geq 0$.

3. The retailer's sales effort can increase its demand directly, but promote the sales of direct channel indirectly. The effort level sensitivity coefficient of demand in retail channel should be greater than that in direct channel, $\lambda_r > \lambda_s$.
4. In this paper, all retailers are symmetric for the convenience to develop our analytical results.

5.3.2 Equilibrium in Centralized Case

As a benchmark, we find out the equilibrium solution of centralized supply chain firstly. In this case, the participants make decision with the goal to maximize the profit of whole supply chain. We assume there is a single decision-maker to choose the prices.

Now, we differentiate partially the total supply chain profit with respect to p_0 and p_i and equal to zero. We can get: $\frac{\partial \pi_T}{\partial p_i} = 0$ and $\frac{\partial \pi_T}{\partial p_0} = 0$.

Solving these equations, we find out the extreme point of π_T. The Hessian matrix is obviously negative definite. Therefore, we can find a maximum of supply chain profit at the extreme point.

Proposition 5.1 *When* $2\beta - \gamma(N - 1) > 0$ *and* $\theta_i a + \lambda_r e_i + (\gamma(N - 1) - \beta)c > 0$*, the centralized supply chain prices and profit are given below.*

$$p_{ic}^* = \frac{\theta_i a + \lambda_r e_i + \beta c}{2\beta - \gamma(N-1)} \tag{5.6}$$

$$p_{0c}^* = \frac{\theta_0 a + N\lambda_s e_i + \beta c}{2\beta} \tag{5.7}$$

$$\pi_{Tc}^* = \frac{(\theta_0 a + N\lambda_s e_i - \beta c)^2}{4\beta} + \frac{N\beta[\theta_i a + \lambda_r e_i - (\beta - \gamma(N-1))c]^2}{[2\beta - \gamma(N-1)]^2} - \frac{N\eta e_i^2}{2} \tag{5.8}$$

Because the price and demand of retail channel need to be positive, the retail channel exists only when $2\beta - \gamma(N - 1) > 0$ and $\theta_i a + \lambda_r e_i + (\gamma(N - 1) - \beta)c > 0$. Otherwise, the suppler sell his products only through the direct channel. In these circumstances, we get $p_{0c}^* = \frac{a + \beta c}{2\beta}, D_{0c}^* = \frac{a - \beta c}{2}, \pi_{Tc}^* = \pi_{0c}^* = \frac{(a - \beta c)^2}{4\beta}$. In this paper, we concentrate on the situation that retailers participate the supply chain.

Proposition 5.2 *When the retailer's effort level increases, the total supply chain profit will increase firstly and then decrease.*

It turns out that total supply chain profit will increase to a maximum as the retailers begin making sales effort to attract customers. But when the retailers continue increasing their effort level, the total profit will decrease. This is because it's too costly as the effort level becomes too high. This is illustrated in Fig. 5.1.

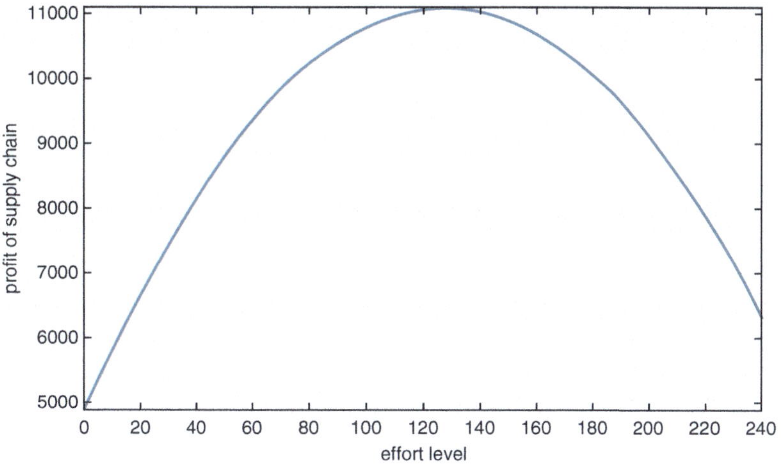

Fig. 5.1 The impacts of total supply chain profit with respect to retailer's effort level, where $N = 10$, $\theta_0 = 0.2$, $a = 50$, $\eta = 0.2$, $\beta = 3$, $\gamma = 0.5$, $\lambda_r = 0.2$, $\lambda_s = 0.1$, $c = 10$

5.3.3 Equilibrium in the Decentralized Case

Then, we consider the decentralized case where all the participants make decisions to maximize their own profit. We use Stackelberg game to find out the optimal pricing strategies. In a Stackelberg game, the leader makes his decision firstly, then the follower observe the leader's decision and decide her actions. We assume that the supplier is Stackelberg leader and the retailers are followers. The order in which events happen is as follows: (1) The supplier chooses his sale price, p_0, and wholesale price, w, anticipating the retailers' corresponding decision. (2) The retailers observe the supplier's decision. (3) The retailers decide their sale prices, p_i, and therefore the demand and profit can be calculated.

Similar to the centralized case, we find the equilibrium solution and reach the following conclusion.

Proposition 5.3 *The decentralized supply chain prices and profits are given below, when* $\frac{\theta_i a+\lambda_r e_i-(\beta-\gamma(N-1))c}{4\beta-3\gamma(N-1)} \geq 0$,

$$p_{id}^* = \frac{3\theta_i a + 3\lambda_r e_i + \beta c}{4\beta - 3\gamma(N-1)} \tag{5.9}$$

$$p_{0d}^* = \frac{\theta_0 a + N\lambda_s e_i + \beta c}{2\beta} \tag{5.10}$$

$$\mathrm{w}_d^* = \frac{(2\beta - \gamma(N-1))c + 2(\theta_i a + \lambda_r e_i)}{4\beta - 3\gamma(N-1)} \tag{5.11}$$

$$\pi_{0d}^{*} = \frac{(\theta_0 a + N\lambda_s e_i - \beta c)^2}{4\beta} + \frac{2N\beta[\theta_i a + \lambda_r e_i - (\beta - \gamma(N-1))c]^2}{[4\beta - 3\gamma(N-1)]^2} \tag{5.12}$$

$$\pi_{id}^{*} = \frac{\beta[\theta_i a + \lambda_r e_i - (\beta - \gamma(N-1))c]^2}{[4\beta - 3\gamma(N-1)]^2} - \frac{\eta e_i^2}{2} \tag{5.13}$$

$$\pi_{Td}^{*} = \frac{(\theta_0 a + N\lambda_s e_i - \beta c)^2}{4\beta} + \frac{3N\beta[\theta_i a + \lambda_r e_i - (\beta - \gamma(N-1))c]^2}{[4\beta - 3\gamma(N-1)]^2} - \frac{N\eta e_i^2}{2} \tag{5.14}$$

As is known to all, the supplier's wholesale price must be greater than unit cost of production, that is to say: $w_d^* - c \geq 0$. Therefore, the supplier operates retail channels only when the constraint (5.15) is binding. Otherwise, supplier sells his products only through direct channel.

$$\frac{\theta_i a + \lambda_r e_i - (\beta - \gamma(N-1))c}{4\beta - 3\gamma(N-1)} \geq 0 \tag{5.15}$$

We define the efficiency of total supply chain, μ, as the ratio of the supply chain profit in decentralized case and centralized case. In other words, we define $\mu = \pi_{Td}^{*}/\pi_{Tc}^{*}$. We can think the efficiency as a supply chain performance indicators. The impact of the number of retailers, N, on supply chain efficiency is shown in Fig. 5.2.

Proposition 5.4 *When the number of retailers increases, the efficiency decreases and then weakly increases.*

As the figure shows, when the quantity of retailers is quite small, reducing the quantity of retailers can enhance the efficiency. Otherwise, if the number of retailers is large enough, the efficiency increases as the number of retailers increases.

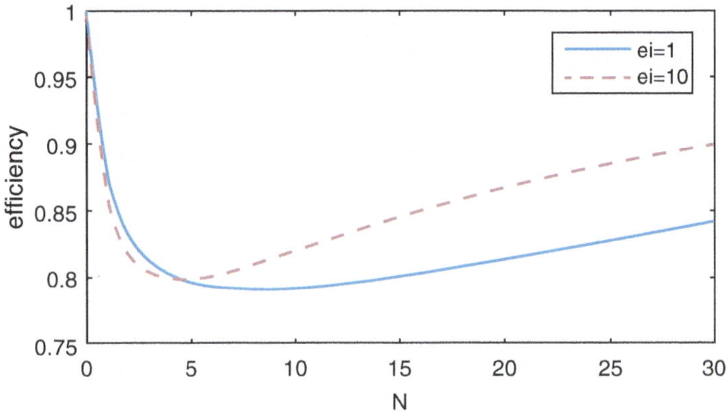

Fig. 5.2 The impacts of supply chain efficiency with respect to N where $\theta_0 = \theta_i = \frac{1}{N}$, $a = 100$, $\eta = 0.02$, $\beta = 3$, $\gamma = 0.01$, $\lambda_r = 2, \lambda_s = 0.5$, $c = 0$

5.3.4 Coordination with Two-Part Tariff Mechanism

For improving the efficiency of the system, we coordinate the supply chain with a two-part tariff mechanism. We regard the wholesale price as the per-unit charge and transfer payments T_i as the lump-sum fee. The participants profit are therefore given as:

$$\pi_0 = D_0(p_0 - c) + \sum_{i=1}^{N} [D_i(w - c) + T_i] \tag{5.16}$$

$$\pi_i = D_i(p_i - w) - \frac{\eta e_i^2}{2} - T_i \tag{5.17}$$

To find out the maximum point of the retailer's profit, we differentiate partially the profit of retailer with respect to p_i and equal to zero: $\frac{\partial \pi_i}{\partial p_i} = 0$.

$$p_{it} = \frac{\theta_i a + \lambda_r e_i + \beta w}{2\beta - \gamma(N-1)}$$

In theory, when the supply chain is coordinated perfectly, the participants should make the same decisions as that in the centralized case. That is to say, p_{it}^* should be equal to p_{ic}^*.

$$p_{it}^* = p_{ic}^* = \frac{\theta_i a + \lambda_r e_i + \beta c}{2\beta - \gamma(N-1)} \tag{5.18}$$

Therefore, we can get the optimal wholesale price, $w_t^* = \text{c}$. Then we solve the optimal problem of the supplier's profit and get:

$$p_{0t}^* = \frac{\theta_0 a + N\lambda_s e_i + \beta c}{2\beta} \tag{5.19}$$

Substituting the prices into demand and profit function, we get corresponding profits. The total supply chain profit should be equal to that in centralized case.

$$\pi_{0t}^* = \frac{(\theta_0 a + N\lambda_s e_i - \beta c)^2}{4\beta} + \sum_{i=1}^{N} T_i \tag{5.20}$$

$$\pi_{it}^* = \frac{\beta[\theta_i a + \lambda_r e_i - (\beta - \gamma(N-1))c]^2}{[2\beta - \gamma(N-1)]^2} - \frac{\eta e_i^2}{2} - T_i \tag{5.21}$$

Proposition 5.5 *When* $T_i \in [T_1, T_2]$*, there exists a two-part tariff mechanism that perfectly coordinates the supply chain.*

Suppliers and retailers will not accept this mechanism unless it can achieve Pareto optimization. Our two-part tariff mechanism need to confirm every participant can

Table 5.1 Numerical example where a = 100, N = 8, β = 3, γ = 0.05, $\lambda_r = 2$, $\lambda_s = 0.5$, $e_i = 10$

	Centralized case	Decentralized case	Two-part tariff
p_i	5.75	8.9	5.75
p_0	8.75	8.75	8.75
D_i	17.26	8.9	17.26
D_0	26.25	26.25	26.25
π_i	89.26	16.43	89.26-T_i
π_0	229.69	652.53	229.69 + $8T_i$
π_T	943.8	783.95	943.8

get no less interest than that in the decentralized case. Thus the transfer payments T_i must satisfy $\pi_{0t}^* - \pi_{0d}^* \geq 0$ and $\pi_{it}^* - \pi_{id}^* \geq 0$. We can get $T_i \in [T_1, T_2]$.

$$T_i \geq \frac{2\beta[\theta_i a + \lambda_r e_i - (\beta - \gamma(N-1))c]^2}{[4\beta - 3\gamma(N-1)]^2} = T_1$$

$$T_i \leq \frac{\beta[\theta_i a + \lambda_r e_i - (\beta - \gamma(N-1))c]^2}{[2\beta - \gamma(N-1)]^2[4\beta - 3\gamma(N-1)]^2} \times \left[12\beta^2 + 8\gamma^2(N-1)^2 - 20\beta\gamma(N-1)\right] = T_2$$

5.4 Numerical Example

To prove our conclusions numerically, we set the coefficients as shown in the following formulas: $a = 100$, $N = 8$, $\beta = 3$, $\gamma = 0.05$, $\lambda_r = 2$, $\lambda_s = 0.5$, $e_i = 10$ and $c = 0$ in appropriate units. For the analytical simplicity, we assume $\theta_0 = \theta_i = 1/N$. For this data set, we obtain the optimal results in each case as given in Table 5.1, where $T_i \in [52.86, 72.83]$. This example has proved the correctness of our conclusions.

From Table 5.1, we can see that total supply chain profit in centralized case is higher than that in decentralized case. The efficiency of supply chain $\mu = 83\%$. Coordinating the supply chain with two-part tariff mechanism, we can get the same result as centralized case.

5.5 Conclusions

We discuss the dual-channel supply chain that is composed of a single supplier and multiple retailers. The retailers compete in price at the terminal market and make sales effort to increase the demands.

In this paper, we make some specific contributions. Firstly, we resume the retailers are symmetric and get the optimal decision. Secondly, the effects of supply chain with respect to effort level and number of retailers are analyzed. Thirdly, we prove that a two-part tariff mechanism coordinates the supply chain.

In this article, we regard the prices as decision variables. In fact, the effort level can also be considered as decision variables. What's more, the supply chain should be tried to coordinate with other contracts and mechanisms. More research is needed in the future.

References

1. Huang, W., & Swaminathan, J. M. (2009). Introduction of a second channel: Implications for pricing and profits. *European Journal of Operational Research, 194*(1), 258–279.
2. Chen, J., Zhang, H., & Sun, Y. (2012). Implementing coordination contracts in a manufacturer Stackelberg dual-channel supply chain. *Omega, 40*(5), 571–583.
3. David, A., & Adida, E. (2015). Competition and coordination in a two-channel supply chain. *Production & Operations Management, 24*(8), 1358–1370.
4. Sinha, S. (2010). Single-vendor multi-buyer discount pricing model: An evolutionary computation-based approach. *International Journal of Operational Research, 8*(1), 1–20.
5. Yao, D. Q., & Liu, J. J. (2005). Competitive pricing of mixed retail and e-tail distribution channels. *Omega, 33*(3), 235–247.
6. Li, Q., & Liu, Z. (2015). Supply chain coordination via a two-part tariff contract with price and sales effort dependent demand. *Decision Science Letters, 4*, 27–34.

Chapter 6
Research on the Construction of Enterprise Brand Competitiveness Evaluation System Based on the Integration of SWOT and AHP Model

Mudan Tao and Zhi Li

Abstract The brand is very important to the enterprise, which determines the enterprise's core competitiveness. It is crucial to assess enterprise brand value based on a series of criteria, and this paper proposed enterprise brand evaluation system based on the SWOT model. It used the SWOT model to analysis and determine the brand competitive strengths, weaknesses, opportunities, threats and other factors of the enterprise brand, and then constructed the enterprise brand evaluation system by integrating SWOT Model and AHP calculation method.

Keywords Enterprise brand competitiveness · Evaluation system · SWOT · AHP

6.1 Introduction

With the progress of internationalization of the market is accelerating, the competition between the state and state, country and country has essentially changed into the competition between enterprises and enterprises, brands and brands. The brand determines the core competitiveness of enterprises. In 2016, The United States occupied 227 seats in the world's top 500 brands, continuing to maintain a brand of great power, while Chinese brands were selected for only 36, and even none was selected into top 100. It can be seen that Chinese enterprises must enhance their brand competitiveness in order to stand out in the international market. So, the evaluation and upgrading of brand competitiveness have become the important development issues for Chinese enterprises. The purpose of this paper is to construct the enterprise brand competitiveness evaluation system by using SWOT model and AHP calculation method, so as to provide a method for enterprise to evaluate its brand strategy and analysis the effect of the brand strategy.

M. Tao · Z. Li (✉)
Sichuan University, Chengdu, China
e-mail: zhil1090@163.com

M. Tavana, S. Patnaik (eds.), *Recent Developments in Data Science and Business Analytics*, Springer Proceedings in Business and Economics,
https://doi.org/10.1007/978-3-319-72745-5_6

6.2 Literature Review

Enterprise brand competitiveness is the unique ability of enterprise product brand which is different from other competitors, and is a form of enterprise comprehensive competitiveness's existence. The research on the brand competitiveness is earlier and the research theory is more mature overseas. There are some typical and excellent theory: The Aaker model proposed by David Aaker (1997), which is derived from five dimensions of brand loyalty, brand awareness, consumer perceived quality, brand associations and other brand equity, and ten indicators to measure the value of brand assets [1]. The global asset model proposed by Motamenti (1988) based on the customer potential, which is measure the brand competitiveness by three indicators of customer potential, competitive potential and global potential [2]. In recent years, domestic scholars also put forward the idea of different perspective for construct the enterprise brand competitiveness evaluation system. Peng Meng, Jinglong Cao, Zhaojun Liu (2014) comprehensively considered the process of brand nurturing and brand performance, and calculated the competitiveness of enterprise brand from three dimensions of "process" and three dimensions of "results" [3]. From the point of view of performance excellence model, Jinrong Xie (2015) constructed an evaluation model which focuses on process index, fusion analysis index and evaluation index [4]. Zhongyuan Ren (2017) bases on the brand value connotation, and constructed the brand evaluation index system from six aspects of brand positioning, brand personality, brand innovation, brand culture, brand communication and customer [5].

In summary, domestic and foreign scholars have rich research results on the evaluation of brand competitiveness. However, in the construction of brand evaluation system, most of the consideration is the enterprise's own brand development capabilities, as well as corporate finance and market performance. However, they overlooked the influence from competitors, competitive environment and other external factors. To solve this problem, this paper will base on a new perspective, and use the SWOT model to analyze and determine the factor index of brand competitive advantage, brand competitive disadvantage, brand competitive weaknesses, brand competitive opportunities and brand competitive threats for enterprise, and use the AHP calculation method to construct the enterprises brand competitiveness evaluation system.

6.3 Enterprise Brand Competitiveness Evaluation System

The SWOT model is a basic mediation model that includes internal factors and external factors [6]. Internal factors include internal strengths and internal weaknesses, and external factors include external opportunities and external threats. Applying SWOT analysis method to the construction of enterprise brand competitiveness evaluation system, and do a comprehensive, systematic and accurate analysis for brand competitiveness of enterprise, is helpful for enterprise to recognize its brand competitive advantages and disadvantages, and actively respond to external opportunities and challenges. It also provide a theoretical basis for enterprise to make scientific brand competiveness evaluate and make reasonable brand development strategy. Based on the above understanding, this paper constructed the brand competitiveness evaluation index system from 4 dimensions and 12 indicators. (As shown in Table 6.1).

Table 6.1 Enterprise brand competitiveness evaluation index system

Dimension	First grade index	Second grade index
Enterprise brand competitive advantage (S)	Brand basic capability	Brand strategy
		Technological innovation capability
		Leadership
		Human capital
		R & D investment
	Brand management capability	Market segmentation
		Brand personality
		Advertisement investment
		Brand image design
		Brand extension capability
	Brand profitability	Sales (operating) profit
		Profit margin
Enterprise brand competitive disadvantage (W)	Brand products (services) quality disadvantages	Unqualified rate of enterprise product
		Complaints rate of enterprise products (services)
	Brand asset inferiority	Enterprise asset scale
		Current asset ratio
		Asset liability ratio
	Brand talent inferiority	Number of brand management staff
		Brand management staff education level
Enterprise brand competitive opportunity (O)	Brand market capability	Market share
		Market coverage
	Brand sustainable development potential	Brand awareness
		Brand reputation
		Brand loyalty
	Brand relationship ability	Customer relationship
		Relationship between brand and supplier
		Relationship between brands and related parties

(continued)

Table 6.1 (continued)

Dimension	First grade index	Second grade index
Enterprise brand competitive threat (T)	Status of negative news	Amount of corporate negative news
		Enterprise negative news propagation path
		The spread speed of negative news
	Competitive power of existing competitors	Competitor capital scale
		Market coincidence degree
		Competitor brand reputation
		Competitor ‘s technological innovation ability
		Competitor's leadership
	Brand competence of new entrant	Market barrier
		New entrant's capital scale
		New entrant's brand image
		New entrant's leadership
		New entrant's technological innovation ability

Due to the enterprise brand competitive advantage and opportunity has positive effect to the brand competitiveness, but brand competitive disadvantage and threat has negative impact on brand competitiveness, this paper will divide the enterprise brand competitiveness evaluation index system into positive enterprise brand competitiveness index table (As shown in Table 6.2) and negative enterprise brand competitiveness index table (As shown in Table 6.3).

6.4 Evaluation of Enterprise Brand Competitiveness

When calculating the score of positive indexes and negative indexes, the influence of each segmentation index on brand competitiveness is different. So, it is necessary to make use of a certain mathematical method to weight its segmentation indicators. In this paper, the analytic hierarchy process (AHP method) is adopted to determine the weight of each index.

6.4.1 Determine Indexes' Weight by AHP

AHP is a good mathematical method to determine weight, its main steps are:

1. Marking the importance of indicators and the construction of judgment matrix

Table 6.2 Positive enterprise brand competitiveness index table

Dimension	Index	Dimension	Index
Brand basic capability (U1)	Brand strategy (U11)	Brand profitability (U3)	Sales (operating) profit (U31)
	Technological innovation capability (U12)		Profit margin (U32)
	Leadership (U13)	Brand market capability (U4)	Market share (U41)
	Human capital (U14)		Market coverage (U42)
	R & D investment (U15)	Brand sustainable development potential (U5)	Brand awareness (U51)
Brand management capability (U2)	Market segmentation (U21)		Brand reputation (U52)
	Brand personality (U22)		Brand loyalty (U53)
	Advertisement investment (U23)	Brand relationship ability (U6)	Customer relationship (U61)
	Brand image design (U24)		Relationship between brand and supplier (U62)
	Brand extension capability (U25)		Relationship between brands and related parties (U63)

Combining with the characteristics of product and industry to choose the method like expert marking method, questionnaire survey method, interview method to compare the importance of indicators. Then, constructing judgment matrix Uij with scoring results according to Satie 1–9 scaling table.

$$\mathbf{Uij} = \begin{bmatrix} u11 & u12 & u13 & u14 & \ldots \\ u21 & u22 & u23 & u24 & \ldots \\ u31 & u32 & u33 & u34 & \ldots \\ u41 & u42 & u43 & u44 & \ldots \\ \ldots & \ldots & \ldots & \ldots & \ldots \end{bmatrix} \tag{6.1}$$

According to the judgment matrix, calculate the eigenvector $\mathcal{W}$ corresponding to the maximum characteristic root λmax based on the weighted mean method of the matrix, which is the corresponding weight of each comparison factor.

2. Consistency test

First, Multiply and add each column in the judgment matrix with the corresponding weight matrix $\mathcal{W}$ which calculated in the first step;

Then, Divide the sum by the corresponding weight coefficient, and calculate the average numerical value A;

Finally, Consistency index: CI = A/N − 1 (N represents the number of comparison items). Carry on the consistency check for the weight distribution's rationality by the formula CR = CI/RI (the numerical value of RI as shown in Table 6.4). When

Table 6.3 Negative enterprise brand competitiveness index table

Dimension	Index	Dimension	Index
Brand products (services) quality disadvantages (u1)	Unqualified rate of enterprise product (u11)	Competitive power of existing competitors (u5)	Competitor capital scale (u51)
	Complaints rate of enterprise products (services) (u12)		Market coincidence degree (u52)
Brand asset inferiority (u2)	Enterprise asset scale (u21)		Competitor brand reputation (u53)
	Current asset ratio (u22)		Competitor's technological innovation ability (u54)
	Asset liability ratio (u23)		Competitor's leadership (u55)
Brand talent inferiority (u3)	Number of brand management staff (u31)	Brand competence of new entrant (u6)	Market barrier (u61)
	Brand management staff education level (u32)		New entrant's capital scale (u62)
Status of negative news (u4)	Amount of corporate negative news (u41)		New entrant's brand image (u63)
	Enterprise negative news propagation path (u42)		New entrant's leadership (u64)
	The spread speed of negative news (u43)		New entrant's technological innovation ability (u65)

Table 6.4 Matrix order n and its corresponding average random consistency index RI

n	1	2	3	4	5	6	7	8	9	10
RI	0	0	0.58	0.90	1.12	1.24	1.32	1.41	1.45	1.49

$CR < 0.1$, it can be considered that the judgment matrix and the weight distribution are reasonable. If there is inconsistency situation happens, return to the first step to re-start to compare the importance of each two indexes in the matrix design cycle.

6.4.2 The Index Weight of the Enterprise Brand Competitiveness Evaluation System

In this paper, we obtain the importance judgment matrix as the source of weight calculation by expert marking. The expert marking carried on three rounds, which lasted 3 weeks, and through the expert group using "1–9 scale method" to mark by comparing the importance of each two indexes. We get the following judgment matrix (take brand basic capacity as an example):

$$\mathbf{U1} = \begin{bmatrix} 1 & 3 & 4 & 5 & 3 \\ 1/3 & 1 & 3 & 4 & 2 \\ 1/4 & 1/3 & 1 & 2 & 1/2 \\ 1/5 & 1/4 & 1/2 & 1 & 1/5 \\ 1/3 & 1/2 & 2 & 5 & 1 \end{bmatrix} \tag{6.2}$$

By the sum calculate method, we get the corresponding weight results as follow: $\mathcal{W}_1 = \{0.4452, 0.2357, 0.0920, 0.0545, 0.1725\}$, $\lambda\text{max} = 5.1893$, consistency check result: CI = 0.0423. As we know, when n = 5, RI = 1.12, therefore, CR = CI/RI = 0.03776 < 0.1, indicates that the judgment matrix's consistency is acceptable. In the same way, we can get the positive and negative indexes' weight of enterprise brand competitiveness evaluation system, as shown in Tables 6.5 and 6.6.

Table 6.5 The positive index weight table

Dimension	Dimension weight	Index	Index weight	Dimension	Dimension weight	Index	Index weight
Brand basic capability (U1)	0.0523	U11	0.4452	Brand profitability (U3)	0.1744	U31	0.3333
		U12	0.2357			U32	0.6667
		U13	0.0920	Brand market capability (U4)	0.3926	U41	0.6667
		U14	0.0545			U42	0.3333
		U15	0.1725	Brand sustainable development potential (U5)	0.2238	U51	0.1047
Brand management capability (U2)	0.0702	U21	0.0641			U52	0.2583
		U22	0.2731			U53	0.6370
		U23	0.0797	Brand relationship ability (U6)	0.0867	U61	0.6483
		U24	0.1711			U62	0.2297
		U25	0.4120			U63	0.1220

Table 6.6 The negative index weight table

Dimension	Dimension weight	Index	Index weight	Dimension	Dimension weight	Index	Index weight
Brand products (services) quality disadvantages (u1)	0.1911	u11	0.5000	Competitive power of existing competitors (u5)	0.1210	u51	0.0554
		u12	0.5000			u52	0.0673
Brand asset inferiority (u2)	0.0781	u21	0.5499			u53	0.5026
		u22	0.2402			u54	0.2572
		u23	0.2098			u55	0.1175
Brand talent inferiority (u3)	0.0488	u31	0.2500	Brand competence of new entrant (u6)	0.0825	u61	0.0774
		u32	0.7500			u62	0.0822
Status of negative news (u4)	0.4785	u41	0.3332			u63	0.3832
		u42	0.0751			u64	0.1264
		u43	0.5917			u65	0.3309

6.5 Conclusion and Discussion

In this paper, we constructed an enterprise brand competitiveness evaluation system which is consist of 12 first grade indexes and 40 second grade indexes from the aspect of enterprise brand competitive advantage dimension, disadvantage dimension, opportunity dimension and threat dimension four dimension based on the SWOT model, and then divided it into positive index system and negative index system two sub-index system. When calculating enterprise brand competitiveness, we calculate the second grade index by expert marking first, and then deduce the first grade index by the weight relation, and calculate the positive index score and negative index score. Finally, the balance of the positive index score and the negative index score is the score of the enterprise brand competitiveness. This enterprise brand competitiveness system and its calculation method not only provided the basis for the evaluation of enterprise brand competitiveness, but also quantitatively calculated the competitive advantage, competitive disadvantage, competitive opportunity and competitive threat of the enterprise brand, moreover, it provided a method for enterprise to make the brand strategy and analysis the effect of the brand strategy, too.

Acknowledgments Thank the financial supports from the Sichuan Science and Technology Project: Study on the Role of Local Government in Regional Brand Formation from the Perspective of Supply-side Reform-Taking Sichuan Province as an Example (SC16TJ001).

References

1. Aaker. (1997). Dimension of brand personality [J]. *Journal of Marketing Research, 34*(8), 347–356.
2. Motameni, R., & Shahrokhi, M. (1988). Brand equity evaluation: A global perspective [J]. *Journal of Product and Brand Management, 7*(4), 275–290.
3. Peng, M., Jinglong, C., & Zhaojun, L. (2014). Research on the evaluation model of brand competitiveness of Chinese industrial enterprises [J]. *The Standard of Science, 1*, 53–57.
4. Xie, J. (2015). *Construction and Application Research of Brand Competitiveness Evaluation Model Based on Performance Excellence Mode [D]*. Zhejiang University.
5. Ren, Z. (2017). Research on brand value connotation and evaluation system [J]. *Business Economy, 4*(1), 10–13.
6. Ilari, K., Kokkonen, M., & Lohteenmoki-Smith, K. (2001). *SWOT-analysis as a basis for regional strategies [J]*. Nord Region Working Paper. 4. 28.

Chapter 7
Reflections on the Training Mode of E-Commerce Professionals with Improved Practical Exercises and Innovative Ability

Shunmin Wang

Abstract Practical teaching is a basic requirement for the cultivation of students' practice and innovation ability. At present, there are some shortcomings in the practice teaching of e-commerce in undergraduate colleges, and it is necessary to cultivate the educational idea of students' autonomous ability in the course of teaching practice. Teaching process should focus on the combination of basic theory and practice teaching, and teaching practice should focus on school-enterprise cooperation and joint training model. Practical teaching should focus on the development of e-commerce professional ability.

Keywords E-commerce · School-enterprise cooperation · Talent cultivation · The talented personnel of e-commerce

7.1 Introduction

In recent years, the e-commerce of our country has the developed rapidly, and e-commerce transaction volume has increased by 10 times in 10 years. At the mean-time, "Internet plus" has become the fastest developing areas, and has affected all areas of the industry, such as retail network, cross-border e-commerce, online services and Internet banking. Such industries have become new bright spot and the new engine of economic upgrade. Among them, the network retail sales increased by 145 times, which surpassed United States from 2013 and become the number one in the world.

S. Wang (✉)
School of Economics and Management, Neijiang Normal University, Neijian, Sichuan, China
e-mail: wsmg1012@163.com

M. Tavana, S. Patnaik (eds.), *Recent Developments in Data Science and Business Analytics*, Springer Proceedings in Business and Economics,
https://doi.org/10.1007/978-3-319-72745-5_7

7.1.1 The Need for Professional Training Practice Patterns of E-Commerce

Based on the rapid development of e-commerce, the teaching model of undergraduate colleges in the "plan 2011" emphasizes a new collaborative model with four type of cooperation, such as: schools cooperated with schools, schools cooperated with institution, schools cooperated with enterprises, and schools integrated in local place. Such new model would contribute to meet the demands of the industry, the regional development and the cultural innovation. Moreover, the education working conference in 2014 further discussed how to push forward vocational education innovation, promote schools and enterprises' combination, enrich the pattern of industry enterprises to participate in the vocational education, and encourage enterprises to participate in the vocational education thoroughly. The e-commerce major itself has the characteristics of strong practicality, and the application type undergraduate graduates mostly work in skillful and operational job. The feature of electronic commerce is that the electronic is the tool while the commerce is the essence. And the electronic commerce is characterized by fast, convenient, and open. Such feature is beneficial to create a favorable environment for e-commerce personnel training. The undergraduate colleges give full play to the function of education service industry. And e-commerce major operation modes also contribute to the economic development and transformation of economic growth mode.

7.1.2 The Status Quo of E-Commerce Talents

According to the "2016 annual Chinese e-commerce talent survey report", released by China Electronic Commerce Research Center (100EC.CN) and domestic professional talents service providers, which is regarded as industrial Internet think-tank, it shows that the proportion of enterprises with stable or large-scale recruitment needs have reached 85%. At the same time, the talents of business enterprise flow frequently, which leads to the growing rate of the loss of talent in the industry.

Further, the report shows that the surveyed enterprises in need of large-scale re-cruitment program account for 45% in year to come, small-scale recruitment program account for 40%, according to the actual development of the enterprise recruitment account for 12%, for the business restructuring, reducing staff size enterprises account for 3%.

Also, the report shows that the electricity supplier industry is a relatively strong field of practical requirements. What values most is the staff's learning ability, sense of responsibility, and executive ability. Data shows that enterprises staff's quality of learning ability account for 48%; responsibility and dedication account for 46%; pro-fessional knowledge and skills account for 38%; work execution capacity account for 38%. The learning speed is greater than the rate of market change, and it is the key to maintain competitiveness. Responsibility is the basic premise of occupation. Professional knowledge and skills are the occupation permission while execution is the biggest security of job performance.

In order to solve the problem of talent, on the one hand, it needs to strengthen the cooperation between schools and enterprises, and build a practical e-commerce training system. On the other hand, it needs for training of middle-level management personnel, so as to enhance the efficiency of enterprises and business performance.

7.2 The Reasons for the Change of Teaching Mode of Existing Talents Training in E-Commerce Major

The teaching of e-commerce major in application-oriented universities is bound to face new opportunities for change. In the past, the courses of e-commerce related courses in each undergraduate college were similar, and the teaching content was relatively backward. At the same time, the teaching technology was relatively backward, and the teaching methods were relatively old. Also, there was a lack of auxiliary materials for teaching materials which kept pace with the times. In terms of employment, many e-commerce enterprises are often reluctant to recruit graduate students, but more willing to recruit experienced talents. The reason is that e-commerce major practical ability is far lower than the requirements of enterprises. The problems existing in practice teaching limit the better and faster development of e-commerce.

7.2.1 Practice Teaching Is Not Enough

First of all, training courses cannot open. College still stay at the primary stage that students only understand the basic operation process of the e-commerce level. Students might read hastily and without thinking. In that way, they have to take a lot of time and effort on learning after the actual work. However, some of them still have a smattering of knowledge. Secondly, those existed training courses could not reach the goal of teaching outline. Because it is hard to find the better teaching methods, those students learning from the training of skills are also less effective. Though the related professional teachers' teaching level is high, yet they are lack of e-commerce experience. They can hardly fully mobilize the interest of students by only teaching knowledge on textbook. Also they can hardly effectively guide the students majoring in electronic commerce on practical activities. Focus on points of textbook while ignoring students' learning curriculum training, then the practice of electronic commerce course will not meet the conditions. Therefore, it is imperative to study the practice teaching model of e-commerce.

7.2.2 *Lack of Systematicness and Pertinence in Curriculum Setting*

Electronic commerce is a new discipline. Undergraduate colleges often simply add the information technology with commerce to build in e-commerce major courses. There have been open too many courses and students have to learn so much that a few years later they still don't know what they are be able to do. In the past 2 years, with the maturity of the e-commerce industry, the demand for talents has become more and more clear. Schools should conform to the current situation and study the demand for talents of various positions in e-commerce enterprises. In view of the characteristics of application-oriented undergraduate students, we should gradually dilute the technical color, and focus on the skills required for "website operation", "online shop operation", and strengthen the cultivation of professionalism and proficiency.

7.2.3 *The Learning Process of Students "Lack of Initiative and Creativity"*

At present, the theory and method of teaching practice in Institutions of higher learning are still in the stage of continuous improvement. The teaching model has not been completely changed from teacher-centered way, and students are dependent on the learning process. As a matter of fact, e-commerce is a highly practical subject, which requires a high level of students' ability. Once been employed, those students would be asked to be qualified with innovative Internet thinking. So during the teaching process, teachers should cultivate students' thinking model. And students should learn to think. Also they should be aware of that business model under the network environment keep changing, and it's useful to grasp the rules and methods of operation of electronic commerce. It's important to train students' initiative, tap and stimulate students' creativity. To do this, teachers have to overcome their inertia of the traditional teaching mode, and actively explore the practice of the new teaching method, undertake the task of leading and motivating the initiative and creativity of students, and excavate the students' potential.

7.3 Reflections on the Teaching Model of Practical Talents Training for E-Commerce Majors in Application-Oriented Universities

7.3.1 *Determination of Practical Teaching Objective of E-Commerce Specialty*

According to the standard and teaching status of e-commerce talents, we should begin with the goal of teaching e-commerce, and cultivate application-oriented talents. First of all, in the business skills aspect, train the students of computer information processing and writing ability, and make the students have the ability to skillfully use the computer information processing business, also be able to do online business communication, which can describe the business platform products etc.. Secondly, in the e-commerce platform, make students understand the network platform operating rules. Also teach them to be familiar with the operation of the platform and related laws and regulations, familiar with warehousing management, SKU and product mix knowledge. Finally, those students need to have good communication skills, negotiation ability, service consciousness, emergency response ability, honest and trustworthy, strong learning ability, sense of responsibility.

7.3.2 Innovation of Practical Teaching in E-Commerce

7.3.2.1 Innovation in School and Enterprise Cooperation, Joint Training Practice

The teaching reform of electronic commerce majors to serve local economic development as the goal, with skill training as the key point, and take the initiative to introduce the e-commerce industry to create a school for students to carry out training, practicing, combat, school productive training base. Set up the idea of "education serves the enterprise". Reform the existing teaching mode, arrange the teaching content according to the actual combat items of the enterprise, train the students' professional ability and job adaptability, and meet the needs of the enterprise. Set up a training base with enterprise culture atmosphere to improve students' professional accomplishment. Combining production with teaching is an important way of innovative vocational education. Promote students to obtain the actual business ability of e-commerce enterprises, both skilled in application and be able to operate practical platform for business activities. You can use the existing Amazon platform and other websites to train students' operating skills. Enterprises are willing to provide students with the company's products and network platform, willing to cooperate with the school to provide its business and related financial handling. Carry out research and establish an integrated model of education, learning and practice. Open the school enterprise cooperation and unite the practice teaching mode. In the enterprise actual position, those experienced ones should guide those green hands. With practical operation, experience comprehensive and true e-commerce platform trade, and grasp the whole process as soon as possible. In the meantime, the student makes full use of the Internet technology and business network platform, also saves the enterprise manpower. The enterprises can reduce the operation cost, trade cost and customer cost management. At the same time, build up efficient, interactive, technical, economic entrepreneurship practical teaching methods to help students realize zero distance employment.

7.3.2.2 The Innovation of Teaching Process Emphasizes the Combination of Basic Theory and Practice Teaching

The construction of e-commerce practice teaching system is reasonable and effective. It is essential to integrate the theory of e-commerce professional teaching content. In the electronic commerce theory and operation method, train the students to solve business problems in the way of thinking and ability. Set up e-commerce experimental training platform. Firstly, integrate all kinds of existing resources, and build a practical teaching platform covering all aspects of e-commerce business curriculum knowledge and skills training. The specific approach is to develop network quality courses, purchase e-commerce laboratory software, build electricity providers entrepreneurship training room, improve e-commerce case teaching resource library. Secondly, education should be oriented in quality, and innovate practice teaching methods. Most of the existing teachers have specialized theoretical teaching, and few of them are engaged in the practice of the enterprise, which lead to lack the actual work experience of e-commerce. The construction of teaching staff should actively consider the selection of "double teacher" and work with the enterprise elite to train students' skills and improve their practical ability. Provide the best choice for enterprise recruitment talents.

7.3.2.3 Training to Pay Attention to the Practical Ability of E-Commerce Major

Because schools often lack the actual environment of e-commerce enterprises, and schools to this part of the training course content development is often due to limited power, the electronic commerce specialized practice teaching course can only stay in the scattered skill points, and it is difficult to form a system. Therefore, the cognition practice, the school curriculum practice, the post-practice, the graduation practice should run through the entire process of education and teaching, and form the practical teaching ability training together. Both internship and graduation practice require students to go deeper into the enterprise. After the ideological and moral education for students, professional knowledge and job skills training and assessment, students should practice in the enterprise through collective or decentralized mode. And also take part in e-commerce activities, learn to accept the real test, so that to lay the foundation for the accumulation of experience. The graduation internship report is based on theoretical study and practical teaching. The report might state students' practice and experience specific problems encountered in the process of elaboration. Students should be under the guidance of teachers in some specific problems and describe their own point of view, select the practical value perspective, try to make full use of professional theory to solve practical problems.

7.4 Summary

In summary, to improve the basic undergraduate colleges and universities autonomy, flexibility and sensitivity of the employment market, it need government guidance, social support, school work. And as in the past, paying attention to the cultivation of e-commerce courses practical teaching, establishing employment oriented training mode and cultivating position real operation ability, are beneficial to the cultivation and development of e-commerce professionals.

Acknowledgments This research was financially supported by the subject of teaching reform project of NeiJiang Normal University (JG201618-333).

References

1. Prahalad, C. K., & Hamel, G. (2005). The core competence of corporation. *Harvard Business Review, 69*(3), 275–292.
2. Chunhua, J., Dongsheng, L., & Yi, Z. (2011). Exploring the training system of classified talents in E-commerce. *Higher Education Research in China, 3*, 87–89.
3. Wu, Q. (2015). Misunderstanding and professional development of Chinese e-commerce personnel training. *University Teaching, China, 2*, 37–41.
4. Lin, B., & Xiaomei, S. (2016). Research on the construction of applied e-commerce course system. *Education and Teaching Forum, 2*, 217–218.
5. Zhiliang, X., Xibo, W., & Zhang, C. E. (2013). Research on training mode of undergraduate e-commerce talents in application-oriented universities. *Education and Profession, 7*, 110–112.
6. Ping, L. (2014). Construction of innovative practical personnel training system for e-commerce majors. *Laboratory Research and Exploration, 3*, 255–258.

Chapter 8
Research on Evaluating Marketing Ability of Traditional Chinese Medicine Enterprises in Gansu Province

Lixin Yun, Zhonghua Luo, and Mingwei Wu

Abstract Marketing ability is the bottleneck of the development of Chinese medicine enterprises in Gansu province. Based on the method of improved gray correlation, with construction of the evaluation index system of marketing ability of Chinese medicine enterprises, this paper analyzes the marketing ability of Chinese medicine enterprises in Gansu province and shows the evaluation process of marketing ability. And then some countermeasures and suggestions are proposed, which can provide reference and guidance in order to promote the marketing ability of Chinese medicine enterprises in Gansu province and promote the rapid development of Gansu traditional Chinese medicine industry.

Keywords Chinese medicine enterprises · Evaluating marketing ability · Gray correlation analysis · Empirical research

8.1 Introduction

Marketing refers to the process of business activities and sales activities for the market [1]. It can bring economic value for customers, partners and society in the process of creating, communicating, disseminating and exchanging products. Not only can marketing drag customers back to the enterprises and open up the market, but also it is the primary way to obtain funds and profits. So marketing is necessary for any enterprise. And enterprise competitiveness is generally determined by marketing ability, as well as the profitability and the pace of development. Since Gansu province is a large one in the field of Chinese medicine, there are more than 3000 different Chinese medicine enterprises at all levels. As a pillar industry in Gansu province, marketing ability of Chinese medicine enterprise not only

L. Yun · Z. Luo (✉) · M. Wu
School of Economics and Business Management, Gansu University of Chinese Medicine, Lanzhou, China
e-mail: luozhpsx@sohu.com

M. Tavana, S. Patnaik (eds.), *Recent Developments in Data Science and Business Analytics*, Springer Proceedings in Business and Economics,
https://doi.org/10.1007/978-3-319-72745-5_8

determines their own development, but also affects the economic and social development of the whole province. However, marketing ability as a whole is still at a low level due to the degree of development of Chinese medicine industry, economic and social environment and the level of management. Products sold mainly in the local market, market development slow, the market size small, the development of traditional Chinese medicine industry is seriously affected and restricted. Therefore, it is very important to guide the management of Chinese medicine enterprises in Gansu province by selecting an appropriate method to evaluate their marketing ability, which help to analyze the marketing ability and also help to find out their problems in marketing activities.

8.2 Methods of Gray Correlation to Evaluate Marketing Ability

The method of gray correlation is a systematic analysis one, which can find the relevance between the sequence of the characteristic data reflecting the behavior of the system and the data sequence of the effective factor affecting the system behavior. With the existing information, the correlation is calculated by a certain amount of data processing to determine the main factors that affect the behavior of the system, and to find out the differences among these factors. All about the factors are quantitatively described, to obtain an evaluation method [2–4]. In the process of evaluating marketing ability of Chinese medicine enterprises, the status and level of marketing ability are analyzed by describing the similarity between marketing ability of each enterprise and the ideal scheme. The formula to calculate the gray correlation is [5, 6]:

$$r\left(X_j X_0\right) = \frac{1}{n}\sum_{k=1}^{n} \xi_i(k) \tag{8.1}$$

where

$$\xi_i(k) = \frac{\min\limits_{j}\min\limits_{k}\left|x_0(k) - x_j(k)\right| + \rho\max\limits_{j}\max\limits_{k}\left|x_0(k) - x_j(k)\right|}{\left|x_0(k) - x_i(k)\right| + \rho\max\limits_{j}\max\limits_{k}\left|x_0(k) - x_j(k)\right|} \tag{8.2}$$

$\rho \in (0, 1)$ is resolution coefficient, and $X_0 = \{x_0(1), x_0(2), \cdots, x_0(n)\}$ is reference series, $X_j = \{x_j(1), x_j(2), \cdots, x_j(n)\}$ is the j-th comparison sequence.

8.3 Constructing the Evaluation Index System of Marketing Ability of Chinese Medicine Enterprises

It is very important to use the method of gray correlation to evaluate the marketing ability of Chinese medicine enterprises and to design a comprehensive evaluation index system. According to the research and summary of different scholars' evaluation system about marketing ability, this study considers that the evaluation of marketing ability of Chinese medicine enterprises should include the following contents.

8.3.1 Marketing Concept

The marketing concept of Chinese medicine enterprises plays a guiding and dominating role in the marketing. The marketing ability is obviously influenced by the features of times and advanced from the marketing concept. With the information era coming, the market is changing rapidly, the demand for Chinese medicine products is also rapidly changing, which forces the enterprises to improve their marketing concepts timely, and to keep up with the pace of the times, therefore advanced marketing concept is always used to guide the specific practice of Chinese medicine marketing [7, 8].

8.3.2 Effective Utilization of Marketing Information

Since effective marketing must be based on comprehensive and correct marketing information, in the marketing activities of Chinese medicine enterprises, only when the correct information become the guide to find out timely the needs of customers and the changes from market, can the scientific marketing strategies be made. And only marketing activities meet the needs of the market development, can product sales be improved and can competitive advantage be obtained. In today's information explosion era, it is the key for Chinese medicine enterprises to timely reorganize and summarize the relevant from the massive information to find out the key information to grasp the needs of customers and the changes from markets, and then to find their opportunities. Therefore, the effective utilization of marketing information becomes important in evaluating the marketing ability of Chinese medicine enterprises.

8.3.3 Marketing Strategy

Marketing strategy means that the Chinese medicine enterprises make their marketing plan, according to the characteristics of their own products, the share in markets and the position in the entire industrial structure. Scientific marketing strategy should make the advantages and competitiveness clear, and highlight the key tasks in marketing activities. Marketing strategy plays a directional role in the marketing of Chinese medicine enterprises, and the irrational marketing strategy will inevitably bring low level of marketing ability, so the quality of marketing strategy is very essential and important in evaluating marketing ability.

8.3.4 Marketing Implementation

In order to achieve marketing goals, marketing strategy and marketing plan must be effectively implemented. In the marketing activities, only when all about marketing programs are implemented, can the definite effect be gotten. The level of marketing execution ability, determining the degree of realization of marketing goals, and also determining the marketing benefit, becomes a key indicator in evaluating marketing ability of traditional Chinese medicine.

8.3.5 Marketing Benefit

Marketing benefit refers to the ratio of profits from markets and marketing costs. The main purpose of marketing is to sell more products, maximize products to market, expand market shares and get the profits. Since marketing needs costs, only when the profits are bigger than the costs, are the marketing activities effective. The better the marketing benefit, the more powerful the marketing ability of Chinese medicine enterprises. Marketing benefit is the core indicator to evaluate the marketing ability of Chinese medicine enterprises.

As the above analysis, evaluating marketing ability of Chinese medicine enterprises mainly involves five aspects: marketing concept (M1), utilization of marketing information (M2), marketing strategy (M3), marketing implementation (M4), and marketing benefit (M5). The index system of marketing evaluation of Chinese medicine enterprise is made of these five factors, which is shown as Fig. 8.1, and EISOMC is the abbreviation of Evaluation Index System of Marketing ability.

Fig. 8.1 Evaluation system of marketing ability of Chinese medicine enterprises

8.4 An Empirical Study on Marketing Ability of Chinese Medicine Enterprises in Gansu Province

8.4.1 Data Sources

In this study, ten Chinese medicine brand enterprises in Gansu province are selected as the research objects, and the marketing ability of ten enterprises is evaluated by hiring the marketing experts, and then the improved gray relational model is used to carry out their marketing ability evaluation. The experts give their scores based on the centesimal system, with the evaluation system of Chinese medicine enterprise. In order to ensure the fair and objective, anonymous evaluation is adopted and the average from group experts becomes the final score. By statistics and calculating the scores from the experts, and ultimately the evaluation information about marketing indicators of the ten Chinese medicine enterprise is as follows:

$$X = \begin{bmatrix} 19.3332 & 49.6562 & 72.7109 & 13.6520 & 79.4819 \\ 68.2199 & 90.0011 & 30.9289 & 21.1760 & 95.6842 \\ 30.2764 & 82.1629 & 83.8496 & 89.3898 & 52.2590 \\ 54.1674 & 64.4910 & 56.8072 & 19.9136 & 88.0142 \\ 15.0874 & 81.7973 & 37.0415 & 29.8723 & 17.2956 \\ 69.7898 & 66.0228 & 70.2740 & 66.1443 & 97.9747 \\ 37.8371 & 34.1970 & 54.6569 & 28.4409 & 27.1447 \\ 86.0013 & 28.9698 & 44.4878 & 46.9224 & 25.2328 \\ 85.3654 & 34.1194 & 69.4563 & 726.478 & 87.5742 \\ 59.3559 & 53.4080 & 62.1308 & 98.8336 & 73.7311 \end{bmatrix}$$

8.4.2 Evaluation of Marketing Ability of Chinese Medicine Enterprises in Gansu Based on Improved Gray Relational Degree

It can be seen from the formula of grey correlation that although the grey correlation can describe marketing ability of each enterprise and the evaluation value is also similar to the ideal scheme, the different resolution parameter $\rho \in (0, 1)$ will get different gray correlation. It may appear different ρ, grey correlation degree appears reverse to disturb the evaluation results. And meanwhile, since the gray correlation does not satisfy the symmetry, discrepancy appears that the sequence of number should satisfy the characteristic of the symmetry. In order to avoid such defects, the improved method to calculate the grey correlation is proposed.

Firstly, the evaluation values are calculated about ideal schema of enterprise marketing ability. The general ideal schema should be the optimal of each evaluation index, and the five evaluation indicators are effective indicators, should be the bigger the better, therefore, when the improved evaluation method is used, the maximum value of each evaluation index is selected to form a reference sequence:

$$X_0 = \{86.0019 \;\; 89.9764 \;\; 83.8498 \;\; 97.9758 \;\; 98.8316\}$$

Gray correlation degree is used to describe the similarity of marketing ability of traditional Chinese medicine enterprises and the ideal schema, which means that the geometric characteristic must be similar. And then it is to require that the slope be as similar as possible. For discrete data, the difference can be seen as the slope of discrete data, therefore, when calculating the gray correlation, the difference of the ideal will be firstly calculated, and the difference is:

$$y_0(1) = 0, y_0(i) = x_0(i-1)(i = 2, 3, \cdots, n) \tag{8.3}$$

So the difference is:

$$Y_0 = \{\, 03.8986 \quad -6.0976 \quad 14.1248 \quad 0.8565 \,\}$$

With the same principle, each evaluation vector is differentiated for marketing ability of each Chinese medicine enterprise, for the ith enterprise, the difference formula is:

$$y_i(1) = 0, y_i(j) = x_i(j-1) - x_i(j-1) \tag{8.4}$$

Finally, the matrix is gotten as follows by differentiating the evaluation value of all the enterprise:

$$Y = \begin{bmatrix} 0 & -35.979 & -17.2654 & -84.3931 & -4.3775 \\ 0 & 3.9849 & -58.9882 & -76.7992 & 11.7952 \\ 0 & -3.8581 & -6.0976 & -8.6017 & -31.6011 \\ 0 & -21.5219 & -33.1595 & -77.9865 & 4.1492 \\ 0 & -4.1929 & -52.9292 & -68.0926 & -66.5537 \\ 0 & -20.0126 & -19.6981 & -31.8410 & 14.1245 \\ 0 & -51.7951 & -35.3209 & -69.5324 & -56.7054 \\ 0 & -57.0305 & -45.4906 & -50.9819 & -58.5993 \\ 0 & -51.8818 & -20.5201 & -71.7959 & 3.7274 \\ 0 & -33.0247 & -28.1688 & 0.9562 & -10.1206 \end{bmatrix}$$

In order to overcome the shortcomings of calculating the original gray correlation degree, with the statistical knowledge, the mean variance is introduced, so it is necessary to calculate the mean value and mean variance. The mean of the ideal scheme is:

$$\bar{X}_0 = \frac{1}{5}\sum_{i=1}^{5} x_0(j) = 90.9854 \tag{8.5}$$

$$\sigma_{x0} = \sqrt{\frac{1}{4}\sum_{j=1}^{5}\left(x_0(j) - \bar{X}_0\right)^2} = 6.7905 \tag{8.6}$$

Similarly, the mean and mean variance of evaluation vector of marketing capacity for each enterprise is as follows respectively:

$$\bar{X} = \begin{Bmatrix} 46.9786 & 61.2174 & 67.6076 & 56.6785 & 36.4201 & 73.9889 & 36.4652 \\ 46.3834 & 60.6286 & 69.5012 \end{Bmatrix}$$

$$\sigma_X = \begin{Bmatrix} 30.00082 & 33.8565 & 25.3927 & 24.4994 & 27.0324 & 13.5249 & 11.0595 \\ 24.1038 & 28.6572 & 18.0125 \end{Bmatrix}$$

After the mean and mean variances of the evaluation vector are determined, the correlation coefficient of evaluation vector between the ith enterprise x_i and the ideal x_O is,

$$\xi\big(x_0(j), x_j(j)\big) = sign(y_0(j) \times y_i(j))\frac{1}{1 + \left|\left|\frac{y_0(j)}{\sigma_{x0}}\right| - \left|\frac{y_i(j)}{\sigma_{xi}}\right|\right|} \tag{8.7}$$

Where $x_i = \{x_i(1), x_i(2), \cdots, x_i(5)\}$, $x_0 = \{x_0(1), x_0(2), \cdots, x_0(5)\}$, $j = 2, 3, \ldots, n$, and $sign\ (x)$ is the sign function

$$sign(x) = \left\{ \begin{array}{c} 1(x > 0) \\ 0(x = 0) \\ -1(x < 0) \end{array} \right\} \tag{8.8}$$

With the sample data, the correlation coefficient is:

$$\xi = \begin{bmatrix} 0 & -0.4802 & 0.7122 & -0.2734 & -0.8631 \\ 0 & 1.0083 & 0.4126 & -0.2985 & 0.9655 \\ 0 & -0.9425 & 0.9308 & -0.7602 & -0.6162 \\ 0 & -0.5610 & 0.4507 & -0.2413 & 0.8795 \\ 0 & -0.9349 & 0.3735 & -0.2914 & -0.3214 \\ 0 & -0.4225 & 0.4276 & -0.3125 & 0.5841 \\ 0 & -0.1835 & 0.2476 & -0.1414 & -0.1810 \\ 0 & -0.3071 & 0.3729 & -0.3222 & -0.3212 \\ 0 & -0.3681 & 0.6295 & -0.2884 & 0.5229 \\ 0 & -0.3672 & 0.4142 & 0.9724 & -0.8021 \end{bmatrix}$$

Finally, calculating the average of the correlation coefficient of each TCM enterprise, the gray correlation is acquired. And the gray correlation between the ith enterprise and the ideal schema is:

$$\eta_i = \frac{1}{4} \sum_{j=2}^{5} \xi(x_0(j), x_j(j)) \tag{8.9}$$

With the sample data, the gray correlation of ten enterprises is as shown in Table 8.1.

The average of gray correlation of each indicator for ten enterprises is as shown in Table 8.2.

Generally, the greater the gray correlation is, the stronger the marketing ability of the enterprises is. With the comparison of the results listed in Table 8.1, the marketing ability of ten companies is sorted as:

$$C2 > C9 > C4 > C6 > C10 > C7 > C8 > C1 > C5 > C3$$

Table 8.1 Improved gray correlation between ten enterprises and the ideal

Enterprise	C1	C2	C3	C4	C5
Correlation	−0.905	1.793	−1.299	0.541	−1.173
Enterprise	C6	C7	C8	C9	C10
Correlation	0.284	−0.240	−0.583	0.829	0.226

Table 8.2 Average of gray correlation of each indicator

Index	M1	M2	M3	M4	M5
Average of correlation	0	−0.35599	0.49725	−0.23589	−0.01542

Shown as the sort sequence, the final evaluation of the marketing ability of ten enterprises is that the enterprise *C*2 is the best and *C*3 is the worst.

As can be seen from Table 8.2, the weakest three factors about the marketing capacity in the ten enterprises are marketing executive ability, utilization efficiency of the marketing information, marketing benefit.

8.5 Discussion and Suggestion

Marketing ability plays a very important role for maintaining and enhancing the competitiveness of Chinese medicine enterprises. The expansion of the market and many activities of marketing in Chinese medicine enterprises become very necessary for the rapid development of Chinese medicine industry in Gansu province. Therefore, both the government and Chinese medicine enterprises should strengthen the marketing evaluation to increase constantly the marketing capacity of enterprises. From this evaluation, the whole marketing ability of Chinese medicine enterprises in Gansu province is still at a low level. There is a huge difference in marketing ability between different enterprises. So it is necessary to support Chinese medicine enterprise with the policy, capital, technology, talent and other aspects to comprehensively improve the marketing ability and marketing level. As the every evaluation index is analyzed, the weakest aspects are the timeliness marketing information, marketing implementation and marketing benefit. So acquiring the information, implementation of marketing schemes, controlling on the management and the cost should be focused to improve the marketing effectiveness of Chinese medicine enterprises and to promote of marketing capacity.

Since evaluation of Chinese medicine marketing ability is a long-term process of sustained activity, the government should evaluate the whole marketing ability regularly of Chinese medicine enterprises in Gansu province and provide some necessary guidance and help based on the evaluation results to promote their development. In order to improve their marketing ability and level so as to improve their competitive advantages, enterprises should regularly and irregularly evaluate their marketing abilities to find out and correct the problems in the process of marketing. The improved method of grey correlation evaluation proposed in this research can effectively overcome the shortcomings of the traditional one. With its simple calculation and easy operating, it provides a reliable method for evaluating the marketing ability of Chinese medicine enterprises.

Acknowledgement This research was financially supported by the FDA of Gansu Province (No. 2017GSFDA016) and also supported by the Health management of key disciplines in Gansu University of Chinese Medicine.

References

1. Yang, S. (2017). Enterprise marketing ability evaluation and empirical research. *Statistics and Decisiong, 6*, 186–188.
2. Lin, Y. (2005). Research on construction the evaluation index system of enterprise marketing ability. *Business Era, 35*, 36–38.
3. Lin, Y., Lan, S., & Yijun, C. (2016). Comprehensive evaluation of enterprise marketing ability. *Science and Technology Information Development and Economy, 19*, 76–79.
4. Yang, R., & Yang, Y. (2009). Research on evaluation index of enterprise marketing capability. *Market Research, 3*, 56–58.
5. Wang, Z., & Fangmei, Q. (2009). The empirical study on enterprise marketing ability and marketing performance. *Commercial Research, 9*, 54–57.
6. Huayou, C., Wu, T., & Xu, Y. (1999). Improvement of gray correlation space and gray correlation degree calculation. *Journal of Anhui University, 23*(4), 39–43.
7. Xiaohong, C., & Yu, T. (2015). Research on the relationship between marketing ability and technological innovation and market performance – Based on the empirical research of small and medium-sized listed enterprises in China. *Journal of Science, 4*, 128–132.
8. Han, S.-p., & Wang, Y.-g. (2013). Research on marketing capability and its performance impact. *Management World, 6*, 67–69.

Part II
Logistics and Operations Analytics

Chapter 9
The Impact of the Relationship Between Operational Cost and Oil Prices on Economic Assessment in Oil and Gas Industry

Lihui Zhu, Dongkun Luo, Xiaoyu Wang, and Rui Guo

Abstract In the long-term, changes in oil prices will certainly have an impact on the estimate of the project income, investment and costs, the variables are not mutually independent, but there is a correlation. Traditional economic evaluation methods tend to ignore the influence of parameter correlation, resulting in project evaluation deviations. In this article, we assume that oil prices obey geometric Brownian motion and mean-reverting stochastic process separately, taking into account the correlation between oil prices and operational costs, using the Monte Carlo model to simulate the project value and risk under different probability. The results show that if the linkage mechanism of the oil prices and operational costs is not considered, it is easy to overestimate the risk of the project, which leads to some feasible projects excluded.

Keywords Oil price · Operating costs · Relevant · NPV · Monte Carlo simulation

9.1 Introduction

Traditional economic evaluation methods generally assumed parameters are independent, but this assumption is not realistic. In recent years, with the rise in oil prices, exploration and development capital investment and operational costs have also increased. On the one hand due to the rising oil prices, industry expansion, increased competition leads to increased investment in equipment and raw material costs. On the other hand, with the exploration and development of the deep-water and unconventional projects, the requirements of technology and environmental protection are becoming higher and higher, and the investment and cost are also increased accordingly. The relationship between oil price and other variables make

L. Zhu · D. Luo · X. Wang · R. Guo (✉)
School of Business Administration, China University of Petroleum, Beijing, China
e-mail: gray000@163.com

M. Tavana, S. Patnaik (eds.), *Recent Developments in Data Science and Business Analytics*, Springer Proceedings in Business and Economics,
https://doi.org/10.1007/978-3-319-72745-5_9

the economic assessment more complex. The dependence between parameters has a little impact on assessment at high oil prices, while the oil price fell down, the relationship between them can make a big difference, even influence the decision-making. With oil prices fell sharp since the second half of 2014, the profitability of oil company decline, the work of economic assessment become more important. Capital investment is mainly influenced by the company's strategy and policy, which is not sensitive to the change of oil prices. In this article, we focus on the relationship between oil prices and operational costs and quantify their impact on the economic assessment.

The remainder of this paper is organized as follows: Sect. 9.2 establishes a NPV model that considered the correlation between oil prices and operational costs. Thereafter, a simple example project is demonstrated to illustrate the effectiveness and practicability of the method in Sect. 9.3. The last section is conclusions and recommendations.

9.2 Methods

9.2.1 Modeling Oil Price

In this article, we consider the Geometric Brownian motion (GBM) and mean-reverting stochastic process to describe the international crude oil prices.

Geometric Brownian motion [1–3] can express as follow:

$$dP_{Oil} = u_P P_{Oil} dt + \sigma_P P_{Oil} dz_P \tag{9.1}$$

Where P_{Oil} is oil price in units of U.S. dollar/Barrel; dz_p is the independent increments of Wiener process $dz_p = \varepsilon_P \sqrt{dt}$, where ε_P is a normally distributed random variable with mean 0 and standard deviation 1; u_P and σ_P represent the drift and variance parameters of the oil price, respectively.

According to Ito's Lemma:

$$d\ln(P_{oil}) = \left(u_p - \frac{1}{2}\sigma_P^2\right) dt + \sigma_P dz_P \tag{9.2}$$

In the simulations, the discrete approximation to oil price process in risk-neutral version is:

$$P_{oil}(t_{i+1}) = P_{oil}(t_i)\exp\left[\left(r - \frac{1}{2}\sigma_p^2\right) \cdot (T - t) + \sigma_p \sqrt{T - t} \cdot \varepsilon_p\right] \tag{9.3}$$

Where r is the risk-free rate.

In addition, we also use a mean reversion process to describe the uncertainty of oil prices which are an alternative type of Markov process. In this article, we use the one factor Ornstein-Uhlenbeck process, which has the form:

$$dP = \eta(m - \ln P)dt + \sigma P dz$$

Where, m is the logarithmic mean of the long-term oil price, η is the magnitude of the speed of adjustment which measures the degree of mean reversion to the long run mean m.

Defining X = lnP and applying Ito's Lemma, this implies that the log price can be characterized by an Ornstein-Uhlenbeck stochastic process:

$$dX = \eta^* \cdot (m^* - X)dt + \sigma dz \tag{9.4}$$

Where $\eta^* = \eta$, $m^* = m - \sigma^2/(2 \cdot \eta)$. lnP obey normal distribution, dz is an increment to a standard Brownian motion. The conditional distribution of X at time T is normal with mean and variance:

$$E[X(T)] = e^{-\eta T} X(0) + \left(1 - e^{-\eta T}\right) m^* \tag{9.5}$$

$$\mathrm{Var}[\mathrm{X(T)}] = \frac{\sigma^2}{2\eta}\left(1 - e^{-2\eta T}\right) \tag{9.6}$$

According to B.J.A. Willigers [4], estimates of low-, base-, high-probability uncertainties are often expressed as P90, P50, P10, respectively, which P means the probability that an uncertain variable has a value less than the specified value. In Swanson's rule, probabilities of 30%, 40% and 30% have been assigned to P10, P50, and P90, respectively [5].

9.2.2 *The Relationship Between Oil Price and Operational Cost*

Previous research most assumes the relationship between oil prices and operational costs is linear. But we all know that, with the rise of the price level and the difficulty of exploration and development, the operating costs also have a ratchet effect. We set up a linear model, exponential model and binomial model fitting oil price index and operational costs index. The results show that, for CNPC companies, the effect of exponential fitting is best, so we have reason to believe that oil price index and operational costs index meet the exponential function. The relationship can express as follows:

$$\mathrm{CNPC}: C = 0.637e^{0.387 \cdot p}$$

where C is operational costs index and p is oil price index.

According to the nonlinear relationship between oil prices and operational costs, we can adjust the operational costs under different oil price assumptions.

9.3 A Case Study

9.3.1 Model Parameters

In case study, we select CNPC (China National Petroleum Corporation) in china as the overseas investor as the case study for the NPV evaluation model discussed in Sect. 9.3. The data related to project were adapted in order to maintain the confidentiality of the information.

The field is expected to produce in 2018, and ends in 2037. The economic value of an exploration target has been made and the basic assumptions of the asset are shown in Table 9.1. We have invested 100 million dollars to explore, and calculated according to 2014 oil price, the project NPV is positive. But as oil prices fell sharply, the profitability of the project affected. The decision in 2017 is whether to drill production wells and develop facilities. The biggest risk to the project from the uncertainty of oil prices, if oil prices do not rebound in 2018, then we have to re-evaluate the economic value of the project. In addition to the uncertainty of oil price, we also analyze the impact of oil price-operational costs correlation and different types of contracts on project evaluation. The relationship between oil price and operational costs refer to the above 3.2, and contract type reference above 3.3. The mining royalty are 10% and 40% respectively under resource royalty systems and production sharing system, income tax is 30%. Depreciation period of 20 years, does not consider the residual value.

In this paper, the operational costs are fixed costs and variable costs, and fixed costs are fixed in the whole production cycle, variable costs change with output. In the short term, the impact of oil prices on the variable costs is large, fixed costs are

Table 9.1 Summary of oil production and expenditures

Year	2017	2018	2019	2020	2021	2022	2023
Production mbbl		1530	5980	8960	8023	6210	4960
Capital $MM	400[a]						
Fix Opex $MM	3.6	12.3	17.8	14.6	11.5	10.2	7.1
Var.Opex oil $/Bbl	11	11	11	11	11	11	11
	2024	2025	2026	2027	2028	2029	2030
Production mbbl	4010	3412	2350	1960	1390	1039	960
Capital $MM							
Fix Opex $MM	5.6	4.1	3.2	2.4	1.8	1.2	0.9
Var.Opex oil $/Bbl	11	11	11	11	11	11	11
	2031	2032	2033	2034	2035	2036	2037
Production mbbl	695	492	356	268	203	114	
Capital $MM							150[b]
Fix Opex $MM	0.7	0.5	0.4	0.2	0.1	0	
Var.Opex oil $/Bbl	11	11	11	11	11	11	

[a]Development drilling and infrastructure costs
[b]Abandonment costs

not sensitive, so we just adjust variable costs. First, we adjust the operational cost index based on the prediction of oil price index. According to the operational cost index to adjust the variable costs, resulting in operational costs associated with the oil price.

9.3.2 *Results and Discussions*

9.3.2.1 Production Sharing Contract

As we can see in Table 9.2, Under the assumption of Geometric Brown motion, the value of the project is significantly greater than that of the mean reversion assumption. When the oil price is assumed to obey the Geometric Brown motion, if the correlation between the price of oil and operational cost is not considered, the volatility of the project value is large, and the error of the expected net present value of the project is 40%. Under the mean reversion hypothesis, the volatility of the project value is small and the correlation between oil price and operation cost has little effect, and the relative error is 9%. In addition, we also found that without considering the correlation between oil prices and operational costs, the breakeven point are basically the same under different oil price assumption. But once considered the relevance, the break-even point becomes lower, and the risk of the project is reduced.

Table 9.2 Oil project values with resource PSC and resource royalty system

PSC			Resource royalty system	
	Consider correlation	Uncorrelated	Consider correlation	Uncorrelated
Scenario	NPV in $MM	NPV in $MM	NPV in $MM	NPV in $MM
GBM				
P90	−155.82	−200.96	24.52	−91.79
P50	125.59	114.57	464.50	437.32
P10	541.16	822.44	908.44	1606.40
ENPV	165.84	232.27	465.69	629.31
Break-even point	69.10	73.92	38.55	50.00
P (NPV > 0) (%)	66.8	63.8	91.9	82.5
Mean reversion				
P90	−25.68	−58.25	231.74	149.43
P50	102.44	87.29	428.51	390.84
P10	260.43	281.27	660.18	712.12
ENPV	111.40	101.82	438.98	414.80
Break-even point	69.10	73.89	38.55	49.94
P (NPV > 0) (%)	82.8	77.0	99.9	99.1

9.3.2.2 Resource Royalty System

Under resource royalty system, cash flow calculation is relatively simple, and the results are show in Table 9.2. When the oil price follows the Geometric Brown motion, the fluctuation range of the project value is relatively large, when the oil price is assumed to be a mean reversion process, the fluctuation of the value of the project is small. Traditional evaluation methods (ignore the correlation between oil price and operational costs) exist error, the relative error is 35% (GBM) and 6%. (mean reversion). For the break-even point, the calculation results of the traditional method are basically the same under different assumption of oil price. When considering the correlation, the break-even point becomes lower.

9.4 Conclusions

In this paper, we research the impact of the relationship between oil prices and operational costs on economic assessment in oil and gas Industry, the main conclusions are as follows:

First, we study the relationship between oil prices and operational costs, and found that the nonlinear exponential relationship can be better fitted to the oil prices and operational costs. Second, the linkage mechanism between the oil prices and the operational costs has an impact on the project evaluation. Whether in resource royalty system or production sharing contract, the linkage mechanism reduced the break-even point, and the risk of the project is relieved. Under different oil price assumptions, the error caused by the correlation between the oil price and the operational costs is also different. Under the assumption of GBM, the error of the traditional evaluation method is relatively large, and the relative error is more than 10%, while under the mean reversion hypothesis, the relative error is less than 10%. The results of two oil price assumptions indicate that ignore the linkage mechanism between oil prices and operational costs will lead to inaccuracies evaluation.

Acknowledgments Funding for this work was supported by International Science and technology support program (2014BAC01B02).

References

1. Dixit, A., Pindyck, R. S., & Sødal, S. (1999). A markup interpretation of optimal investment rules. *The Economic Journal, 109*(455), 179–189.
2. Willigers, B. J., Begg, S., & Bratvold, R. B. (2014). Combining geostatistics with bayesian updating to continually optimize drilling strategy in shale-gas plays. *SPE Reservoir Evaluation & Engineering, 17*(04), 507–519.
3. Zhu, L., Zhang, Z., & Fan, Y. (2015). Overseas oil investment projects under uncertainty: How to make informed decisions? *Journal of Policy Modeling, 37*(5), 742–762.

4. Willigers, B. J. (2009). Enhanced economic modeling by correlated stochastic models of E&P costs and hydrocarbon prices: The limitations of fixed price decks and the versatility of least-squares Monte Carlo simulation. In EUROPEC/EAGE Conference and Exhibition. Society of Petroleum Engineers.
5. Hurst, A., Brown, G. C., & Swanson, R. I. (2000). Swanson's 30-40-30 rule. *AAPG Bulletin, 84*(12), 1883–1891.

Chapter 10
The Construction of University Students' Entrepreneurship Competency Model in Application-Oriented Universities

Chunxiao Chen and Jinjian Wang

Abstract As a special entrepreneurial group, university student enterprises have different psychological features, behavior characteristics and entrepreneurship environment, so they have different entrepreneurship competency relative to others groups. Based on competency model, by means of personnel assessment, the article established an entrepreneurship competency hypothetical model for application-oriented Universities' students, and modified the model through statistical analysis and test.

Keywords Entrepreneurship competency · Quotient · Factor

10.1 Introduction

At present, under the innovation driven development strategy, China's economy kept transformation and upgrading and entered a period of new normal. Innovation and entrepreneurship have become the themes in the context of the present era, and elevate to the position of national strategy. As the precious human resources and innovation resources, university students are the vital new force and main force of popular entrepreneurship and innovation [1]. In the process of cultivating creative talents and promoting the construction of innovative society, colleges and universities face heavy tasks and shoulder unshirkable responsibility. Entrepreneur is the main body of entrepreneurship. Entrepreneurial success is a result of the effect of entrepreneur capability. As a special entrepreneurial group, university student enterprises have different psychological features, behavior characteristics and entrepreneurship environment, so they have different entrepreneurship competency relative to other groups [2]. Therefore, we

C. Chen (✉) · J. Wang
Taishan University, Taian, P. R. China
e-mail: ch7766@163.com; 15153878668@163.com

M. Tavana, S. Patnaik (eds.), *Recent Developments in Data Science and Business Analytics*, Springer Proceedings in Business and Economics,
https://doi.org/10.1007/978-3-319-72745-5_10

should explore the diathesis characteristic of university students' entrepreneurship competency according to their conditions, environment and growth law, using scientific methods.

10.2 Research Process

10.2.1 The Determination of University Students' Entrepreneurial Competence

Literature search.The factors of undergraduates' entrepreneurship competency.Literature search. According to searching journal articles and relative books, we acquire some concepts of entrepreneurship competency, and arranged them from high frequency to low frequency. We take the top 60 as the alternatives of the factors of undergraduates' entrepreneurship competency [3].

Behavioral event interview. Among the application-oriented universities, We select nine entrepreneur teams with representative sample characteristics related to finance, accounting, business management, logistics management, hotel management, clothing design, food science and engineering, electronic commerce and marketing. And we carry out deeply behavior event interviews with 25 undergraduates which entrepreneur succeeded.

Base on collecting the entrepreneur behavior and results, we have screened and modified the competency factors which involved in the interview process according to the criterion of accuracy simplicity. Then, we ranked them according to frequency and compared them to the top 60 factors in the above. Finally, we got the 38 competency factors (Table 10.1).

10.2.2 Build the Hypothetical Model of the University Students' Entrepreneurship Competency

Entrepreneurship competency factors classification, hypothetical model construction. We selected 38 university students' entrepreneurship competency factors through behavioral event interview. We divided these factors into ten major categories, there are learning ability, decision making ability, theoretical knowledge and skills, market ability, sense of social responsibility, interpersonal quotient, self-management skills, innovation ability, anti-pressure ability and problem-solving skill. The ten major categories can be classified into three concepts, Intelligence quotient (IQ), Emotional quotient (EQ) and Adversity quotient (AQ). Based on the above analysis, we constructed a hypothetical model university students' entrepreneurship competency (Table 10.2).

Table 10.1 The factors of undergraduates' entrepreneurship competency

1 Interpersonal skills	11 Coordinate ability	21 Communication skill	31 Psychological diathesis
2 Contemplative faculties	12 Spirit of dedication	22 Foreseeability	32 Market consciousness
3 Sanguine nature	13 Well-connected	23 Strategic foresight	33 Positivity
4 Not conservative	14 Responsibility	24 Decision-making capacity	34 Understanding of national laws and regulations
5 Marketing ability	15 Perseverance	25 Jump at the chance of learning	35 Professional knowledge and skills
6 Confidence	16 Moral character	26 Insight	36 Asset management ability
7 Face up to challenge	17 Firmness of will	27 Creative thinking	37 Sense of time
8 Dauntless	18 Team work	28 Knowledge application ability	38 Creation inspiration
9 Strong belief	19 Possessiveness	29 Market analysis	
10 Wonderful thinking	20 Memory	30 Attention	

Table 10.2 The classification of competency factors

Category	Classification of entrepreneurship competency	Factor description
Intelligence quotient (IQ)	I1 Learning ability	20 Memory 22 Foreseeability 33 Positivity 25 Jump at the chance of learning
	I2 Decision making ability	23 Strategic foresight 26 Insight 30 Attention
	I3 Theoretical knowledge and skills	34 Understanding of national laws and regulations 35 Professional knowledge and skills 36 Asset management ability
	I4 Market ability	5 Marketing ability 29 Market analysis 32 Market consciousness
Emotional quotient (EQ)	E1 Sense of social responsibility	12 Spirit of dedication 14 Responsibility 16 Moral character
	E2 Interpersonal skills	1 Interpersonal skills 11 Coordinate ability 13 Well-connected 18 Team work 21 Communication skill
	E3 Self-management skills	2 Contemplative faculties 37 Sense of time 19 Possessiveness 31 Psychological diathesis
Adversity quotient (AQ)	Al Innovation ability	4 Not conservative 10 Wonderful thinking 7 Creative thinking 38 Creation inspiration
	A2 Anti-pressure ability	3 Sanguine nature 6 Confidence 7 Face up to challenge 9 Strong belief 15 Perseverance 17 Firmness of will
	A3 Problem-solving skills	8 Dauntless 24 Decision-making capacity 28 Knowledge application ability

Table 10.3 The results of KMO test and Bartley sphere test of each dimension

		Intelligence quotient (IQ)	Emotional quotient (EQ)	Adversity quotient (AQ)
Kaiser-Meyer-Olkin measure of sampling adequacy		0.715	0.707	0.778
Bartlett's test of sphericity	Approx. Chi-square	383.100	244.242	146.062
	df	105	66	36
	Sig.	0.000	0.000	0.000

Questionnaire survey. According to the 38 competence factors, we established university students' entrepreneurship competence questionnaire, using Likert's Five Scaling Method. Taking their experience in entrepreneurship process into account, the respondents will be required to grade the competence factors considering the factors' significance in entrepreneurship process. The 55 respondents were selected from ten application-oriented universities in Shandong Province. We distributed 55 questionnaires, and recycled 55. There are 52 effective questionnaires in the end.

Factor analysis. Basing on the effective questionnaires, we used Cronbach alpha coefficient to do reliability analysis. The conclusion will be shown as Table 10.3. The total Cronbach alpha coefficient is 0.934, and the reliability coefficient of Intelligence quotient (IQ), Emotional quotient (EQ) and Adversity quotient (AQ) are 0.860, 0.826 and 0.795, respectively.

We did KMO test and Bartley sphere test before factor analysis, and the conclusion will be shown as Table 10.3. The average value of KMO test is 0.7 which can be accepted. The significance level of Bartley sphere test is 0.000(<0.01), so there is a common factor in the correlation matrix, which is suitable for factor analysis.

Because there are too many elements in the questionnaire, we divided the competency factors into three groups, that is, intelligence quotient group, emotional quotient group and adversity quotient group while doing factor analysis.The results of factor analysis will be shown as follows.

Intelligence quotient (IQ) group. The 13 competency factors included in the original hypothesis model can be attributed to 4 common factors, we gave the name of I1–I4 to them. These factors explained 72.527% of the variance (Table 10.4).

The rotated component matrix of intelligence quotient will be shown in Table 10.5. The data showed that the hypothesis of competency factors (including I1, I2, I3, I4) was consistent with the proposition of the original hypothetical model.

I1 is designated as learning ability in the original hypothetical model. The competency factors of I1 contains 20 Memory, 22 Foreseeability, 25 Jump at the chance of learning and 33 Positivity, so we can rename I1 to independent learning ability considering the connotation of active learning. The others factors' name in the original hypothetical model are no need to rename.

Table 10.4 Partitioning of total variance

	Initial eigenvalues			Extraction sums of squared loadings		
Factor	Total	% of variance	Cumulative %	Total	% of variance	Cumulative %
I1	10.872	55.966	55.966	10.872	55.966	55.966
I2	2.905	8.299	64.265	2.905	8.299	64.265
I3	1.658	4.737	69.103	1.658	4.737	69.103
I4	1.198	3.424	72.527	1.198	3.424	72.527
E1	1.673	56.467	56.467	1.673	56.467	56.467
E2	1.513	9.061	65.528	1.513	9.061	65.528
E3	1.251	3.575	69.103	1.251	3.575	69.103
Al	2.429	57.300	57.300	2.429	57.300	57.300
A2	1.885	8.386	65.686	1.885	8.386	65.686

Emotional quotient (EQ) group. Based on the data in Table 10.4, the 12 competency factors can be attributed to 3 common factor, that is, E1, E2, E3. These factors explained 69.103% of the variance.

The rotated factor loading matrix of Emotional quotient (EQ) is in Table 10.5. From the data it is shown that the hypothesis of competency factors (including E1, E2, E3) was consistent with the proposition of the original hypothetical model.

Because the factor loading of 21 (Communication skill) and 31 (Psychological diathesis) are under 0.5, we winkled out these two factors. The others factors' name in the original hypothetical model are no need to rename.

Adversity quotient (AQ) group. In the process of factor analysis, we winkled out four factors, that is 4 (Not conservative), 10 (Wonderful thinking), 27 (Creative thinking) and 38 (Creation inspiration), because their factor loading is too small for factor analysis. The value of KMO is 0.676(>0.5) after deleting.

In Table 10.4, we selected Anti-pressure ability and Problem-solving skills as the explanatory factors. Anti-pressure ability is named for Al, and Problem-solving skills is named for A2. According to data in Table 10.10, after factor analysis, the 13 competence factors can be attributed to two common factors (A1 and A2). These factors explained 65.686% of the variance.

The rotated factor loading matrix of Adversity quotient (AQ) is in Table 10.5. The data showed that the hypothesis of competency factors (including A1, A2) was consistent with the proposition of the original hypothetical model. The factors' name in the original hypothetical model are no need to rename.

Correlation analysis. In order to investigate the internal relations of the competence factors, we carried out correlation analysis based on the method of Pearson correlation analysis, and tested the significance level by two-tailed test [4]. The results will be shown as Table 10.6.

Through the hypothesis test of correlation, we can find that the correlation among most of the common factor is significant. We believe that there are positive relations among the university students' entrepreneurship competency factors with 95% certainty while the significance level $p = 0.05$ (2-tailed).

Table 10.5 Rotated component matrix

Factor	The Factors of Entrepreneurship Competency	Factor Loadings			
		1	2	3	4
I1 Learning ability	20 Memory	0.571			
	22 Foreseeability	0.607			
	25 Jump at the chance of learning	0.630			
	33 Positivity	0.795			
I2 Decision making ability	23 Strategic foresight		0.637		
	26 Insight		0.745		
	30 Attention		0.675		
I3 Theoretical knowledge and skills	34 Understanding of national laws and regulations			0.523	
	35 Professional knowledge and skills			0.707	
	36 Asset management ability			0.745	
I4 Market ability	5 Marketing ability				0.838
	29 Market analysis				0.720
	32 Market consciousness				0.496
E1 Sense of social responsibility	12 Spirit of dedication	0.637			
	14 Responsibility	0.730			
	16 Moral character	0.720			
E2 Interpersonal skills	1 Interpersonal skills		0.830		
	11 Coordinate ability		0.718		
	13 Well-connected		0.632		
	18 Team work		0.747		
	21 Communication skill		0.438		
E3 Self-anagement skills	2 Contemplative faculties			0.571	
	19 Possessiveness			0.596	
	31 Psychological diathesis			0.424	
	37 Sense of time			0.585	
Al Innovation ability	3 Sanguine nature	0.581			
	6 Confidence	0.667			
	7 Face up to challenge	0.597			
	9 Strong belief	0.779			
	15 Perseverance	0.720			
	17 Firmness of will	0.592			
A2 Anti-pressure ability	8 Dauntless		0.523		
	24 Decision-making capacity		0.809		
	28 Knowledge application ability		0.622		

Table 10.6 The results of correlation analysis of the competence factors

Component score covariance matrix									
Component	1	2	3	4	5	6	7	8	9
I1 Learning ability	1								
I2 Decision making ability	0.385**	1							
I3 Theoretical knowledge and skills	0.385**	0.696**	1						
I4 Market ability	0.333**	0.517**	0.514**	1					
E1 Sense of social responsibility	0.362**	0.643**	0.573**	0.582**	1				
E2 Interpersonal skills	0.349**	0.340**	0.426**	0.467**	0.395**	1			
E3 Self-managementskills	0.471**	0.408**	0.420**	0.363**	0.478**	0.645**	1		
A1 Anti-Pressure ability	0.294*	0.348**	0.162	0.291*	0.296*	0.480**	0.478**	1	
A2 Problem-solving skills	0.428**	0.424**	0.401**	0.332**	0.398**	0.345**	0.319*	0.358**	1

Correlation is significant at the 0.05 level (2-tailed)

Table 10.7 The revised model of the university students' entrepreneurship competency

Category	Classification of entrepreneurship competency	Factor description
Intelligence quotient (IQ)	I1 Learning ability	20 Memory 22 Foreseeability 25 Jump at the chance of learning 33 Positivity
	I2 Decision making ability	23 Strategic foresight 26 Insight 30 Attention
	I3 Theoretical knowledge and skills	34 Understanding of national laws and regulations 35 Professional knowledge and skills 36 Asset management ability
	I4 Market ability	5 Marketing ability 29 Market analysis 32 Market consciousness
Emotional quotient (EQ)	E1 Sense of social responsibility	12 Spirit of dedication 14 Responsibility 16 Moral character
	E2 Interpersonal skills	1 Interpersonal skills 11 Coordinate ability 13 Well-connected 18 Team work
	E3 Self-management skills	2 Contemplative faculties 19 Possessiveness 37 Sense of time
Adversity quotient (AQ)	A1 Anti-pressure ability	3 Sanguine nature 6 Confidence 7 Face up to challenge 9 Strong belief 15 Perseverance 17 Firmness of will
	A2 Problem-solving skills	8 Dauntless 24 Decision-making capacity 28 Knowledge application ability

10.2.3 Modify the Hypothetical Model of the University Students' Entrepreneurship Competency

Considering the factor composition of Adversity quotient (AQ), Emotional quotient (EQ) and Adversity quotient (AQ), We Modified the hypothetical model of the university students' entrepreneurship competency. The revised model of the university students' entrepreneurship competency is shown in Table 10.7.

10.2.4 The Correlation Between the University Students' Entrepreneurship Competency and Entrepreneurial Performance

Let's suppose that there are positive relation between the university students' entrepreneurship competency and entrepreneurial performance [5]. Entrepreneurial performance consists of operating period, living crisis, operational condition, financial distress, market share and competitive power. Using the questionnaire survey method, we require the respondents to evaluate the entrepreneurship competency

Table 10.8 The results of KMO test and Bartley sphere test of each dimension

Entrepreneurial performance		
Kaiser-Meyer-Olkin measure of sampling adequacy		0.748
Bartlett's test of sphericity	Approx. Chi-square	48.897
	df	10
	Sig.	0.000

factors', that is, Independent learning ability, Decision making ability, Theoretical knowledge and skills, Market ability, Sense of social responsibility, Interpersonal skills, Self-management skills, Anti-pressure ability and Problem-solving skills, effects on entrepreneurial performance.

Firstly, we test the reliability of the questionnaire. All the Cronbach alpha coefficient are greater than 0.6, so the reliability of our questionnaire is fairly good.

According to the data in Table 10.8, the average value of KMO test is greater than 0.7 which can be accepted. The significance level of Bartley sphere test is 0.000 (<0.01), so there is a common factor in the correlation matrix, which is suitable for factor analysis.

There is a simple correlation between the entrepreneurship competency factors and entrepreneurial performance. Table 10.9 shows that the nine factor all have correlation with entrepreneurial performance, but the strength of correlation is different.

The results of regression analysis will be shown in Table 10.10. The data shows that Self-management skills and Problem-solving skills have no significant effects on entrepreneurial performance. The other factors have positive effects on entrepreneurial performance in varying degrees. On the whole, the entrepreneurship competency has positive effect on entrepreneurial performance.

Because the entrepreneurship competency factors are no independent mutually, the positive relationship does not show in the testing results. The competency system is indispensable to successful entrepreneurship, so we retain Self-management skills and Problem-solving skills in the entrepreneurship competency model.

10.3 The Construction of University Students' Entrepreneurship Competency Model in Application-Oriented Universities

Base on the relationship analysis, we test the reliability of the composition and relationship of the entrepreneurship competency factors. So we can establish the entrepreneurship competency model as shown in Fig. 10.1. The relationship of the variables will be shown as Table 10.7.

Table 10.9 The results of correlation analysis

Component score covariance matrix										
Component	1	2	3	4	5	6	7	8	9	10
I1 Learning ability	1									
I2 Decision making ability	0.385**	1								
I3 Theoretical knowledge and skills	0.385**	0.696**	1							
I4 Market ability	0.333**	0.517**	0.514**	1						
E1 Sense of social responsibility	0.362**	0.643**	0.573**	0.582**	1					
E2 Interpersonal skills	0.349**	0.340**	0.426**	0.467**	0.395**	1				
E Self-management skills	0.471**	0.408**	0.420**	0.363**	0.478**	0.645**	1			
A1 Anti-pressure ability	0.294*	0.348**	0.162	0.291*	0.296*	0.480**	0.47**	1		
A2 Problem-solving skills	0.428**	0.424**	0.401**	0.332**	0.398**	0.345**	0.319*	0.358**	1	
Entrepreneurial Performance	0.370**	0.293*	0.037	0.371*	0.153*	0.330**	0.250*	0.269*	0.391**	1

Correlation is significant at the 0.05 level (2-tailed)

Table 10.10 The results of regression analysis

Factor	Regression Coefficient	The standard regression coefficient	t	Sig.
I1 Learning ability	0.246	0.246	3.679	0.001
I2 Decision making ability	0.242	0.242	3.561	0.001
I3 Theoretical knowledge and skills	0.295	0.295	−4.913	0.000
I4 Market ability	0.378	0.378	4.466	0.000
E1 Sense of social responsibility	0.378	0.378	6.983	0.000
E2 Interpersonal skills	0.390	0.390	4.768	0.000
E3 Self-management skills	0.008	0.008	0.210	0.834
A1 Anti-pressure ability	0.741	0.741	18.648	0.000
A2 Problem-solving skills	0.055	0.055	1.376	0.175

Correlation is significant at the 0.05 level (2-tailed)

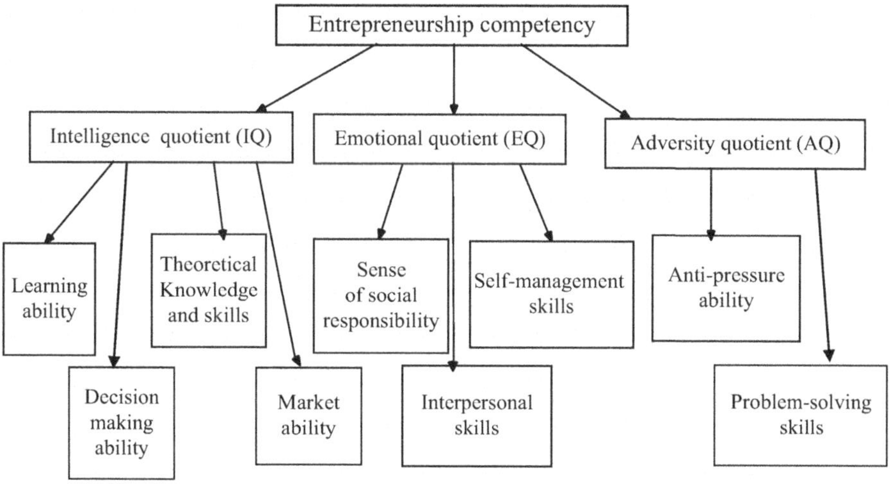

Fig. 10.1 University students' entrepreneurship competency model

References

1. Jie, L., & He, Y. (2012). Research on the competency of college graduates' success in entrepreneurship. *Technology and Economy, 4*, 81–85.
2. Weiling, F., & Yingchun, L. (2011). The construction of contemporary college students' entrepreneurial competency mode. *Scientific Information, 23*, 178–179.
3. Zongtong, C., & Juan, L. (2011). The construction of college students' entrepreneurial competency model. *Economic Research Guide, 16*, 292–294.

4. Xiaoying, W., Exiang, L., & Yi, Z. (2010). Research on the relationship between college students' entrepreneurial competence and entrepreneurial performance. *Technology and Management, 5*, 126–128.
5. Zhenghua, Z. (2009). A study on the relationship between entrepreneurial team competence structure and entrepreneurial performance. *Contemporary Economic Research, 12*, 22–25.

Chapter 11
Improving Airport Security Screening System in Terms of Efficiency and Fairness Via Network Model

Tan Liming, Cai Xiaohang, and He Yuandi

Abstract In this paper, we focus on a dual-objective task of maximizing the airport throughput, efficiency, and reducing the inconvenience to travelers, fairness. Based on the real screening data collected, we construct a network model to make a precise analysis aiming to help figure out a set of efficient strategies. After evaluating the performance of these methods and taking feasibility into consideration, we find that modifying the group security process can improve the overall efficiency of the airport screening system while maintaining the good experience for each passenger.

Keywords Screening security · Passenger throughput · Network model · Multi-objective optimization

11.1 Introduction

The role of aviation security checkpoint screening is important in protecting airport and passengers from terrorist attacks. However, airlines have a vested interest in maintain a positive flying for passengers by minimizing the time they spend waiting in line at security checkpoint. Therefore, there is a tension between desires to maximize security while minimizing inconvenience to passengers.

Nevertheless, this contradiction is intensified recently. In the first year in American airports, most of passengers wait for more than 2 h before their departure and still missed their flights. Apart from that, 10% drop in staffing which cannot afford the 15% increase in passenger volume makes the situation worse still [1].

There is a trade-off between maximizing the passenger throughput and minimizing variance in wait time given the same standard of security and safety. And there are a lot of method to deal with this problem.

T. Liming (✉) · C. Xiaohang · H. Yuandi
School of Information Science and Technology, Jinan University, Guangzhou, China
e-mail: 903247863@qq.com

M. Tavana, S. Patnaik (eds.), *Recent Developments in Data Science and Business Analytics*, Springer Proceedings in Business and Economics,
https://doi.org/10.1007/978-3-319-72745-5_11

For example, assigning passengers to a set of security classes, when passenger perceived risk levels are known. Babu [2] and Nie [3] use linear programming models to investigate the benefit of diving passengers into different groups.

In addition, the allocation of customers to a queuing system with multiple servers has been studied extensively, through either dynamic flow models [4] or static flow models [5].

This paper focuses on using quantitative models to figure out the problem facing the airport security in America now. In particular, the most challenging part is that we are not only required to determine the passenger throughput macroscopically but also to obtain the wait time individually. And in order to search for the appropriate policy and procedural recommendations, we also need to take the practical constraints into consideration due to that our assumption may not always match the actual. Therefore, the potential modifications should be evaluated based on our model and knowledge of the real world.

11.2 Fundamental Assumptions

- **There is only one Pre-check lane open for every three regular lanes.** Based on the given information and illustration, we only consider four lanes (three regular lanes and one pre-check lane) to simplify the model.
- **Under a certain observation time, the queue length is unlimited.** That is to say, passengers who arrive later will continue to line up regardless the queue length.
- **Average 45% of total passengers line up in the Pre-check lane over a long time.** We have already learned that approximately 45% of passengers enroll in Pre-check program. According to the law of large number, we believe that the average number of pre-check lane takers is around 45%.

11.3 The Passenger Flow Network Model

Airport security screening often consist of lots of tedious processes like documents checking, belongings removing, scans (for both people and required items), baggage collecting and so on. We construct a quantitative model here to explore the flow of passengers and figure out the bottlenecks.

Based on our researches, there is often one Pre-check lane open for every three regular lanes, despite the fact that more passengers use the Pre-check process. According to this fact, we build our model just for four lanes, three for regular lanes and one for Pre-check lane.

Every node shown in Fig. 11.1 represents different processes. We presume that every passenger goes through these nodes consecutively until he or she departs the checkpoint area. Node 1 is the point from which passengers start to enter the

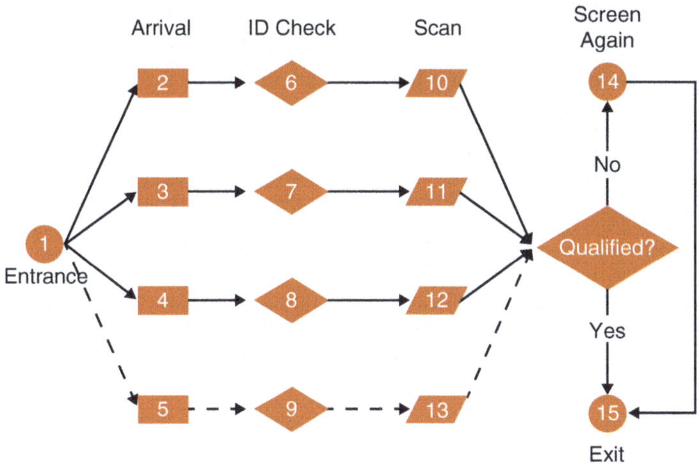

Fig. 11.1 The passenger flow network model

different entrances randomly. Arc (1, 3), (1, 4), (1, 5) stands for the regular lanes while arc (1, 6) presents the pre-check lane.

11.3.1 Throughput Estimation

Airport throughput refers to the number of passengers transported in a certain period of time. According to some definitions and equations of the word "Throughput", we have

$$Q = Q_{in} + Q_{out} \tag{11.1}$$

Q is the airport throughput, Q_{in} is the number of inflow passengers, Q_{out} is the number of outflow passengers.

We assume that Q_{in} is not equal to Q_{out} in our model. It is reasonable for that there are always some visitors waiting in the queue for screening. The time taken to finish all security process is much longer than that of inflow. And it is also the problem that we need to figure out in this paper. However, in the long run, this two values matches each other.

Based on our network model, we can easily obtain the following equations

$$Q_{in} = \sum_{k=2}^{5} Q(1, k), \quad Q_{out} = \sum_{k=10}^{13} Q(k, 15) \tag{11.2}$$

$Q(i,j)$ is the number of inflow passengers in arc (i,j). It is constrained by the flow rate of the previous arc and the velocity in node i. Therefore, we build a passenger flow network model by subjecting to this constraint.

Time required to pass each node

$$t(i) = \begin{cases} \varepsilon & ,i = 2,3,4,5 \\ E(t_{ID}) & ,i = 6,7,8,9 \\ E\left(t_{scan-reg}\right) & ,i = 10,11,12 \\ E\left(t_{scan-pre}\right) & ,i = 13 \end{cases} \quad (11.3)$$

ε is an infinite quantity. In our model, the mean value of each channel entry is used as the inflow. Then, based on the data from the COMAP [6], we can get the following data

$$Q(1,2) = Q(1,3) = Q(1,4) = 0.08, Q(1,5) = 0.11 \quad (11.4)$$

Flow rate in the process of ID check

$$Q(i,i+4) = \min\{Q(1,i), 1/t(i)\}, i = 2,3,4,5 \quad (11.5)$$

Flow rate in the process of scan

$$Q(i,i+4) = \min\{Q(i-4,i), 1/t(i)\}, i = 6,7,8,9 \quad (11.6)$$

Flow rate in the process of departure

$$Q(i,15) = \min\{Q(i-4,i), 1/t(i)\}, i = 10,11,12,13 \quad (11.7)$$

Through MATLAB, we analyze that the annual airport throughput is about 15.97 million which is almost equal to the previous result 15.54 million. That is to say, our model building is reasonable and valid.

However, we think this result is still reasonable because there are only four channels in our model, while in real life, there are far more than four channels. In addition, the throughput actually includes the number of passengers on the airplanes which make a stopover in these airports.

11.3.2 Identify the Potential Bottlenecks

By comparing the data during the calculation, we identify the potential bottlenecks in the screening process (Table 11.1).

When the number of inflow passengers of the node is smaller than the flow rate of the arc, the subsequent flow rate is limited, and the passengers this arc cannot be transferred to the next arc. That is to say, the marked area in Table 11.1 is the place where bottleneck may exist.

Table 11.1 Comparison of node flow rate and arc flow rate

Node Flow Rate			Arc Flow Rate		
2	1.00	>	(1,2)	0.0	Arrival
3	1.00	>	(1,3)	0.0	
4	1.00	>	(1,4)	0.0	
5	1.00	>	(1,5)	0.1	
6	0.09	>	(2,6)	0.0	ID Check
7	0.09	>	(3,7)	0.0	
8	0.09	>	(4,8)	0.0	
9	0.09	<	(5,9)	0.1	
10	0.03	<	(6,10)	0.0	Scan
11	0.03	<	(7,11)	0.0	
12	0.03	<	(8,12)	0.0	
13	0.06	<	(9,13)	0.0	

One bottleneck is the ID check process for the pre-check takers, another bottleneck is the scan process of all channels.

11.4 Modified Model: Focus on Individuals

In the previous Passenger Flow Network Model, we only care about the flow of people in each link. As a result, we cannot accurately figure out the individual wait time, so we modify our first model by introducing matrix to meet the needs of recording passengers' wait time separately.

11.4.1 Model Modification

- **Combination of macroscopic and microscopic method**

Matrix P introduced here helps us to record individual wait time more accurately.

$$\mathbf{P}_{pre}(t) = \begin{bmatrix} x_{11}(t) & x_{12}(t)\cdots x_{1p}(t) & \\ x_{21}(t) & x_{22}(t)\cdots x_{2p}(t) & \\ \vdots & \vdots \quad \ddots & \vdots \\ x_{n1}(t) & x_{n2}(t)\cdots x_{np}(t) & \end{bmatrix}, \mathbf{P}_{reg}(t) = \begin{bmatrix} x_{11}(t) & x_{12}(t)\cdots & x_{1p}(t) & \\ x_{21}(t) & x_{22}(t)\cdots & x_{2p}(t) & \\ \vdots & \vdots & \ddots & \vdots \\ x_{m1}(t) & x_{m2}(t)\cdots & x_{mp}(t) & \end{bmatrix} \quad (11.8)$$

$\mathbf{P}_{pre}$(t) and $\mathbf{P}_{reg}(t)$ record the security state at time t in the same lane. Each row is the representative of one passenger and shows which position he or she is in at time t. n and p is set based on our observation time T (normally set to a larger number). p is the total number of security steps. $x_{ij}(t)$ is described as follows

$$x_{ij}(t) = \begin{cases} 1, & \text{passenger i is at the j} - \text{th stage} \\ 0, & \text{passenger i is not at the j} - \text{th stage} \end{cases} \tag{11.9}$$

Of course, one passenger cannot take more than one security process simultaneously.

$$\sum_{j=1}^{p} x_{ij}(t) = 0, 1 \quad i = 1, 2, \cdots, n \tag{11.10}$$

In addition, based on previous assumption, when $x_{ij}(t) \le x_{kj}(t) \quad j = 1, 2, \cdots, p$, we write

$$x_{ij}(t) \le x_{kj}(t) \quad j = 1, 2, \cdots, p \tag{11.11}$$

- **Modification of time used in the process of scan**

In the first model, we use the mean value of processed data to represent the time spent by each individual in each process step. However, it is not so accurate due to that we leave out the individual difference (gender, baggage, etc.). According to the data, we consider that the time consumed is normally distributed.

$$\begin{array}{l} t_{scan-pre}(i) \sim N(\mu_{pre}, \sigma_{pre}) \\ t_{scan-reg}(i) \sim N(\mu_{reg}, \sigma_{reg}) \end{array} \tag{11.12}$$

- **Calculation of the average and variance of the wait time**

During $[0, t]$, the number of the passenger flowing out from the node can be calculated as follows:

$$Q(10, 15) = Q(11, 15) = Q(12, 15) = \sum_{i=1}^{m} \mathbf{P}_{reg}[x_{ip}(t)] \tag{11.13}$$

$$Q(13, 15) = \sum_{i=1}^{n} \mathbf{P}_{pre}[x_{ip}(t)] \tag{11.14}$$

Total number of outflows:

$$Q_{out}(t) = \sum_{k=10}^{13} Q(k, 15) \tag{11.15}$$

Average wait time and variance of wait time are as follows

$$E(t) = \frac{\sum_{i=1}^{Q_{out}} (t_{outi} - t_{ini})}{Q_{out}}, Var(t) = \frac{\sum_{i=1}^{Q_{out}} [(t_{outi} - t_{ini}) - E(t)]^2}{Q_{out}} \tag{11.16}$$

Solutions are shown in the Table 11.2.

Table 11.2 The average and variance of wait time

Time stamps	Average wait time	Variance of wait time
100	51.2	98.8
150	82.4	704.3
200	101.0	1204.3
250	107.1	1599.0

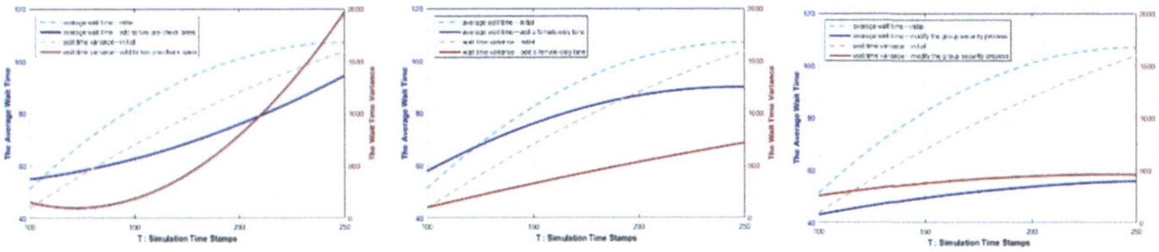

Fig. 11.2 The effect of three strategies in waiting time

11.4.2 Three Recommended Strategies

Aiming to reduce the variance of wait time, we offer three strategies here:

- **Strategy 1: Add to two pre-check lanes**. If one more pre-check lane is added, it may share the pressure with original pre-check lane.
- **Strategy 2: Add a female-only lane**. If the female fail the millimeter wave scan, they will be inspected only by female officers, while the male can be screened by both female and male officers. Therefore, adding a female-only lane can improve the overall efficiency.
- **Strategy 3: Modify the group security process**. This policy allows the passengers who buy group tickets to start removing their belongings simultaneously. In addition, we limit the number of groups not more than 5.

The blue line represents the average of wait time and the red one represents the variance of wait time in Fig. 11.2. Also, the dotted line represents the initial situation and the solid line represents the modification.

As shown above, we can easily find that all the strategies have positive effect on reducing the average waiting time especially the third one. In another aspect, the attribution of the three strategies to reducing the waiting time variance is decreasing successively.

- **The analysis of throughput after optimizing**

By comparison, all the three policies have a positive impact on the passenger throughput. Among them, strategy 1 has a relatively weak effect. For the other two, there is no such big difference. However, taking the practical situation into account, adding female-only lane is much harder to conduct for more operating costs. In a word, it is better to modify the group security process (Fig. 11.3).

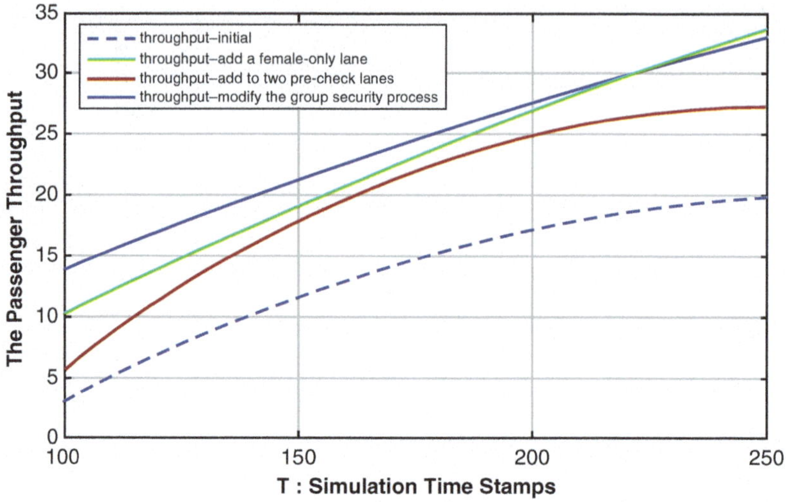

Fig. 11.3 The effect of three strategies in throughput

Table 11.3 Multiple strategies are used simultaneously

i	Modifications	Comprehensive evaluation index
I	1 + 2	18.98
II	1 + 3	24.73
III	2 + 3	31.45 (max)
IV	1 + 2 + 3	27.02

11.4.3 Comprehensive Evaluation Index

Our model is introduced to realize the two-objective task of maximizing the airport passenger throughput and minimizing the variance of wait time. We carry out a comprehensive evaluation index as follows:

$$CV_i = Q_i - \frac{Var(i)}{\alpha} \tag{11.17}$$

Q_i and $Var(i)$ are the throughput and the variance of wait time under different modifications, α is the coefficient which makes throughput and variance comparable due to the different magnitude (Table 11.3).

As indicated in the table, we find that the third combination methods (add a female-only lane and modify the group security process both) is the most optimal solution to the dual-objective task with the index value of 31.45. This is also in accordance with the previous analysis.

11.5 Conclusions

As our team set out to build a quantitative mathematical model to explore the flow of passengers, we figure out that the bottlenecks exist in the process of ID check and scan. Then, modified model is introduced to determine the effect of bottlenecks more accurately. And, in order to fulfill the ambition of maximizing the throughput and minimizing the variance of wait time, we offer some ways and examine their impact and feasibility.

The result shows that the combination of the policies we choose is influenced by some factors like operating costs and the ratio of the pre-check lane takers. When the ratio is between 45% and 61.25%, we pick the policies of adding female-only lane and modify the group security process.

References

1. (2016, May). *TSA security line waits inevitable, DHS secretary says.* CNN. http://edition.cnn.com/2016/05/13/aviation/tsa-long-lines-us-airports/.
2. Babu, V. L. L., Batta, R., & Lin, L. (2006). Passenger grouping under constant threat probability in an airport security system. *European Journal of Operational Research, 168*, 633–644.
3. Nie, X., Batta, R., Drury, C. G., & Lin, L. (2009). Passenger grouping with risk levels in an airport security system. *European Journal of Operational Research, 194*(2), 574–584.
4. Kumar, P. R. (1985). Individually optimal routing in parallel systems. *Journal of Applied Probability, 22*, 989–995.
5. Jain, M. (2005). Finite capacity M/M/r queueing system with queue-dependent servers. *Computers & Mathematics with Applications, 50*(1), 187–199.
6. (2017). *MCM/ICM problems.* http://www.comap.com/undergraduate/contests/mcm/contests/2017/problems/.

Chapter 12
Study on the Choice of Strategic Emerging Industries in Gansu Province Based on Multi-level Grey Model

Zhonghua Luo, Qi Men, and Lixin Yun

Abstract In the economic transformation period, it is an important realistic choice for local government to accelerate developing its strategic emerging industries and to cultivation them into leading industries for revitalizing local economy and transforming economic increasing mode. In the context of "One Belt And One Road" strategy, how to choose strategic emerging industries will determine its future economic trends and development speed. This paper uses multi-level grey correlation evaluation method, and studies how to select its future strategic emerging industries relying on date of four industries such as health services, processing and manufacturing industry in 2015 Gansu statistical yearbook. This paper can provide the necessary theoretical reference for Gansu province to determine its future emerging industries.

Keywords Strategic emerging industries · A multi-level gray model · Industry choice · The empirical research

12.1 Introduction

Scientific and rational selection of emerging strategic industries is critical. If you choose the right one, you can leap the development, however the reverse is not. So a question is what kind of surveying indicators should be used, what kind of selection methods should be choose for potential strategic emerging industries. Therefore, it is the important content of the current economic research to avoid the phenomenon of blindly disordering and convergent development in the new round of industry development. Gansu province is located in the northwest of China, the economic society is backward. In the western development strategic and one Belt and one Road

Z. Luo · Q. Men · L. Yun (✉)
School of Economics and Business Management, Gansu University of Chinese Medicine, Lanzhou, China
e-mail: luozhpsx@sohu.com

M. Tavana, S. Patnaik (eds.), *Recent Developments in Data Science and Business Analytics*, Springer Proceedings in Business and Economics,
https://doi.org/10.1007/978-3-319-72745-5_12

strategic background, It is a great importance for Gansu to seize the opportunity, utilize the local advantage and resources to improve the economic development speed of Gansu province [1].

12.2 Strategic Emerging Industries and Their Choices

Strategic emerging industries are based on breakthroughs in major technological development, and represent the new direction of science and technology and industry development, and reflect the world knowledge economy, circular economy and low carbon economy development trend. Strategic emerging industries are still in its early growth, so their future development potential is huge. Strategic emerging industries have obvious competitive advantage, and may play an significant leading role in driving regional economic and social development and leading the development of other industries. Strategic emerging industries are characterized by strategic, competitive, leading, mobility and development. The industries play an important role to promote regional economic, social development and regional economic competitiveness. Strategic emerging industry choice refers to the government combining with the regional resource advantages and characteristics, on the basis of the scientific and reasonable evaluation, reasonable choose and determine the strategic emerging industries, and management activities to promote the development of the process. Correct selection of strategic emerging industries will benefit local effective use of regional resources and competitive advantage, to speed up the regional economic development [2]. On the other hand, the strategic emerging industry selection error will affect the development of local economy and society. So using the scientific method, evaluate the related industries from a host of economic and social development needs of industry, select the suitable strategic emerging industries, is the important content of government macroeconomic management, and is also the basic duty of the government [3].

12.3 Create a Grey Three-Tiered Strategic Emerging Industries Evaluation Index System

In order to objectively and effectively evaluate which industry can be used as a potential strategic emerging industry, adhere to the principles of scientific nature and operability, can be universal, choose some which can response characteristics of strategic emerging industries, and to obtain objective data from the current statistics index to evaluate as grey correlation theory, is the precondition of scientific choose strategic emerging industries [4]. According to the different scholars research on evaluation of strategic emerging industries, and combining with the characteristics of strategic emerging industries, this research will be strategic emerging industries as competitiveness index and economic benefit evaluation index design and sustainability indicators, financial indicators, innovative five level indicators. The five secondary indexes were designed under each category, and 25 secondary indexes were put together to form the secondary industry evaluation index system [5].

1. Competitiveness index, Z1, the index is mainly industry anti-risk ability, and in the economic and social position, competitive advantage, and the means of development of the enterprise ability. In combination with the target requirements, the five sub-indices of U1, total output value, U2, total value U3, total assets U4 and current assets U5, are taken.
2. The index of economic performance, Z2, measures the performance of industry price and the overall economic performance. One, based on the characteristics of the strategic emerging industries, it must be someone who has a good economic benefit, to grow as the leading and mainstay industry of industry is the appropriate choice of strategic emerging industry; Reflect the economic efficiency of enterprises. Two, Reflect the economic efficiency of enterprises many indicators, each indicator can't fully reflect its connotation, considering the acquired data, this study selected total assets contribution U7, U6, the asset-liability ratio, current assets turnover U8, cost, profit marginsU9and product sales U10 of five subsystems refers to as a secondary indicator of economic benefit index.
3. The financial indicator Z3 mainly examines and evaluates the profits of enterprises in the industry. The growth and development of an enterprise, needs to constantly invest, if without the support of surplus profit, enterprise development is unsustainable, it will bring great influence to industry development, industry is difficult to become a pillar industry in the economy. This paper examines the five sub-indexes of the average individual enterprise owners' equity U11, the total liabilities of U12, the main business revenue U13, the main business cost U14 and the total profits total U15.
4. The innovation indicator Z4 mainly measures the space of future development of the industry, including technological innovation potential, business model innovation and so on. Strategic emerging industry to highlight the industry pioneer and grasp core technology competence, therefore, must be those who dare to investment in the research and development, investment, and possess the core technology advantage industry, can in the face of competition, finally have the ability to deal with, and get development. This study from the R&D expenses within budget spending U17, test development U16, scientific research personnel for stepping, count U19 and owns a patent for invention patent application U20 is the index to evaluate.
5. The sustainability indicator Z5 focuses on industry and environmental compatibility. According to the international evaluation indicators related to the study on sustainable development, this article selects raw coal U22 consumption power U21, consumption of gasoline U23, U24, "three wastes" emissions and energy consumption per unit output value of U25 five indicators as a secondary indicator of the sustainability evaluation of strategic emerging industries.

12.4 The Multi-level Grey Evaluation Process of Strategic Emerging Industries

On the basis of the rational evaluation index system, the paper evaluates the strategic emerging industries using the multi-level grey evaluation method, and the basic evaluation process is as follows:

12.4.1 Select the Reference Number Column

Let's say that Vij is the actual value of the i(i = 1,2,. . ., n) of the industry for the j(j = 1,2,. . ., n). Take the various best values of Vij as the specific value of Vo which is:

Vo = (V01, V02,. . ., V0n)

Among them: Voj = Optimum(V_{ij}), i = 1,2,. . ., m, j = 1,2,..., n is a model of an m evaluation project, n indicators.

$$V = \left(V_{ij}\right)_{m\times n} = \begin{bmatrix} V_{11}\ V_{12}\ldots V_{1n} \\ V_{21}\ V_{22}\ldots V_{2n} \\ \cdot \\ \cdot \\ \cdot \\ V_{m1}\ V_{m2}\ldots V_{mn} \end{bmatrix} \tag{12.1}$$

The selected reference number is listed as: $V_o = (V_{01}, V_{02}, \ldots, V_{0n})$

12.4.2 The Metric Is Normalized

In order to ensure that the indexes are comparable, the formula is processed through normalization

$$X_{ij} = \frac{V_{ij} - \min_i V_{ij}}{\max_i V_{ij} - \min_i V_{ij}} \tag{12.2}$$

$$X = \left(X_{ij}\right)_{m\times n} = \begin{bmatrix} X_{11} & X_{12} & \cdots & X_{1n} \\ X_{21} & X_{22} & \cdots & X_{2n} \\ & \vdots & & \\ X_{m1} & X_{m2} & \cdots & X_{mn} \end{bmatrix} \tag{12.3}$$

12.4.3 Evaluate the Correlation Values ξij

After the normalized$X_0 = (x_{01}, x_{02}, \ldots, x_{0n})$ as a reference value,

$X_i = (x_{i1}, x_{i2}, \ldots, x_{in})(i = 1, 2, \ldots, m)$as a comparative sequence, deduced ξ_{ij} calculation formula is:

$$\xi ij = \frac{\min_i \min_j X_{oj} - X_{ij} + \xi \max_i \max_j X_{oj} - X_{ij}}{X_{oj} - X_{ij} + \xi \max_i \max_j X_{oj} - X_{ij}} \quad (\xi = 0.5) \tag{12.4}$$

By the formula above, $\xi ij = (i = 1, 2, m; j = 1, 2, \ldots, n)$ to obtain the following matrix:

$$E = (\xi_{ij})_{m \times n} = \begin{bmatrix} \xi_{11} & \xi_{12} & \cdots & \xi_{1n} \\ \xi_{21} & \xi_{22} & \cdots & \xi_{2n} \\ & \vdots & & \\ \xi_{m1} & \xi_{m2} & \cdots & \xi_{mn} \end{bmatrix} \tag{12.5}$$

In the formula, ξij is the jth index of the ith industry and the associated value of the jth reference index.

12.4.4 Calculates the Lowest Correlation Values

Because each metric plays a different role in the target layer, the calculation of the associated values is obtained by the product of the weighting and correlation coefficients. Using the AHP method, get the weight $W = (w_1, w_2, \ldots, w_n)$.

Among them:$\sum_{j=1}^{a} w_j = 1$, a represents the total number of indices.

The associated values $R = (r_i)_{1 \times m} = (r1, \ r2, \ldots, \ rm) - WE^{T.}$

12.4.5 The Final Correlation Value of the Model Is Chosen, and the Appropriate Strategic Emerging Industries Are Selected

Based on the method in 3.4, the correlation value of the target layer is obtained by layer by layer $r_i(i = 1, 2, \ldots, m)$. According to the size of r_i, the bigger the r_i is more suitable to cultivate and support development as a strategic emerging industry

12.5 Research on the Evaluation and Selection of Strategic Emerging Industries in Gansu Province

Based on Gansu province statistical yearbook 2015, selection of processing and manufacturing, health services, information industry and cultural tourism four characteristic industry figures as raw data, the empirical test on the above model, to analysis and study of Gansu province should choose single industry as the best for cultivating and developing strategic emerging industry. From Gansu province statistical yearbook from 2015 finishing get four industries (V_1, V_2, $V_{3,}V_4$) the index data of the original, as shown in Table 12.1:

According to Table 12.1, select the reference sequence

$$V_{ij}(i = 1,2,3,4, j = 1,2,3,\ldots,25)$$

$V_0 = (V_{01}, V_{02},\ldots, V_{0n}) =$ (629,971.27,932.71,871.92,421.74,60.25,49.1,6.25, 10.98.87, 1.23,46.3933.72,299.08,75.01,90567.2,89006.6,0.39,1155.59,401.35,7247, 5326,84,713.78,6.64)

According to the method of the model, the correlation value of 25 points and v0 is obtained by using the relevant software $\xi_{ij} = (i = 1,2,\ldots,m; j = 1,2,\ldots,n)$

As shown in Table 12.2.

Through the analytic hierarchy process (ahp), invite the industry experts hierarchy in Table 12.1, according to 1–9 scale method to construct double two judgment matrix, it is concluded that the weight of each level values are as follows:

$$\begin{gathered}\mathrm{W}_{\mathrm{YZ}}(0.1249, 0.3751, 0.0615, 0.3748, 0.0631)\\ \mathrm{W}_{\mathrm{Z1U}}(0.0579, 0.4698, 0.2401, 0.1167, 0.1202)\\ \mathrm{W}_{\mathrm{Z2U}}(0.3531, 0.1169, 0.1169, 0.3528, 0.0591)\\ \mathrm{W}_{\mathrm{Z3U}}(0.2222, 0.0563, 0.2222, 0.0563, 0.4444)\\ \mathrm{W}_{\mathrm{Z4U}}(0.3333, 0.1671, 0.0833, 0.3333, 0.0833)\\ \mathrm{W}_{\mathrm{Z5U}}(0.0833, 0.0833, 0.0833, 0.2498, 0.4978)\end{gathered}$$

Calculated by $\mathrm{R} = \mathrm{WE}^{\mathrm{T}}$, U associated values:

$$\begin{aligned} R_{Z1} &= W_{Z1}E_{Z1}^{T}(0.9989, 0.9964, 0.9960, 0.9989)\\ R_{Z2} &= W_{Z2}E_{Z2}^{T}(0.9969, 0.9989, 1.00000.9999,)\\ R_{Z3} &= W_{Z3}E_{Z3}^{T}(0.9996, 0.9992, 0.9991, 0.9989)\\ R_{Z4} &= W_{Z4}E_{Z4}^{T}(1.0000, 0.8628, 0.8203, 0.8712)\\ R_{Z5} &= W_{Z5}E_{52}^{T}(0.9548, 0.9140, 0.9661, 0.9558)\end{aligned}$$

The $\mathrm{E}_{\mathrm{B1C}}^{\mathrm{T}}$,$\mathrm{E}_{\mathrm{B2C}}^{\mathrm{T}}$,$\mathrm{E}_{\mathrm{B3C}}^{\mathrm{T}}$, $\mathrm{E}_{\mathrm{B4C}}^{\mathrm{T}}$, $\mathrm{E}_{\mathrm{B5C}}^{\mathrm{T}}$ respectively corresponding to the lines of data in Table 12.2 of matrix. Repeat the process to calculate the associated value of the Y layer:

Table 12.1 The initial data for four industry indicators

Indicator/unit	Health services V1	Processing manufacturing V2	Cultural tourism V4	Information industry V3
U1/numer	629	266	105	327
U2/100 million yuan	970.27	439.19	438.12	802.61
U3/100 million yuan	942.71	421.24	334.29	768.99
U4/100 million yuan	678.97	296.57	81.86	868.92
U5/100 million yuan	387.89	130.62	83.51	421.74
U6/%	67.09	21.82	60.25	32.37
U7/%	46.69	59.1	36.56	43.28
U8/frequency/ year	6.43	1.69	7.25	4.87
U9/%	7.66	10.7	5.01	9.83
U10/%	97.16	90.91	98.87	93.81
U11/million	0.86	0.46	0.62	1.03
U12/100 million yuan	284.93	245.62	46.3	162.96
U13/million	933.72	457.23	332.99	675.08
U14/100 million yuan	780.61	360.7	299.08	664.23
U15/100 million yuan	75.01	24.11	20.43	42.24
U16/10 thousand yuan	19006.6	7640.3	627.9	16272.1
U17/10 thousand yuan	70587.2	48340.3	28238.8	51083.4
U18/ten thousand people	0.39	0.17	0.03	0.31
U19/piece	1125	488	156	803
U20/piece	399	143	51	206
U21/gw	96,940	149,456	7387	86,898
U22/ton	65,591	286,988	79,710	7326
U23/ton	1928	9952	10,986	1065
U24/ton	4282	6233.72	713.78	1045.97
U25/ton	116.07	653.45	24.9	16.64

$RY(r_1, r_2, r_3, r_4) = W_{YZ}, R_{Z1}, R_{Z2}, R_{Z3}, R_{Z4}, R_{Z5} = (0.9971, 0.9229, 0.9501, 0.9530)$

According to Rr, if choosing right from the above four industry development in Gansu province to cultivate strategic emerging industries, in this research model validation, health services will be the best selection of industry, can be used as to nurture and support the development of strategic emerging industries.

Table 12.2 classifies index correlation tables

Refers to the standard	Health servicesV1	Processing and manufacturing V2	Cultural tourism V4	Information industry V3
ξi1	0.9998	0.9963	0.9962	0.9979
ξi2	1.0000	0.9971	0.9959	0.9988
ξi4	0.9987	0.9959	0.9944	1.0000
ξi5	1.0000	0.9979	0.9974	1.0000
ξi6	0.9997	0.9987	1.0000	0.9997
ξi7	0.9999	1.0000	0.9999	1.0000
ξi8	1.0000	0.9983	1.0000	1.0000
ξi9	1.0000	1.0000	1.0000	1.0000
ξi10	1.0000	0.9691	0.9997	1.0000
ξi11	1.0000	1.0000	1.0000	1.0000
ξi12	0.9986	0.9993	1.0000	0.9971
ξi13	1.0000	0.9966	0.9958	0.9989
ξi14	0.9996	0.9976	1.0000	0.9974
ξi15	1.0000	0.9998	0.9996	1.0000
ξi16	0.9200	0.6293	0.6426	0.7641
ξi17	1.0000	0.6327	0.7466	0.7672
ξi18	1.0000	1.0000	1.0000	1.0000
ξi19	1.0000	0.9953	0.9930	0.9961
ξi20	1.0000	0.9984	0.9976	0.9986
ξi21	0.7387	0.6937	1.0000	0.5207
ξi22	0.9121	0.3333	0.5987	0.8030
ξi23	0.9735	0.9939	1.0000	0.9931
ξi24	0.9686	0.9793	0.9871	0.9858
ξi25	0.9996	0.9342	0.9920	0.9820

12.6 Conclusion

The rational choice of strategic emerging industries is vital to the development of local economic and social development. The academic community has not able to find a feasible scientific method to effectively evaluate the industry as a strategic emerging industry. And, in the real economy and society, the statistical methods, statistical rules and other limitations, made some information about the strategic emerging industry is in a state of "grey", that is part of the state of information is not clear, even more difficult for industry to evaluate. A multilevel grey evaluation model, this study builds on the system data requires less, at the same time just have a small amount of calculation, on the characteristics of less sample size requirements, very suitable for use in the selection of strategic emerging industries. Evaluation results of this study strategic emerging industry in Gansu province, and determining the health services industry as a strategic emerging industry in Gansu province is suitable for resources characteristic and the economic and social situation in Gansu province. Gansu has the development of health service resources advantage, competitive advantage and good market prospects, it is reasonable for Gansu province to regard health service industry as a strategic emerging industry. This proves that the multi-level grey evaluation model is an effective method for industry evaluation and selection of strategic emerging industries.

Acknowledgement This research was financially supported by the key research project of Gansu university of Chinese medicine (No. XZD-01). and also supported by the Health management of key disciplines in Gansu University of Chinese Medicine.

References

1. Li, D. L. (2016). Selection of regional strategic emerging industries. *Development Research, 1*, 73–77.
2. Zhang, Y. (2013). Research on the development strategy of strategic emerging industries in Henan province based on multi-level grey evaluation. *Technology Management Research, 2004*, 47–50.
3. Zhang, L. Q., He, C. C., & Wu, Y. (2010). Based on the grey relation analysis evaluation of strategic emerging industries in biological medicine, for example. *Journal of Economic Mathematics, 03*, 79–84.
4. He, Z. C., & Zhang, X. (2014). The selection and evaluation of strategic emerging industries and the empirical analysis. *Science and Technology Management, 31*(12), 62–66.
5. Hao, M. L. (2011). The study of selection of regional strategic emerging industries. *Journal of North China Water Conservancy and Hydropower University, 2004*, 38–42.

Chapter 13
The Study of "Big Quality" Satisfaction Evaluation

Lizhong Tong and Taoyu Jia

Abstract The concept of "high quality" became more and more concerned in the world, the study of the "big quality" satisfaction evaluation is necessary, which is a basic work of macro quality management. This paper drew on the widely accepted ACSI model, expanded its range of application, and successfully constructed a new "big quality" satisfaction evaluation model. Based on this model, a "big quality" satisfaction evaluation system has been established. And then this study used Analytic Hierarchy Process to give weight to each evaluation index, so that the system can be used in empirical study in the further.

Keywords Quality · Satisfaction evaluation · Evaluation index system · AHP

13.1 Introduction

The International Standardization Organization defines "quality" as "the ability to meet requirements with a set of inherent characteristics". From the comprehensive definition of quality we can see that the concept of quality is not only about product or service, but also including engineering and environment, and so on. This is the concept of "big quality" that is popular all over the world. But in the research of the evaluation of quality satisfaction, most scholars still carry on the satisfaction evaluation of a product, a service or regional environment. Such as, Ramdass conducted a case study that the proposed company faced various problems with regards to product quality and customer satisfaction. The most fundamental definition of a quality product is one that meets the expectations of the customer. In order to use quality as a strategic advantage, quality dimensions relevant within the specific industry needs to be specified [1]. Kang combined the customer satisfaction,

L. Tong · T. Jia (✉)
Business School, Sichuan University, No.24 South Section 1, Yihuan Road, Chengdu 610065, P. R. China
e-mail: jiataoyu_scu@163.com

M. Tavana, S. Patnaik (eds.), *Recent Developments in Data Science and Business Analytics*, Springer Proceedings in Business and Economics,
https://doi.org/10.1007/978-3-319-72745-5_13

customer demand structure and modern products theories to study the satisfaction evaluation of products, and used AHP to establish evaluation system [6]. Park, Kim et al. adopted a Delphi method to identify suitable criteria for evaluating the farm accommodation quality and applied the analytic hierarchy process to determine the relative weights of those evaluation criteria [2]. Ghiasi's study focused on satisfaction with the Plant Clinics services with the objective to examine farmers' familiarity, willingness to use, and willingness to pay for services offered by Plant Clinics in Guilan Province of northern Iran [4]. Hsu, Hung et al.'s study aims to propose a hybrid analytic network process (ANP) model as an improved method to evaluate electronic service quality with the interdependence perspective [3].

There are also some scholars noticed the evaluation of "big quality" satisfaction from the macro perspective, Cheng 's study based on "big quality" category, put forward the construction of regional Total quality index (TQI), which is to measure the overall quality of region, including products, services, engineering and environment [7]. Feng 's study is based on the "big quality" concept, which is divided into four dimensions of products, services, engineering and environment. According to the geographical seven division of China, a total of 15,000 samples was selected to conduct a questionnaire survey of satisfaction. But the study failed to design the evaluation system on the basis of customer satisfaction theory, and weight the evaluation indicators [8].

This paper is also about the evaluation of quality satisfaction. However, the research object is not limited to certain aspects, but about "big quality" status of an area which including products and services, engineering, environment as four aspects. The focus of this research is to establish the evaluation system of the "big quality" satisfaction. The difference from the previous research is no longer simply divided the "big quality" into products, services, engineering, and environment to evaluate the public satisfaction respectively but combined a customer satisfaction model and transformed it to establish a new evaluation system.

At present, there are some related models to assess the public satisfaction, such as the Swedish SCSB model, the German DK model, the Korean KCSI model, the Malaysia McsI model, etc.. But the most widely accepted and used one is the American ACSI model. Such as, Chai&Fan's study analyzed the relations between user motivations, attempted to establish a user satisfaction evaluation model for mobile SNSs based on the Uses and Gratifications Theory on media communication and the American Customer Satisfaction Index [5]. Du exceeded the scope of application of the ACSI model, combined with the characteristics of the retail industry, and used the method of fuzzy judgment to extend and improve the ACSI model. [9].

Based on the model, this study puts forward a new quality satisfaction evaluation system, with quality expectation, quality consciousness, quality perception and quality participation and weights each of it.

Analytic Hierarchy Process (AHP) is a method of combining the subjectivity and objectivity by comparing the indexes of the same level by experts and establishing the judgment matrix. After a series of mathematical treatments, the weights of each index are obtained. Therefore, this paper chooses to use Analytic Hierarchy Process to assign weight to each index.

13.2 Quality Satisfaction Evaluation Model

The American Customer Satisfaction (ACSI) evaluation model is a customer-based evaluation system used to evaluate and improve organizational performance. The structural model of ACSI consists of the above six structure variables. The final target variable is customer satisfaction, the causal variables are consumer's expectation, customer expectation and perceived value and the outcome variables are customer complains and customer loyalty. Of these six structural variables, each structural variable contains one or more observed variable(s) and the data of the observed variables are generally obtained by actual investigation.

ACSI system is applicable to all different business, industry and departmental customer satisfaction whose evaluation results are intercomparable. Not only does it make customer satisfaction comparable in different products and industry, but also in different customers using the same product.

Therefore, customer satisfaction measurement can also be extended from the product and service quality areas to the field of engineering and environmental quality.

Inspired by the ACSI model variables, the evaluation of the satisfaction of the four quality areas can derive the following quality satisfaction evaluation model on the original basis (Fig. 13.1):

This model only focuses on the evaluation of the satisfaction of quality, so the variables of "perceived value" and "customer loyalty" in ACSI model are removed, the "customer complaint" is extended to "quality participation" and the "quality awareness" variable is created. The details as follows:

Quality perception: It is equivalent to the "perceived quality" in the ACSI model, which refers to the subjective feeling and psychological experience of the quality of the public. Quality expectations and quality participation can affect quality perception while quality perception can also affect social quality satisfaction.

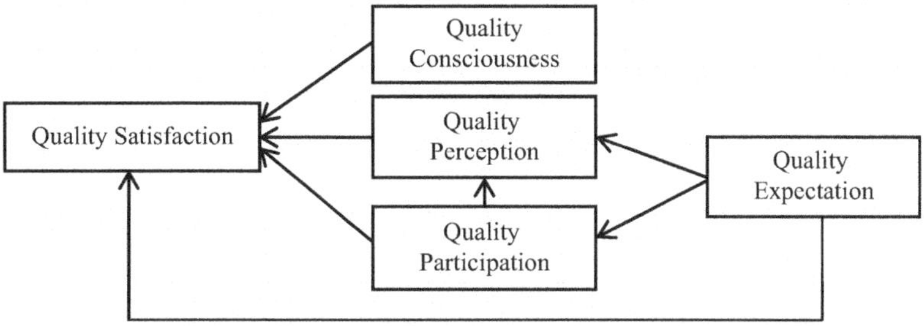

Fig. 13.1 Quality satisfaction evaluation model

Quality expectation: It is equivalent to the "customer expectation" in the ACSI model, which refers to the quality level expected by the public and the quality of construction requirements based on past and present experience.

Quality awareness: As a new variable, it mainly refers to whether the public has quality awareness. Social public awareness of quality is the cornerstone of social quality satisfaction.

Quality participation: It refers to the public's own initiatives on the quality issues, including the initiative of supervision, participation and the variable "customer complaints" in ACSI model which include encouraging the public to report fake and shoddy products and participation in product quality supervision. "Customer complaints" model makes the public feel the quality construction is closely related to their own lives which leaves a feeling of self-realization and enhances satisfaction.

13.3 Establishment of Quality Satisfaction Evaluation System

Refine the first level index including product, service, engineering and environment to 12 second level index on the basis of quality evaluation model so as to determine 33 observable variable as follows (Table 13.1):

The quality consciousness indicator is reflected by quality concerns, quality activities, and quality complaints. Specifically speaking, the quality concerns refers to the public attention to the four quality and the quality complaints refers to the public awareness of complaints and will.

The survey content of quality consciousness is what the "customer perception" is in the ACSI model, which directly investigates the public perception of the four quality.

Quality participation is divided into three parts: quality supervision, quality promotion and quality activities which investigates whether the public has conducted quality supervision through rights protection, carried out quality promotion and participated in quality related activities.

The quality expectation indicator is explained by quality expectations, which, namely, is investigation the expectations of the public about the quality of the four quality.

Table 13.1 Quality satisfaction evaluation system

Objective	First level index	Second level index	Observed variables
Quality satisfaction	Quality consciousness	Quality concerns	Concerns about product quality
			Concerns about service quality
			Concerns about engineering quality
			Concerns about environment quality
		Quality complaints	Complaint consciousness
			Complaint will
	Quality perception	Perception of product quality	Perception of agricultural product quality
			Perception of consumer goods quality
			Perception of import and export product quality
			Perception of food product quality
			Perception of drug & medical equipment quality
			Perception of dangerous chemicals quality
		Perception of Service Quality	Perception of tourism & hotel quality
			Perception of transport services quality
			Perception of communication services quality
			Perception of electrical utility services quality
			Perception of express mail services quality
			Perception of health services quality
		Perception of engineering quality	Perception of architectural engineering quality
			Perception of transport engineering quality
			Perception of hydrographic engineering quality
		Perception of Environment Quality	Perception of air quality
			Perception of water quality
			Perception of acoustic environment quality
	Quality participation	Quality supervision	Frequency of safeguarding rights
		Quality publicity	Frequency of publicizing quality
		Quality activity participation	Frequency of quality activity participation

(continued)

Table 13.1 (continued)

Objective	First level index	Second level index	Observed variables
	Quality expectation	Quality expectation	Product quality expectation level
			Service quality expectation level
			Engineering quality expectation level
			Environment quality expectation level
		Quality improvement expectation	Product quality improvement expectation degree
			Service quality improvement expectation degree
			Engineering quality improvement expectation degree
			Environment quality improvement expectation degree

Table 13.2 Evaluation index

Objective	First level index	Second level index
Quality satisfaction (U)	Quality consciousness (U1)	1. Quality concerns
		2. Quality complaints
	Quality perception (U2)	1. Perception of product quality
		2. Perception of service quality
		3. Perception of engineering quality
		4. Perception of environment quality
	Quality participation (U3)	1. Quality supervision
		2. Quality publicity
		3. Quality activity participation
	Quality expectation (U4)	1. Quality expectation
		1. Quality improvement expectation

13.4 Weighting Quality Satisfaction Evaluation Index

13.4.1 Determination of Evaluation Index Set

Through the interview with scholars in the field of quality research, government officials of quality departments and managers of enterprise market departments, the evaluation index set have formed as shown in Table 13.2.

Table 13.3 Quality satisfaction system index weight table

Dimension	The weight of dimension	Index	The weight of index
Quality consciousness U1	0.1061	U11	0.2
		U12	0.8
Quality perception U2	0.5536	U21	0.25
		U22	0.25
		U23	0.25
		U24	0.25
Quality participation U3	0.0513	U31	0.6833
		U32	0.1998
		U33	0.1168
Quality expectation U4	0.2890	U41	0.8
		U42	0.2

13.4.2 *The Calculation of the Weight of the Index*

In this experiment, expert group scoring method is used to obtain the importance judgment matrix as the source of weight calculation. Being different from the standard of regular selection of experts. This experiment invited six experts (quality professors, quality-related government officials and enterprise market managers) and four civilians who are concerning about products, services, engineering and environment quality to participate in the process of scoring to the judgment matrix .

The expert scoring is a total of three rounds, which lasted 3 weeks, After the expert group compared and corrected scores using "1–9 scale method" several times, the following judgment matrix was achieved:

(U as an example here)

$$\mathrm{U} = \begin{bmatrix} 1 & 0.2 & 3 & 0.25 \\ 5 & 1 & 7 & 3 \\ 0.33 & 0.14 & 1 & 0.17 \\ 4 & 0.33 & 6 & 1 \end{bmatrix} \tag{13.1}$$

Corresponding weighting results:

$w_2 = \{0.1061, 0.5536, 0.0513, 0.2890\}$, λ max $= 4.1725$, $CR = CI/RI = 0.0646 < 0.1$, That means the uniformity of the judgement matrix is acceptable. Similarly, the weight of other indexes is shown in Table 13.3.

13.5 Conclusion and Further Work

The issues of quality satisfaction evaluation have attracted the interest of researchers since 1980s. Continuing the previous works in this area, the work has successfully achieved its objectives.

The first contribution of the work was the establishment of "big quality" satisfaction evaluation model and index system. Based on the ACSI model, this paper expanded its application scope and successfully built a new "big quality" satisfaction evaluation model and system, which provides a reference for the scholars of relevant fields and government department manager. The second, instead of studying the satisfaction of one kind of product or service, this research focused in the category of "big quality" including product, service, engineering and environment, which is a rare perspective in the field. Finally, this paper used AHP to give weight to each evaluation index, so that the system can be used in empirical study, which is the further work of this research.

Acknowledgments The fund project: Study on the Role of Local Government in Regional Brand Formation from the Perspective of Supply-side Reform—Taking Sichuan Province as an Example (SC16TJ001).

References

1. Ramdass, K. (2012). *Quality dimensions relevant to a first tier automotive supplier: Case study at an automotive seat cover supplier.* 2012 I.E. International Conference on Industrial Engineering and Engineering Management (Ieem) (pp. 1835–1839).
2. Park, D. B., et al. (2017). The development of quality standards for rural farm accommodations: A case study in South Korea. *Journal of Hospitality & Tourism Research, 41*(6), 673–695.
3. Hsu, T. H., et al. (2012). A hybrid ANP evaluation model for electronic service quality. *Applied Soft Computing, 12*(1), 72–81.
4. Ghiasi, R., et al. (2017). Crop protection services by plant clinics in Iran: An evaluation through rice farmers' satisfaction. *Crop Protection, 98*, 191–197.
5. Chai, J. X., & Fan, K. K. 2016. *User satisfaction and user loyalty in mobile SNSs: Wechat in China.* Proceedings of 2016 International Conference on Applied System Innovation (ICASI).
6. Kang, D., & Xumei, Z. (2003). Research on evaluation index system and method of product customer satisfaction. *Computer integrated manufacturing system, -CIMS, 05*, 407–411H.
7. Lei, F., & Yu, W. (2016). Analysis on the survey of Residents' satisfaction degree in China from 2014 to 2015. *Standard Science, 05*, 88–93.
8. Cheng, H., & Qingquan, L. (2009). Research on regional mass index model system and evaluation in China. *Management World, 01*, 2–9.
9. Xuan, D., Wei, L., & Xiaojun, T. (2007). Construction of the retail customer satisfaction evaluation model and application based on extended ACSI model. *statistics and decision, 10*, 113–115.

Chapter 14
Research on Evaluation Index System of Enterprise Brand Competitiveness: Taking Liquor Industry as an Example

Yan Fan and Lizhong Tong

Abstract With the increasingly fierce market competition, the competition among enterprises has risen to brand competition. The brand competitiveness is essentially a different advantage beyond the competitors, which enable the enterprises to gain a sustainable competitiveness advantage and a stable development. So, in this paper, building a scientific and reasonable evaluation index system of brand competitiveness for enterprises, taking the liquor industry as an example, which helps to make a scientific and accurate evaluation for the enterprise brand competitiveness and provide a decision-making basis for building brand and enhancing the brand competitiveness for the enterprise.

Keywords Enterprise brand · Brand competitiveness · Evaluation index system

14.1 Introduction

China's liquor industry is the traditional industries which have a long history and cultural heritage, and Sichuan wine is well-known for Wuliangye, Luzhou Laojiao and other famous brands throughout the country. However, due to many foreign wine brands entering the Chinese market in recent years, many Chinese liquor enterprises have been facing the stress and challenge increasingly. The brand is the source of enterprise development, determining the core competitiveness of enterprises. Therefore, the enterprises must build excellent brands and make unremitting efforts to continuously improve their brand competitiveness if they want to stand out in the fierce competition. Under the background, building the brand competitiveness evaluation index system of liquor industry enterprises, through drawing lessons from the research results of domestic and foreign scholars on the evaluation index system of brand competitiveness and fully considering the liquor industry inheritance and

Y. Fan · L. Tong (✉)
Business School, Sichuan University, Chengdu, China
e-mail: tonglifh@126.com

M. Tavana, S. Patnaik (eds.), *Recent Developments in Data Science and Business Analytics*, Springer Proceedings in Business and Economics,
https://doi.org/10.1007/978-3-319-72745-5_14

development of traditional culture and the sustainable development ability of brand, can objectively evaluate brand competitiveness of the liquor industry enterprises, but also as the basis of comparison of the liquor brand competitiveness with industry applicability.

14.2 Literature Review

On the definition of brand competitiveness, the domestic and foreign scholars have different views. The following views are the current understanding of the brand competitiveness.

The foreign scholar Philip Kotler thought that the essence of the brand is the invisible contract between enterprises and consumers [1]. They reached an invisible contract that the enterprises make commercial commitments with the quality of products or services and other projects and the consumers pay a "brand premium"to companies as leverage, which formed the brand competitiveness. And David A Aaker argues that the brand competitiveness is the ability to shape and support a strong brand in a given market environment.

The domestic scholar Xu Jinan defined the brand competitiveness: the enterprise can better meet the needs of consumers than the competitor's brand through the effective allocation and the use of resources, thus expanding market share and obtaining the high profits and the comparative brand competitive ability [2]. Liu Yingqiu thought that: the brand competitiveness refers to reflect the capacity of production organization, technological innovation, marketing service and the brand market developing and capturing of the enterprise, industry, regional and national [3]. He Amao believed that: the brand competitiveness is the result of brand competition and is a comprehensive ability to participate in market competition, specifically reflecting the particularity and difficulty to imitate in the brand, image, personality, service and other aspects, which makes the enterprises occupy the market, access to dynamic competitive advantage and long-term profits.

This paper establish the comprehensive evaluation index system of brand competitiveness of liquor industry enterprises based on the understanding of brand competitiveness and the previous brand competitiveness evaluation index system research results, according to the degree of attention of the traditional cultural heritage and the lasting vitality of the liquor industry.

14.3 Research Methods

The main research methods adopted in this paper are:

1. Delphi method

The relevant information of the brand competitiveness evaluation index system of the liquor industry enterprises are sent to the university and representatives of the six experts and senior users 4, through the expert scoring in the form of the judgment matrix to build the various elements of the score [4], to determine the importance of the indicators and get a more satisfactory answer in the constant feedback.

2. Analytic Hierarchy Process (AHP)

It is the combination of the quantitative and qualitative analysis that expressing and dealing with subjective judgments in quantity [5]. In this paper, the judgement matrix is determined by the comparison of the two factors of AHP, and the eigenvector is obtained by the 1–9 scale to determine the weight coefficient of the index which is an important basis for calculating the final target value [6].

14.4 The Construction of the Enterprise Brand Competitiveness Evaluation Index System in the Liquor Industry

On the basis of the study of the brand competitiveness literature, this paper focuses on exploring the construction of the relevant brand competitiveness evaluation system in the research object of the liquor industry enterprises [7]. The three principles of validity, practicability and instructiveness need to follow in the construction of enterprise brand competitiveness evaluation index system, to build scientific and rational construction of the index system and ensure the collectability of the data. So it can help enterprises to scientifically assess the competitiveness of brands and improve the weak aspects of the system through the evaluation results, thereby enhancing the brand's competitiveness.

14.4.1 Determination of the Evaluation Index System of Brand Competitiveness in Liquor Industry

The competitiveness index is the basis of reflection of the accurate and objective competitiveness of enterprises. Based on the connotation of brand competitiveness, this paper analyzes the indicators of brand competitiveness of liquor industry in terms of resources, market, profitability and development, combining the feasibility of data collection, and builds the evaluation index system from the five aspects of the brand management ability, the brand market ability, the brand financial ability and the brand development potential containing 5 first level index and 28 second level index (specific content as shown in Table 14.1).

Table 14.1 The brand competitiveness evaluation index system of the liquor industry enterprise

Target	First level index	Second level index
The brand competitiveness of the liquor industry enterprise	The brand base ability A1	The management expense profit ratio B1
		The number of technical personnel B2
		The human capital development costs B3
		The investment in brand culture construction B4
		The brand association degree B5
		The number of brand patents B6
		The R & D innovation investment B7
	The brand management ability A2	The number of market segments B8
		The advertising costs B9
		The brand distribution ability B10
		The brand image B11
		The brand scale expansion ability B12
	The brand market ability A3	The market share B13
		The market coverage B14
		The number of new market B15
		The investment of product development B16
	The brand financial ability A4	The main business income B17
		The total profit B18
		The rate of return on total assets B19
		The profit margin of the cost and expense B20
		The sales growth rate B21
		The profit growth rate B22
		The operating cost profit ratio B23
	The brand development potentiality A5	The brand awareness B24
		The brand reputation B25
		The brand loyalty B26
		The customer satisfaction B27
		The technology leadership B28

14.4.2 Determination of the Weight of Evaluation Index System of Brand Competitiveness in Liquor Industry

On the basis of the evaluation index system of brand competitiveness of the liquor industry enterprises, it is necessary to determine the weight vector W_n of each level index in the evaluation index system of brand competitiveness of liquor industry due to the existence of complex hierarchical relationship among the evaluation indexes, which is to quantitatively describe the importance of each indicator in the overall index system. The weight of each index refers to its importance to the system for each indicator according to the impact of the various indicators to corporate brand competitiveness.

To determine the steps of the weight of the evaluation index system:

1. The importance scoring of each level index and the construction of each judgment matrix

In this paper, we use the Analytic Hierarchy Process (AHP) to form the weight distribution scheme, the experts grading and correcting the evaluation index system by using the "1–9 scaling method", to get judgement matrix of every level index Uij and calculate and determine the evaluation index weight system according to the specific mathematical method.

$$\mathbf{Uij} = \begin{bmatrix} u11 & u12 & u13 & u14 & \ldots \\ u21 & u22 & u23 & u24 & \ldots \\ u31 & u32 & u33 & u34 & \ldots \\ u41 & u42 & u43 & u44 & \ldots \\ \ldots & \ldots & \ldots & \ldots & \ldots \end{bmatrix} \tag{14.1}$$

According to the judgment matrix, the eigenvector W corresponding to the maximum eigenvalue λmax is obtained by using the matrix weighted average method, that is, the corresponding weight of each level index.

2. Consistency check

In view of the contradiction between the two dimensions of multi-dimension and multi-factor, it is necessary to verify the consistency of the importance of the process before and after the comparison of the weight. When the weight is obtained, the comparison of the two comparisons should be compared with the calculation of the consistency index.

By using the "1–9 scale method" of the evaluation index system by the expert group, the following judgment matrix was obtained (here take the brand development potential as an example):

$$\mathbf{U5} = \begin{bmatrix} 1 & 2 & 3 & 2 & 4 \\ \frac{1}{2} & 1 & 2 & \frac{1}{2} & 3 \\ \frac{1}{3} & \frac{1}{2} & 1 & \frac{1}{2} & 2 \\ \frac{1}{2} & 2 & 2 & 1 & 2 \\ \frac{1}{4} & \frac{1}{2} & \frac{1}{3} & \frac{1}{2} & 1 \end{bmatrix} \tag{14.2}$$

The corresponding weights are calculate and obtained by the method:

W_5 = {0.0662, 0.0339, 0.0215, 0.0418, 0.0144}, consistency check available CI = 0.0293, when n = 5, RI = 1.12, so CR = CI / RI = 0.0262 < 0.1, describing that judgment matrix consistency is acceptable. Similarly, other indicators available right levels of weight are obtained (as shown in Table 14.2), which lay the foundation for the liquor industry brand competitiveness measurement and comparison.

Table 14.2 The brand competitiveness evaluation index weight table of the liquor industry enterprise

Target	First level index	Second level index	The index weight Wn (n = 1,2,3,. . .,28)
The brand competitiveness of the liquor industry enterprise	The brand base ability A1	The management expense profit ratio B1	0.0101
		The number of technical personnel B2	0.0203
		The human capital development costs B3	0.0166
		The investment in brand culture construction B4	0.0166
		The brand association degree B5	0.0101
		The number of brand patents B6	0.0207
		The R & D innovation investment B7	0.0332
	The brand management ability A2	The number of market segments B8	0.0126
		The advertising costs B9	0.0117
		The brand distribution ability B10	0.0226
		The brand image B11	0.0153
		The brand scale expansion ability B12	0.0325
	The brand market ability A3	The market share B13	0.1234
		The market coverage B14	0.0670
		The number of new market B15	0.0430
		The investment of product development B16	0.0282
	The brand financial ability A4	The main business income B17	0.0438
		The total profit B18	0.0680
		The rate of return on total assets B19	0.0208
		The profit margin of the cost and expense B20	0.0317
		The sales growth rate B21	0.0557
		The profit growth rate B22	0.0896
		The operating cost profit ratio B23	0.0288
	The brand development potentiality A5	The brand awareness B24	0.0662
		The brand reputation B25	0.0339
		The brand loyalty B26	0.0215
		The customer satisfaction B27	0.0418
		The technology leadership B28	0.0144

Table 14.3 The brand competitiveness evaluation criteria of the liquor industry enterprise

THE comprehensive evaluation value E	<30	30~55	55~75	75~90	>90
The brand competitiveness	Very weak	Weak	Medium	Strong	Very strong

14.4.3 *Quantitative Analysis of Target Value of Enterprise Brand Competitiveness in Liquor Industry*

The liquor industry enterprises brand competitiveness evaluation index system, some of which indicators are qualitative, and others are quantitative. In the conversion process, the evaluation method of the different types of indicators can also be different. These qualitative indexes which are very fuzzy are quantified by grade evaluation method. In the index system of this article, the score of the qualitative evaluation index through the questionnaire survey can be obtained by computer processing. (the score is full of 100 points, divided into five grades: excellent 100 points, good 80 points, medium 60 points, low 40 points and bad 20 points), the rest of the indicators are the quantitative indicators whose value can be directly or indirectly calculated.

To calculate the total score of the brand competitiveness evaluation system of the liquor industry, only the weight value and the dimensionless score of each evaluation index are substituted into the mathematical model $E = \Sigma wi*Pi$, where E is the total score and wi is the weight of the index i, Pi is the score of the evaluation index, and n is the number of the evaluation index. Drawing the level of evaluation method, the brand competitiveness can be divided into five grades (as shown in Table 14.3) with reference to the existing industry brand competitiveness evaluation criteria. The enterprises can identify their position and determine the next goal according to the comprehensive evaluation of the value, so as to enhance their own brand competitiveness.

14.5 Conclusions and Remarks

Based on the comprehensive analysis of the existing relevant theories and achievements, this paper constructs the comprehensive evaluation index system of brand competitiveness of liquor industry enterprises that can scientifically and accurately evaluate the brand competitiveness of enterprises and provide the basis for decision-making for enterprises to comprehensively understand and improve their own ability and better participate in the market competition. The brand competitiveness evaluation index system, mainly through the way of the mathematical statistics to calculate the comprehensive evaluation value E, is intuitive and strong persuasive. It can be used for the evaluation of enterprise competitiveness and analysis, to provide evaluation basis and reference.

Many questions need to carry on the thorough research and the analysis because of the limited time and the difficulties of data acquisition, the following are some prospects for further research: (1) the use of empirical research to prove the scientific and feasibility of the index system, and according to the specific single enterprise brand five modules of the respective scores to determine the module layer on the enterprise brand competitiveness E contribution (that is the brand weight), to identify gaps and unreasonable places to determine the next phase of enterprise brand building the focus of the direction. (2) Further analysis of the brand base ability, the brand management ability, the brand market ability, the brand financial ability and the brand development potentiality, and to find the fundamental problems and core issues in the current enterprise brand building from their respective secondary indicators level, and to take targeted countermeasures for related issues to further enhance the brand competitiveness of enterprises.

Acknowledgements The authors are grateful to the financial supports from the Sichuan Science and Technology Project: Study on the Role of Local Government in Regional Brand Formation from the Perspective of Supply-side Reform-Taking Sichuan Province as an Example (SC16TJ001).

References

1. Kotler, P. (2003). *Marketing management* (11th ed.). First Edition. Shanghai: Shanghai People's, Publishing House.
2. Nan, X. J. (2005). *Rresearch on brand competitiveness*. Beijing: Economic Management Press.
3. Liu, Y., & Xu, Z. (2005). *China's private enterprise competitiveness report*. Beijing: Social Science Literature Publishing House.
4. Yaping, Z. (2010). *Study on hotel brand competitiveness evaluation based on analytic hierarchy process* (pp.33–39). Suzhou University.
5. Wei, S., Tao, C., & Hu, S. (2008). Evaluation of brand competitiveness: Domestic research review. *Marketing Guide, 6*, 55–60.
6. Haiyan, G. (2012). *The evaluation model construction and application research of the brand competitiveness index*. Liaoning: Liaoning University.
7. Yaqi, J., & Yaping, Z. (2011). The evaluation and measurement of enterprise brand competitiveness based on AHP. *Economic Research Guide*. Serial No. 118. pp. 139–141.

Chapter 15
Simulation of Stochastic Volatility Variance Swap

Shican Liu, Yanli Zhou, Yonghong Wu, and Xiangyu Ge

Abstract This paper aims to propose efficient mathematical model of variance swap to study the effect of stochastic volatility in different time-scales on the option pricing. Two types of stochastic volatility, including Ornstein-Uhlenbeck (OU) process and Cox-Ingersoll-Ross (CIR) process are considered. Analytical solution of CIR model is presented. For the OU process, a numerical algorithm based on the finite element approach is established for solution of the model.

Keywords Variance swaps · Time-scale · Stochastic volatility · Finite element method

15.1 Introduction

A variance swap is a financial instrument which allows investors to speculate on the spread between future volatility and implied volatility. Variance swap provides us a straightforward method to cover the exposure risk of the volatility of the underlying asset. Recently, many researchers have investigated the variance pricing based on the classical Greek option with constant volatility which lead to the underlying process of a fat-tailed distribution. The stochastic volatility model is one of the

S. Liu
Department of Mathematics and Statistics Curtin University, Bentley, Australia

School of Statistics and Mathematics, Zhongnan University of Economics and Law, Wuhan, China

Y. Zhou
School of Finance, Zhongnan University of Economics and Law, Wuhan, China

Y. Wu
Department of Mathematics and Statistics Curtin University, Bentley, Australia

X. Ge (✉)
School of Statistics and Mathematics, Zhongnan University of Economics and Law, Wuhan, China

M. Tavana, S. Patnaik (eds.), *Recent Developments in Data Science and Business Analytics*, Springer Proceedings in Business and Economics,
https://doi.org/10.1007/978-3-319-72745-5_15

approaches to overcome the shortcomings of the constant volatility models. Carr and Madan [1] briefly reviewed three different methods for trading realized volatility, including static replication, delta hedge, and volatility contract, and connected their work with the stochastic volatility. However, little work has been done to study the variance swap pricing problem under stochastic volatility.

The concept of time scales in Finance was first introduced by Fouque et al. in 1998, in Ref. [2] in which the option pricing model with fast-scale stochastic volatility is proposed. In practice, variations of data, include high frequency data always appears only in the short period, while low frequency data appears in the long period.

Many numerical algorithms have been proposed to study the time-scale option pricing problem. Little and Pant [3] applied the finite difference method (FDM) the variance swaps problem based on constant volatility [4], in which a two-dimensional (2D) problem was reduced to a one-dimensional one, and the price of variance swap was obtained as an average of the 2D solutions. As well known, the stochastic volatility emerges as a solution to the constant volatility, which has been studied for years. Zhu and Lian [5] applied the Fourier transformation to price variance swaps with discrete sampling times and found a closed-form solution of the Heston's two-factor stochastic volatility model [6].

In this paper, we extend Zhu and Lian's work to study the variance swap based on the fast-scale model. A little attempt has been done on using the OU process for the variance swap problem. Numerical approximation is carried out using Finite Element Method. By introducing the technique implanted by Little and Pant [3], the 3D model reduces to a 2D model. The model is then split into two stages at $t_{i-1} \leq t \leq t_i$, and $t_i \leq t \leq T$ respectively. The solution of the second stage at $t_i \leq t \leq T$ is first carried out and is then implemented to the initial solution of the first state. The solution in Zhu and Lian [5] is used as a benchmark to show the validity of our algorithm. The effect of maturity time and different time-scale rates on strike price is investigated, also the long level convergence of the strike is showed in our numerical results.

The rest of this paper is as follows. In Sect. 15.2, we set up the model. Section 15.3 concerns the numerical study of the problem. Section 15.4 is the conclusion of this paper.

15.2 Model Setup

This section concerns dynamics of the underlying asset, which can be described by the following stochastic differential equation:

$$dS_t = \mu S_t dt + \sigma_t S_t dW_t^{(0)}, \tag{15.1}$$

where S_t denotes the stock price, $dW_t^{(0)}$ is a Geometric Brownian motion, σ_t represents volatility which is driven by a factor diffusion process and is determined by $\sigma_t = f\left(Y_t^{(k)}\right)$, with the factor $Y_t^{(k)}$ obtained from a Gaussian OU process [7]:

$$dY_t^{(k)} = k\left(m - Y_t^{(k)}\right)dt + v\sqrt{2k}dW_t^{(1)}, \tag{15.2}$$

for the drift scale $k > 0$ and the diffusion coefficient scale $\sqrt{k}$. From (15.1) and (15.2), $W_t^{(1)}$ and $W_t^{(0)}$ is correlated with the covariation $\mathrm{cov}\left(W_t^{(t)}, W_t^{(0)}\right) = \rho_1$.

The scale of k affects the volatility process significantly. When k is large, the process is referred to the fast scale process; otherwise, is known as the slow scale process. The value of a variance swap at the expiration date can be written as $V_T = L^*$ $(\sigma^2 - K)$, where σ^2 is the final realized volatility, K is the strike price and L is the variance amount. In the risk-neutral world, the value of variance swap at time t is denoted by $V_T = E\left[e^{-r(T-t)}\left(\sigma_R^2 - K\right)L\right]$. We let $V_0 = 0$, because there is no cost to enter a swap at the right beginning. Based on the martingale property, $K = E[\sigma^2]$ is obtained. The problem thus becomes to calculate the realized volatility σ^2. According to Little and Pant [3], the final realized volatility is defined as

$$\sigma_R^2 = \frac{AF}{N}\sum_{i=0}^{N-1}((S_{i+1} - S_i)/S_i)^2, \tag{15.3}$$

where AF is an animalization factor and N is number of the expected scheduled trading days in the observation period. The AF value of 252 is used when the sampling frequency is every trading day, 52 for everyweek and 12 for everymonth. As it is shown in (15.3), there are two underlying processes in the final payoff function, which makes the problem difficult to deal with. In this work, the method used by Zhu and Lian [5], and Little and Pant [3] are implemented. Firstly, a new variable I_t is introduced as

$$I_t = \int_0^t \delta(t_{i-1} - \tau)S_\tau d\tau, \tag{15.4}$$

where δ is the Dirac-delta function, which means $I_t = 0$ if $t < t_{i-1}$, and $I_t = S_{i-1}$ if $t \geq t_{i-1}$. By the usual no-arbitrage argument, we rewrite (15.1), and (15.2) into the forms:

$$\begin{cases} dS_t = rS_t dt + \sqrt{Y_t}S_t dW_t^{(0)}, \\ dY_t = \left(k(m - Y_t) - \lambda v\sqrt{2k}\right)dt + v\sqrt{2k}dW_t^{(1)}, \end{cases} \tag{15.5}$$

Where λ denotes the Risk price which is the same as Heston [6]. Letting $U(t, S, Y, I)$ be the price of a derivative whose payoff at time point t_{i+1} from t_i is $((S_{i+1} - S_i)/S_i)^2$, and according to Fouque and Sircar [8] and Feyman-Kac Theorem [5], we obtain:

$$U_t + \frac{1}{2}YS^2 U_{SS} + \rho v S\sqrt{2kY_t}U_{SY} + v^2 kU_{YY} + r(U_S - U) \\ + \big(k(m - Y) - \lambda v\sqrt{2k}\big)U_Y + \delta(t_{i-1} - t)U_I = 0 \tag{15.6}$$

with the terminal condition $U(t, S, Y, I) = (S/I - 1)^2$. According to the definition of $U(t, S, Y, I)$, we have $e^{r(t_i - t)}U(t, S_t, Y_t, I_t) = E_0^Q\left[(S_i/I_i - 1)^2\right]$. Based on the definition (15.4), I_t is deterministic and is only related to the previous state of S_{i-1}, which means that even though S is a stochastic process, I_t can be determined if we fix the previous stage of S_{i-1}. The variation of a deterministic process equals to zero, and the proof of this argument can be found in Klebaner et al. [9]. Because the discounted price of U is a martingale, by $V(t, S, Y, I) = e^{-rt}U(t, S, Y, I)$, we obtain.

$$dV = e^{-rt}(-rUdt + dU) \tag{15.7}$$

From Ito's formula,

$$dU = U_t dt + U_S dS + U_Y dY + U_I dI + \frac{1}{2}U_{SS}[dS, dS] + \frac{1}{2}U_{YY}[dY, dY] + U_{XY}[dX, dY] \\ = \Big(U_t + \frac{1}{2}YS^2 U_{SS} + \rho v S\sqrt{2kY}U_{SY} + v^2 kU_{YY} + r(U_S - U) \\ + \big(k(m - Y) - \lambda v\sqrt{2k}\big)U_Y + \delta(t_{i-1} - t)U_I\Big)dt + \phi(X, Y, I, t)dW \tag{15.8}$$

where ϕ is a complicated function of X, Y, I and t. By substituting (15.8) into (15.7), and using the Martingale Representation Theorem, (15.6) is obtained by taking expectation of (15.7). Let $x = \ln S$, $y = \ln Y$ and $\gamma = \ln I$, lEq. (15.6) becomes

$$U_t + \frac{1}{2}yU_{xx} + \rho v\sqrt{2ky}U_{xy} + v^2 kU_{yy} + r(U_x - U) \\ + \big(k(m - y) - \lambda v\sqrt{2k}\big)U_y + \delta(t_{i-1} - t)U_\gamma = 0 \tag{15.9}$$

with the terminal condition $U(T, x, y, \gamma) = (e^{x-\gamma} - 1)^2$. Based on the property of Dirac-delta function, Eq. (15.9) can be represented by two-stages PDEs.

Stage one for $0 \le t < t_{i-1}$,

$$U_t + \frac{1}{2}yU_{xx} + \rho v\sqrt{2ky}U_{xy} + v^2 kU_{yy} + r(U_x - U) \\ + \big(k(m - y) - \lambda v\sqrt{2k}\big)U_y = 0, \lim_{t \uparrow t_{i-1}} U_i = \lim_{t \downarrow t_{i-1}} U_i. \tag{15.10}$$

Stage two for $t_{i-1} \le t \le Y$,

$$U_t + \frac{1}{2}yU_{xx} + \rho v\sqrt{2ky}U_{xy} + v^2 kU_{yy} + r(U_x - U) \\ + \big(k(m - y) - \lambda v\sqrt{2k}\big)U_y = 0, U_i(x, y, \gamma, T) = (e^{x-\gamma} - 1)^2. \tag{15.11}$$

Let $t_{i-1} = T - \Delta t$, and $\Delta t = T/N$, $N = 1, 2, \ldots, \tilde{N}$ stage one and stage two should be solved by backward algorithm.

15.3 Numerical Analysis

In Subsect. 15.3.1, the finite element method (FEM) is applied to solve the model with the CIR process, and our approach will be benchmarked by making a comparison between the approximate solution and the semi-analytical solution at each point. Also, for the reason that it is no analytic solution for the model with the OU process, we investigate the relationship between the time scale rate k, the maturity time T and the expected value of strike price through numerical solution based on the FEM method in Subsect. 15.3.2.

15.3.1 Validity Study

In order to show that the proposed model is applicable, we apply the model (15.12) and assumption as used by Zhu and Lian [5], and compare the approximated solution with the closed form solution. The Heston Model:

$$\begin{cases} dS_t = rS_t dt + \sqrt{v_t} S_t d\tilde{W}_t^S, \\ dY_t = k^*(\theta^* - v_t)dt + \sigma_V \sqrt{v_t} d\tilde{W}_t^V, \end{cases} \tag{15.12}$$

where $k^* = k + \lambda$ and $\theta^* = k\theta/(k + \lambda)$ are the risk-neutral parameters, λ is the premium of volatility risk [6]. The parameters used here is the same as those in Zhu and Lian [5], namely $k^* = 11.35$, $\theta^* = 0.022$, $\sigma_V = 0.618$ while we choose $v_0 = 0.5$ in this paper. We also apply the same assumption, as in Zhu and Lian [5] that the strike price is defined by

$$K_{\text{var}} = \kappa^* 10^4 = e^{r\Delta t} E[(S_{t_i} - S_{t_{i-1}})/S_{t_{i-1}}] * 10^4/T \tag{15.13}$$

which is only related with the time step size and value of v_t. Figure 15.1 shows the comparison of the FEM approximation and the exact solution. Clearly, in Fig. 15.1, the strike price falls as maturity time increases. Even though the difference between the two methods becomes larger and larger as the maturity time increases, we can apply this method for the reason that exact solution does not always exist, and even in Zhu and Lian's paper, they derived the semi-analytic solution with the integral form instead of the exact solution. The FEM method is accurate to describe the tendency and in most cases the maturity time cannot be that long. In order to increase the absolute accuracy and make our method more persuasive, we calculate the κ in Eq. (15.13) instead of K_{var} with the mesh size of 100, and then obtain Fig. 15.2.

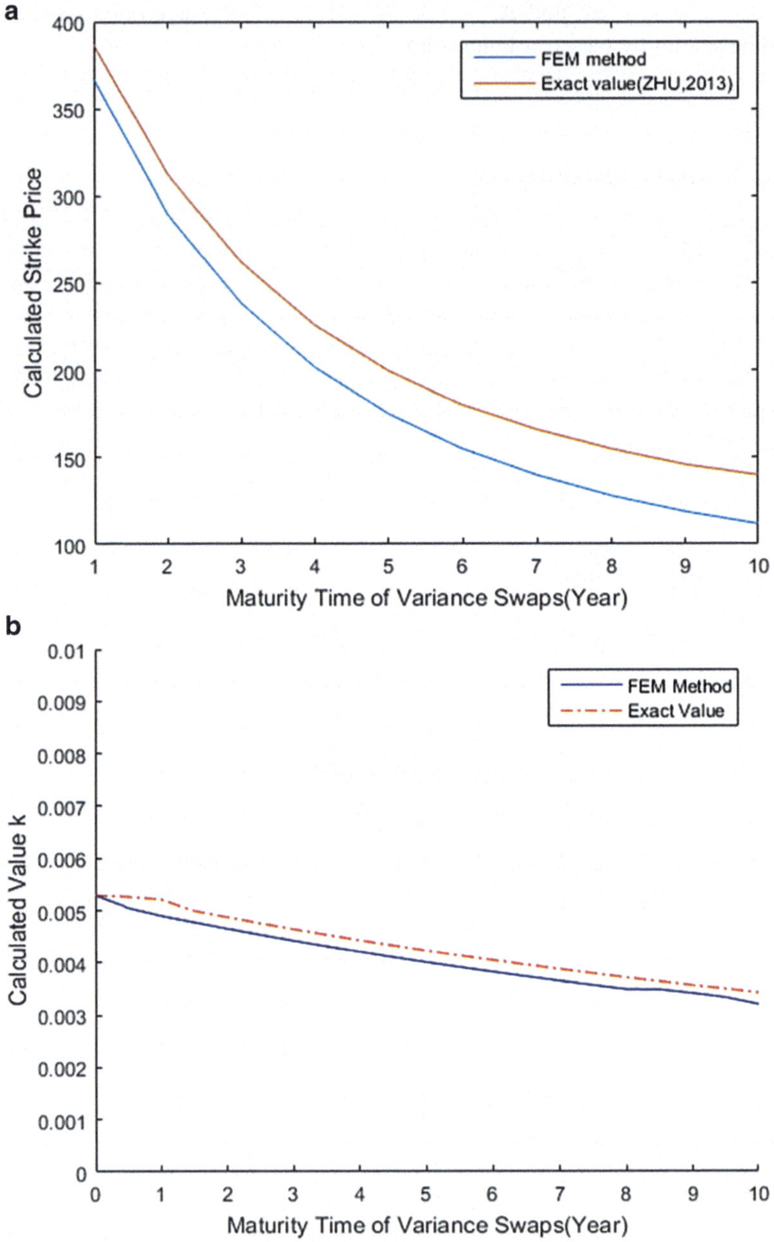

Fig. 15.1 Calculated strike values as a function of maturity time

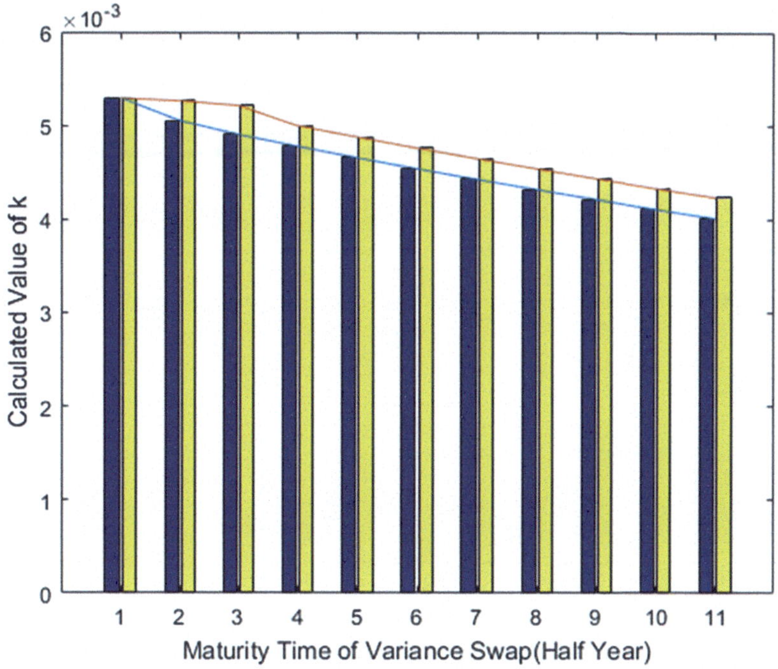

Fig. 15.2 Relationship between maturity time and strike price

15.3.2 Numerical Results of Our Model

Zhu and Lian's paper is based on Heston's two-factor stochastic volatility model. In Heston's model, the stochastic volatility process is a CIR process, from which it is easy to construct a closed form solution by using Heston's Scheme [6]. However, if the stochastic volatility process is the OU process instead of the CIR process, it is not easy to obtain analytical solution by simply constructing a specified form. The approximate solution is thus obtained instead.

As it is shown in Fig. 15.3a, the strike price is inverted anti-correlated with the time scale rate. There is a mechanism behind the phenomenon: Lager k brings more risk exposure, which contributes more to the strike price. But this effect will not go to infinity, when k is larger than one, the strike price experiences a slightly decrease and converges to zero.

Also, Fig. 15.3b shows the relationship between the maturity time T and the strike price. Obviously, the strike price is anti-correlated with the maturity time. It decreases sharply when the maturity time T is less than 1.5 years and approaches a steady level when time goes by. This agrees with the result proved by Zhu and Lian [5]. The result verifies that volatility provides a measure of risk exposure. The longer the investors hold the contract, the higher risk they have to take.

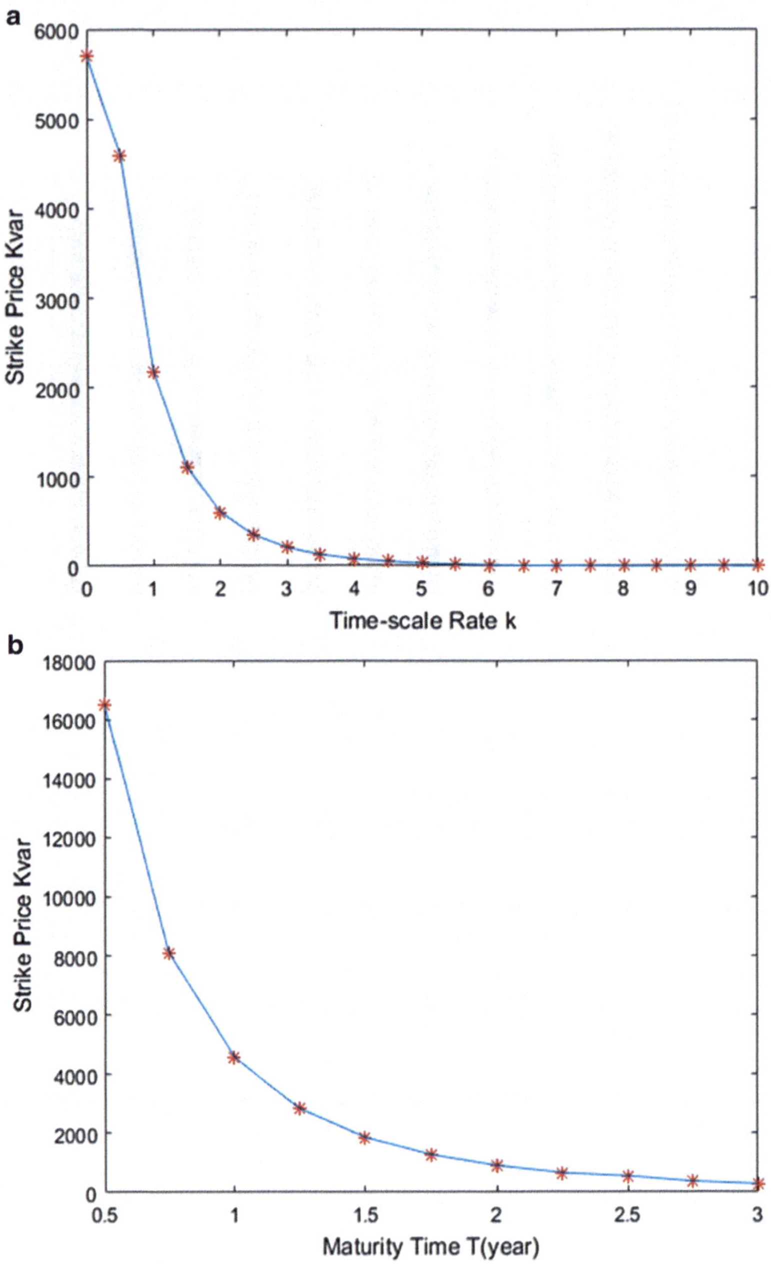

Fig. 15.3 Relationship between the maturity time and strike price

15.4 Conclusion

In this paper, we apply the finite element method to obtain the approximate solution of variance swaps under stochastic volatility. The time scale rate of stochastic volatility is considered to describe the long term and short term perturbation and draw the conclusion that the strike price of variance swap is anti-correlated with the time scale rate, especially when k is less than one. Also, for the reason that the volatility is a measure of risk, the strike price falls when the maturity time increases. We have also compared the results produced by the FEM method with the model with the CIR process for describing the volatility and found that our approximate solution agrees with the exact solution. The significance of this work can be illustrated in two aspects. First of all, the exact solution can only be obtained for specified models. For most PDE, we cannot derive the closed form solution, which makes the numerical approach necessary. Besides, even though most work has considered the stochastic volatility, they do not study the property of the stochastic volatility, we apply the time scale rate to describe our model and show how it works on the variance swaps pricing.

Acknowledgments This research work is supported by Humanities and Social Science fund of Chinese Ministry of Education (17YJC630236).

References

1. Carr, P., & Madan, D. (1998). Towards a theory of volatility trading. *Volatility: New Estimation Techniques for Pricing Derivatives, 29*, 417–427.
2. Fouque, J. P., Papanicolaou, G., & Sircar, K. R. (1998). Asymptotics of a two-scale stochastic volatility model. In *Comparative Biochemistry & Physiology Part B Comparative Biochemistry*, 53(2), 187–190.
3. Little, T., & Pant, V. (2001). *A finite-difference method for the valuation of variance swaps.* Quantitative Analysis in Financial Markets: Collected Papers of the New York University Mathematical Finance Seminar (Vol. III), pp. 275–295.
4. Black, F., & Scholes, M. (1973). The pricing of options and corporate liabilities. *The Journal of Political Economy, 81*(3), 637–654.
5. Zhu, S. P., & Lian, G. H. (2011). A closed form exact solution for pricing variance swaps with stochastic volatility. *Mathematical Finance, 21*(2), 233–256.
6. Heston, S. L. (1993). A closed-form solution for options with stochastic volatility with applications to bond and currency options. *Review of Financial Studies, 6*(2), 327–343.
7. Schobel, R., & Zhu, J. W. (1999). Stochastic volatility with an Ornstein--Uhlenbeck process: An extension. *European Finance Review, 3*(1), 23–46.
8. Fouque, J. P., Papanicolaou, G., & Sircar, K. R. (2000). *Derivatives in financial markets with stochastic volatility*. Cambridge: Cambridge University Press.
9. Klebaner, F. C., et al. (2005). *Introduction to stochastic calculus with applications*(Vol. 57). London: World Scientific\Imperial College Press.

Chapter 16
Empirical Research of the Contribution Rate of University Science and Technology to the Regional Economic Development

Tianhang Wang and Fengge Yao

Abstract In this paper, the development of the times, scientific and technological progress has become the decisive factor of regional economic growth. As an important part of scientific and technological progress, University Science and technology has made important contributions to the regional economic development. The research shows that the rate of scientific and technological progress in Colleges and universities is 1.46%, the contribution rate of regional economic development is 10.09%, and its scientific research efficiency is about 3 times the efficiency of scientific research institutions and enterprises. Therefore, increasing the investment of scientific research funds in universities has a very important role in improving the efficiency of scientific research and promoting the rapid development of regional economy.

Keywords University science and technology · Technological progress rate · Contribution rate

16.1 Introduction

Since the twenty-first century, the world economy has rapid development, human society gradually from the industrial age into the era of knowledge economy, the growing status and role of science and technology, have become the fresh power to promote the economic development, improving the comprehensive national strength important factor. As an important part of our country's scientific and technological innovation in colleges and universities, has been charged in the forefront of scientific

T. Wang (✉)
The Institute of Finance, Harbin University of Commerce, Harbin, China
e-mail: 15046005471@163.com

F. Yao
The Institute of Commercial Economy Research, Harbin University of Commerce, Harbin, China

M. Tavana, S. Patnaik (eds.), *Recent Developments in Data Science and Business Analytics*, Springer Proceedings in Business and Economics,
https://doi.org/10.1007/978-3-319-72745-5_16

research front, in order to promote our country's scientific and technological progress, promote national and regional economic development has made outstanding contributions. In the process of research, this article uses Eviews8 as auxiliary software, set up multiple regression model to explore the productivity and the intrinsic relation between scientific research spending, then calculate the technological progress's contribution.

16.2 Model Selection

University of science and technology is a part of social science and technology, to explore the progress of science and technology's influence on the economic development of colleges and universities, you first need to understand the productivity contribution to economic output level of science and technology, measurement of university science and technology progress in the overall ratio of science and technology progress, finally get the university science and technology progress contribution rate to economic growth. Therefore, in the measurement of the contribution rate of science and technology in Liaoning province [1].

16.2.1 Technical Productivity Model Selection

At present, there are many ways to calculate the contribution rate of science and technology at home and abroad, such as sermon glass production function method, the elasticity of substitution production function, solo residual method, growth factor analysis, DEA method, logarithmic production function method, etc. Among them, solo's residual method is favored by researchers and is widely used in the research field of technology contribution rate [2].

In this article, using solo residual method for scientific and technological progress rate using the most commonly used C-D production function, according to Thoreau improvements on C-D production function, technology, productivity can be divided into initial science and technology and scientific and technological progress, scientific and technological progress as a function of time t.

In order to avoid the influence of multiple co-linear, it is assumed that the scale compensation is constant, that is, a + b = 1, and the model as follows:

$$\mathrm{Y} = Ae^{mt}K^{a}L^{b} \tag{16.1}$$

Y = output, A = the initial technical level, m = the rate of technological progress, $\mathrm{Ae^{mt}}$ = A period of technology productivity level, K = capital input, L = the labor input, A and b = the output elasticity of capital and labor. We can divide the Eq. (16.1) by both sides by L.

$$\mathrm{Y/L} = Ae^{mt}(K/L)^{a} \tag{16.2}$$

Take the logarithm of both sides:

$$\ln(Y/L) = \ln A + a\ln(K/L) + mt \tag{16.3}$$

16.2.2 *The Model Construction of the Contribution Rate of Scientific and Technological Progress in Colleges and Universities*

On the basis of estimating the contribution rate of scientific and technological progress to regional economic development, we can estimate the contribution rate of science and technology in colleges and universities [3]. The progress of regional science and technology is generally derived from the research of universities, research institutes and research departments.

In our country, colleges and universities, scientific research institutions and enterprises is the main research and development spending, T can establish technology productivity and the relationship between departments of scientific research funds, in the case of omitted distractions, model is as follows:

$$T = e^{A}G^{c}Q^{d} \tag{16.4}$$

In formula (16.4), A is a constant term, and G is the internal expenditure of research and development funds of universities, Q is the internal expenditure of research and development funds of other scientific research institutions and enterprises [4].

To estimate the parameters, you need to take the logarithm of the Eq. (16.4) to make the model:

$$\ln T = A + c\ln G + d\ln Q \tag{16.5}$$

The relationship between technical productivity and r&d expenditure can be obtained by estimating the values of A, c and d by parameter estimation of Eq. (16.5).

The technical progress rate is obtained by using the whole differential Eq. (16.5):

$$\lambda = dT/T = c{*}dG/G + dQ/Q \tag{16.6}$$

16.3 Data Selection and Processing

This paper, taking Liaoning province as an example, chooses and processes the data of the above model variables.

16.3.1 Technical Productivity Model Data Selection and Processing

In the measurement of scientific and technological productivity, the basic model of choice involves total output, capital input and labor input into three variables. Technical progress at home and abroad the research about the relation between economic growth, gross domestic product (GDP) is chosen mostly as output indicators, in here, this paper also made the same choice, with GDP of Liaoning province [5]. According to the total assets of three economic census in Liaoning province and the new calendar year (removal of first industry) fixed assets value, can get a new liquid assets and the proportion of new fixed assets is about 0.84:1, according to the new can calculate the new calendar year total assets, fixed assets and total capital throughout the years.

16.3.2 The Selection and Disposal of College Funds Data

When selecting data for scientific research funds, there are two kinds of data in the yearbook, which are the internal expenditures of research and development funds and the internal expenditure of science and technology activities.

The activities of science and technology in recent years due to the internal spending this measure, all the choice of research and development expenses within budget as required for variable data, and a total funds for research in Liaoning in 1997 before the yearbook of does not exist, in order to guarantee the consistency of the data, this article choose in Liaoning province from 1998 to 2013 in research and development spending data as the variable of the model for numerical data from Liaoning statistical yearbook and China statistical yearbook of science and technology.

16.4 An Empirical Study on the Contribution Rate of Scientific

16.4.1 Measurement of Technical Productivity

The data in Eviews8 and Table 16.1 are used to analyze the Eq. (16.3) and get the following results:

$$\ln Y/L = 0.5135 \ln K/L + 0.069t - 138.3946$$
$$(3.8650) \quad (5.0381) \quad (-5.1307)$$

Table 16.1 The value of total output, total assets, labor and ratio of Liaoning province

Year	Output	Capital	Labor	Ln(Y/L)	Ln(K/L)
2003	5386.70	36,977.50	1318.10	1.4077	3.3341
2004	5873.60	40,572.57	1376.10	1.4512	3.3838
2005	7164.90	44,898.76	1398.20	1.6340	3.4692
2006	8365.10	49,920.52	1411.90	1.7791	3.5655
2007	10,030.90	56,516.07	1475.00	1.9170	3.6459
2008	12,366.60	71,286.01	1497.50	2.1112	3.8629
2009	13,797.60	83,693.27	1579.60	2.1673	3.9700
2010	16,826.20	1,00,791.34	1613.90	2.3443	4.1344
2011	20,311.13	1,16,992.89	1665.01	2.5013	4.2523
2012	22,690.58	1,37,979.83	1729.16	2.5743	4.3795
2013	24,997.00	1,56,553.40	1835.07	2.6117	4.4463

Two coefficient and t test value of the constant term 3.8650, 5.0381 and 5.0381, respectively, F = 1376.526, variables are visible through the t test and F test, at the same time, the coefficient of determination R2 = 0.9953, the goodness of fit is very good also, but the DW = 1.0660 < DL = 1.1, shows that residual autocorrelation sequence, need to eliminate residual autocorrelation, after using the generalized difference method is used to eliminate the first-order autocorrelation, regression analysis for the model are as follows:

$$\begin{aligned} \ln Y/L = 0.532 \ln K/L + 0.0661t - 132.6987 \\ (2.0860) \quad (2.3040) \quad (-2.3441) \end{aligned}$$

Eliminate autocorrelation, DW = 1.4865, the model coefficients and constant term t test value of 2.0860, 2.3040 and 2.3040 respectively, F = 417.7831, the explanation of the variables in the model through the t test and F test, and R2 = 0.9858.The model shows that the technological progress rate m = 0.0661 = 6.61%, and the elasticity of capital output a = 0.532, according to a + b = 1, the elasticity of labor output is equal to 0.468.

According to formula (16.5) and a and b, the model can be obtained by moving items:

$$y = \lambda + 0.532 + 0.4681$$

The model of economic growth obtained from the above shows the relationship between economic growth rate and technological progress rate, capital growth rate and labor growth rate. After the progress of science and technology, according to formula (16.6), the contribution rate of scientific and technological progress can be calculated, as shown in Table 16.2.

Table 16.2 Expenditure on research and development of Liaoning province in 2003–2013

Year	G	Q	Ln G	Ln Q	g(%)	q(%)
2003	7.6545	75.3154	2.0353	4.3217	25.89	15.02
2004	10.2049	96.7093	2.3229	4.5717	33.32	28.41
2005	12.2408	112.4678	2.5048	4.7227	19.95	16.29
2006	15.6078	120.1779	2.7478	4.7890	27.51	6.86
2007	17.3274	148.0715	2.8523	4.9977	11.02	23.21
2008	21.1333	168.9329	3.0508	5.1295	21.96	14.09
2009	23.8643	208.5044	3.1724	5.3400	12.92	23.42
2010	24.4612	263.0091	3.1971	5.5722	2.50	26.14
2011	32.6010	331.2338	3.4843	5.8028	33.28	25.94
2012	37.2531	353.6149	3.6177	5.8682	14.27	6.76
2013	41.3586	404.5736	3.7223	6.0028	11.02	14.41
Ave	–	–	–	–	21.59	21.77

16.4.2 The Calculation of the Contribution Rate of Science and Technology in Colleges and Universities

According to type (16.6), must carry on the university of science and technology contribution rate is calculated to estimate the first university of science and technology progress rate and growth rate of GDP, GDP growth rate has been calculated in Table 16.2, so the most important thing is to measure the rate of scientific and technological progress.

Using the data in Table 16.2, the combination Eq. (16.6), after eliminating the autocorrelation, makes the parameter estimation, which can get linear regression results:

$$\ln T = 0.0777 \ln G + 0.2712 \ln Q - 1.6499$$
$$(2.1588) \quad (7.4226) \quad (-13.7144)$$

The numerical value of the upper brackets is the t statistic, and the coefficients and constants are all passed the t test. In the regression results, F = 274.1481, R2 = 0.9856, the model also passed the F test, and the fitting is good.

The regression results showed that every 1% increase in expenditure in colleges and universities increased by 0.0752% in scientific and technological productivity, 1% in research institutions and enterprises and 0.2529% in scientific and technological productivity.

So I'm going to take the logarithm of the upper and the right hand side,

$$dT/T = 0.777dG/G + 0.2712dQ/Q$$

The growth rate of scientific and technological progress and the growth rate of college funds is equal to the growth rate of scientific research institutions, which can be:

$$\lambda = 0.0777g + 0.2717q$$

According to the college in Table 16.2 average growth rate of funds, funds for scientific research institutions and enterprises average growth rate can be calculated lambda Liaoning province scientific and technological progress rate of 7.57%, 1.67% for university science and technology progress, the other 5.90% for scientific research institutions and enterprises of science and technology progress, the regression model of scientific and technological progress rate values and the average rate of scientific and technological progress in Table 16.2 numerical similar, smaller error exists.

Acknowledgments Therefore, increasing the investment in science and technology and encouraging the innovation of science and technology to drive economic growth will lead to higher efficiency and better results.

Research and development spending growth rate of 21.59% in colleges and universities, scientific research institutions and companies fund growth rate of 21.77%, the growth rates are similar, but under the condition of the same research and development spending, the efficiency of scientific research in colleges and universities more than scientific research institutions and enterprises, so that increase the funds to promote economic growth plays an important role in Liaoning [6]. To establish scientific research evaluation system based on market survey, according to the law of value to judge the real value of scientific research achievements, replace the number of scientific research with scientific research value, improve the scientific research achievements of actual use value, the number of output less practical transformation and application of poor or hard to the number of research projects, increase the proportion of applied research in scientific research in colleges and universities.

For generic technology and proprietary technology, realize the conversion of scientific research, generally uses the school-run enterprises, investment in technology, scientific research personnel self-employed, technology to sell or transfer such as form, at the same time encourage interdisciplinary cross-industry cooperation of science and technology, to maximize the use scientific research achievements.

References

1. Cui, Y. (2001). China's education contribution to economic growth. *Education and the Economy*, (01), 1–5.
2. Mao, S. (2010). Higher education workforce and China's economic growth. *Statistical Study*, (05), 34–39.
3. Zhou, G. (2014). The empirical study on the influence of technological innovation ability on the regional economy of Jiangsu Province. *China's Trade*, (28), 174–176.
4. Li, X. (2015). From a statistical perspective, the relationship between science and technology development and economic development in Zhejiang Province is seen. *Journal of Shaoxing University of Arts and Sciences*, (04), 64–68.
5. Jiaming, L. (2004). Research on the contribution rate of Chinese science and technology to economic growth. *Research on Natural Dialectics*, (02), 42–52.
6. Lingling, C. (2006). The empirical analysis on the contribution rate of economic growth in Anhui Province. *Anhui Science*, (01), 46–47.

Chapter 17
Research on Incentive Strategy of Logistics Outsourcing About Manufacturing Enterprises

Bo Yang and Wenwen Jin

Abstract Facing of China's manufacturing industry development status quo, this paper based on evolutionary game theory foundation and relationship between manufacturing enterprises and TPL to build game model of logistics outsourcing. Then this paper analyzes evolutionary game to get optimal strategies. It is in order to make game both sides to achieve maximum benefits. Finally, according to model results, this paper proposes strategies to enhance enthusiasm of outsourcing cooperation and promotes common development of manufacturing enterprises and TPL. It makes manufacturing companies respond quickly to markets and with high quality services meet individual needs of consumers.

Keywords Manufacturing enterprises · TPL · Outsourcing · Evolutionary game

17.1 Introduction

Since the reform and opening up, although China's manufacturing industries have made rapid development, there is a big gap in technological innovation with developed countries [2]. Facing these statuses, advantages that establish in the past are difficult to bring huge profits for enterprises [6, 7]. At present, the study of manufacturing enterprise logistics outsourcing has been concerned many scholars. In the literature, the authors from different areas use game theory to study manufacturing enterprise logistics outsourcing. In these documents, they are based on that game players are completely rational preconditions, but in real life, it is difficult to achieve. Therefore, from reality of the situation, we should be from a limited rational point to analyze manufacturing enterprise logistics outsourcing [5]. In addition, there are few studies on its incentive mechanism [3, 4]. Facing manufacturing enterprises present situation, the study on logistics outsourcing incentive mechanism that has become a

B. Yang · W. Jin (✉)
JiangXi University of Finance and Economics of Information Management, Nanchang, China
e-mail: 635709019@qq.com

M. Tavana, S. Patnaik (eds.), *Recent Developments in Data Science and Business Analytics*, Springer Proceedings in Business and Economics,
https://doi.org/10.1007/978-3-319-72745-5_17

hot. Therefore, this paper researches logistics outsourcing incentive mechanisms of manufacturing enterprises and TPL. Using evolutionary game theory, establishing evolutionary game model of encourage behavior and analyzing the game's balance under different conditions results, then drawing the appropriate conclusions.

17.2 Constructing Evolutionary Game Model of Manufacturing Enterprises and Tpl Enterprise

Evolutionary game theory is based on incomplete rational and incomplete information [1], combined with game theory analysis and dynamic evolution process. In the game processes, both sides improve their strategy by learning and exploring. And ultimately tend to a stable strategy. Compared with the classical game theory, the evolutionary game theory makes up its shortcomings. So it is widely used to various fields, especially in the process of enterprise cooperation competition.

17.2.1 Set of Model Parameters

In the evolutionary game model of logistics outsourcing include manufacturing enterprises(c) and TPL(v).θ_i: normal benefits; α_i: information shares in outsourcing; β_i: business matching in outsourcing; γ_i: the level of corporate confidence; ϑ_i: information gain coefficient;μ_i: synergies coefficient; π_i: additional revenue coefficient; ρ_i: risk; C_i: cost; φ: encourage cooperation coefficient; ω_i: punishment.

17.2.2 Replication Dynamic Equation and Stability Strategy Analysis of Model

Each participant has two options: "Outsourcing", "No Outsourcing"; "Cooperation", "Non-cooperation". The income matrix of two sides can be divided into four cases and based on analysis, the income matrix of both sides is shown in Table 17.1:

The probability of manufacturing enterprises choosing outsourcing is x, the probability of TPL choosing cooperation is y, which $o \leq x \leq 1, 0 \leq y \leq 1$. Also suppose the time variable t, with the passage of time t, the x and y will eventually remain stable.

The average income of manufacturer is: $\overline{u_c} = xu_{c_1} + (1 - x)u_{c_2.}$

$$= \theta_c + y(x\vartheta_c + x\mu_c + \pi_c - x\pi_c)\gamma_v\beta_v\alpha_v + x(\varphi - \rho_c)\gamma_c\beta_c\alpha_c + (x - 1)\omega_c - xC_c \quad (17.1)$$

Then the average income of TPL is: $\overline{u_v} = yu_{v_1} + (1 - y)u_{v_2}$

Table 17.1 Income matrix for both sides of the game

		TPL	
		Cooperation	Non-cooperation
Manufacturing enterprises	**Outsourcing**	$\theta_c + (\vartheta_c + \mu_c)\gamma_v\beta_v\alpha_v + (\varphi - \rho_c)\gamma_c\beta_c\alpha_c - C_c$	$\theta_c + (\varphi - \rho_c)\gamma_c\beta_c\alpha_c - C_c$
		$\theta_v + (\vartheta_v + \mu_v)\gamma_c\beta_c\alpha_c + (\varphi - \rho_v)\gamma_v\beta_v\alpha_v - C_v$	$\theta_v + \pi_v\gamma_c\beta_c\alpha_c - \omega_v$
	No Outsourcing	$\theta_c + \pi_c\gamma_v\beta_v\alpha_v - \omega_c$	$\theta_c - \omega_c$
		$\theta_v + (\varphi - \rho_v)\gamma_v\beta_v\alpha_v - C_v$	$\theta_v - \omega_v$

$$= \theta_v + x(y\vartheta_v + y\mu_v + \pi_v - y\pi_v)\gamma_c\beta_c\alpha_c + y(\varphi - \rho_v)\gamma_v\beta_v\alpha_v + (y-1)\omega_v - yC_v \quad (17.2)$$

The dynamic equation of manufacturer replication is: $F(x) = dx/dt = x\left(u_{c_1} - \overline{u_c}\right)$

$$= x(1-x)\left(y(\theta_c + \mu_c - \pi_c)\gamma_v\beta_v\alpha_v + (\varphi - \rho_c)\gamma_c\beta_c\alpha_c + \omega_c - C_c \right. \quad (17.3)$$

Order $dx/dt = 0$, available $x^* = 0, x^* = 1$ or $y^* = (\varphi - \rho_c)\gamma_c\beta_c\alpha_c + \omega_c - C_c/(\theta_c + \mu_c - \pi_c)\gamma_v\beta_v\alpha_v$

Similarly, TPL replication dynamic equation is: $F(y) = dy/dt = y\left(u_{v_1} - \overline{u_v}\right)$

$$= y(1-y)\left(x(\theta_v + \mu_v - \pi_v)\gamma_c\beta_c\alpha_c + (\varphi - \rho_v)\gamma_v\beta_v\alpha_v + \omega_v - C_v\right) \quad (17.4)$$

Order $dy/dt = 0$, available $y^* = 0$, $y^* = 1$or $x^* = (\varphi - \rho_v)\gamma_v\beta_v\alpha_v + \omega_v - C_v/(\theta_v + \mu_v - \pi_v)\gamma_c\beta_c\alpha_c$

17.3 Solving the Evolutionary Game Model of Manufacturing Enterprises and Tpl

In logistics outsourcing game, the payoff function main body should be interest and cost, incentive and punishment play tutoring effect. So the sum of the incentives and penalties should be less than the sum of costs and risk. In evolutionary game, there are five local equilibrium points of game: (0,0),(1,0),(0,1), (1,1),(x^*,y^*). According to the local stability analysis of Jacobi's proof to analyze the stability of the equilibrium point.

1. $(\vartheta_c + \mu_c)\gamma_v\beta_v\alpha_v + (\varphi - \rho_c)\gamma_c\beta_c\alpha_c - C_c > \pi_c\gamma_v\beta_v\alpha_v - \omega_c$ and $(\vartheta_v + \mu_v)\gamma_c\beta_c * \alpha_c + (\varphi - \rho_v)\gamma_v\beta_v\alpha_v - C_v > \pi_v\gamma_c\beta_c\alpha_c - \omega_v$, the results are shown in Table 17.2:

As shown in Fig. 17.1, the stability strategy of the evolution game is $x^* = 0$, $y^* = 0$ and $x^* = 1$, $y^* = 1$. It because two sides of game share less information and two sides are weak in absorption and conversion capacity of information. So eventually tend to: manufacturing enterprises choose outsourcing strategy, TPL choose cooperation strategy or manufacturing enterprises choose no outsourcing strategy, TPL choose non-cooperation strategy. It is a Pareto policy balance that manufacturing enterprises adopt outsourcing strategy and TPL adopts cooperative strategy.

2. $(\vartheta_c + \mu_c)\gamma_v\beta_v\alpha_v + (\varphi - \rho_c)\gamma_c\beta_c\alpha_c - C_c < \pi_c\gamma_v\beta_v\alpha_v - \omega_c$ and $(\vartheta_v + \mu_v)\gamma_c\beta_c * \alpha_c + (\varphi - \rho_v) * \gamma_v\beta_v\alpha_v - C_v < \pi_v\gamma_c\beta_c\alpha_c - \omega_v$, the results are shown in Table 17.3:

As shown in Fig. 17.2, the evolution strategy of the evolution game is $x^* = 0$, $y^* = 0$. Case 2 of the initial state of game is manufacturing enterprises adopt

Table 17.2 Case 1 the Jacobi matrix local stability analysis results

Equilibrium point	J's determinant sign	J's trace symbol	Result
x = 0, y = 0	+	−	ESS
x = 1, y = 0	+	+	Unstable point
x = 0, y = 1	+	+	Unstable point
x = 1, y = 1	+	−	ESS
$x = x^*$, $y = y^*$	−	0	Saddle point

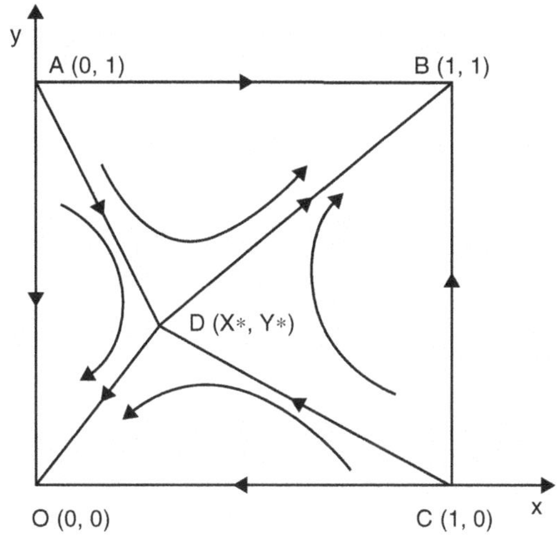

Fig. 17.1 Case 1 evolutionary game replication dynamic phase diagram

Table 17.3 Case 2 the Jacobi matrix local stability analysis results

Equilibrium point	J's determinant sign	J's trace symbol	Result
x = 0, y = 0	+	−	ESS
x = 1, y = 0	−	Uncertain	Saddle point
x = 0, y = 1	−	Uncertain	Saddle point
x = 1, y= 1	+	+	Unstable point

outsourcing strategy and TPL adopt cooperation strategy. But in the process of long-term game, the two sides found that they don't choice teamwork strategy is greater than teamwork strategy, eventually both sides don't choice teamwork strategy. The final equilibrium result is that manufacturing enterprises adopt no outsourcing strategy and TPL adopt non-cooperation strategy.

3. $(\vartheta_c+\mu_c)\gamma_v\beta_v\alpha_v+(\varphi-\rho_c)\gamma_c\beta_c\alpha_c-C_c > \pi_c\gamma_v\beta_v\alpha_v-\omega_c$ and $(\vartheta_v+\mu_v)\gamma_c\beta_c\alpha_c+(-\varphi-\rho_v)\gamma_v\beta_v\alpha_v-C_v < \pi_v\gamma_c\beta_c\alpha_c-\omega_v$, the results of are shown in Table 17.4:

As shown in Fig. 17.3, the evolution strategy of the evolution game is $x^* = 0$, $y^* = 0$. Case 3, the original state is that TPL chosen cooperation strategy. In the course of the game, TPL found that chosen non-cooperative strategy is better than

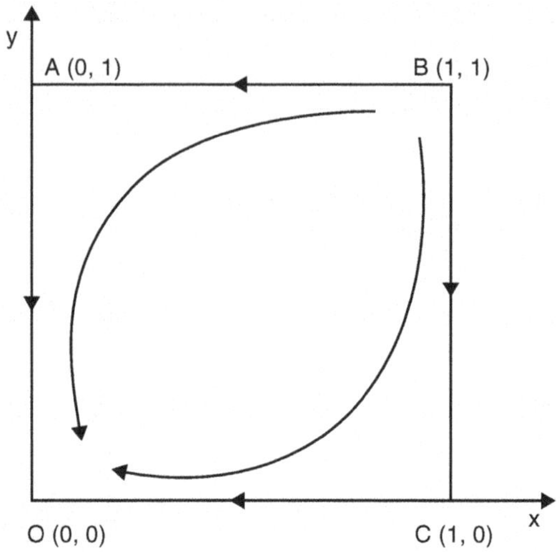

Fig. 17.2 Case 2 evolutionary game replication dynamic phase diagram

Table 17.4 Case 3 the Jacobi matrix local stability analysis results

Equilibrium point	J's determinant sign	J's trace symbol	Result
x = 0, y = 0	+	−	ESS
x = 1, y = 0	−	Uncertain	Saddle point
x = 0, y = 1	+	+	Unstable point
x = 1, y= 1	−	Uncertain	Saddle point

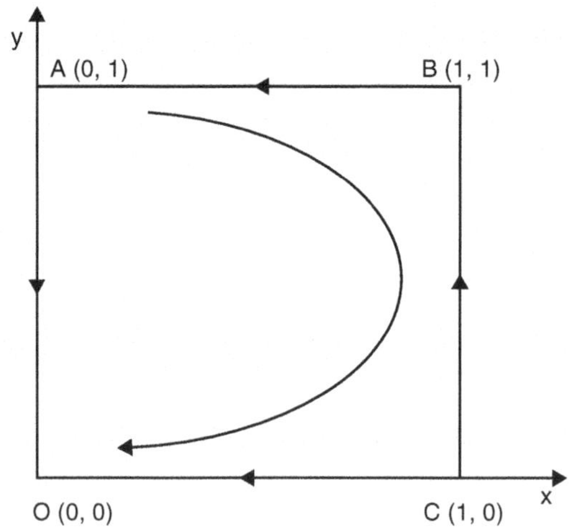

Fig. 17.3 Case 3 evolutionary game replication dynamic phase diagram

chosen cooperation strategy. At the same time, manufacturing enterprises research and study found that adopted no outsourcing strategy is better than outsourcing strategy. Long-term evolutionary game final result is manufacturing enterprises adopt no outsourcing strategy and TPL adopt non-cooperation strategy.

Table 17.5 Case 4 the Jacobi matrix local stability analysis results

Equilibrium point	J's determinant sign	J's trace symbol	Result
x = 0, y = 0	+	–	ESS
x = 1, y = 0	+	+	Unstable point
x = 0, y = 1	–	Uncertain	Saddle point
x = 1, y= 1	–	Uncertain	Saddle point

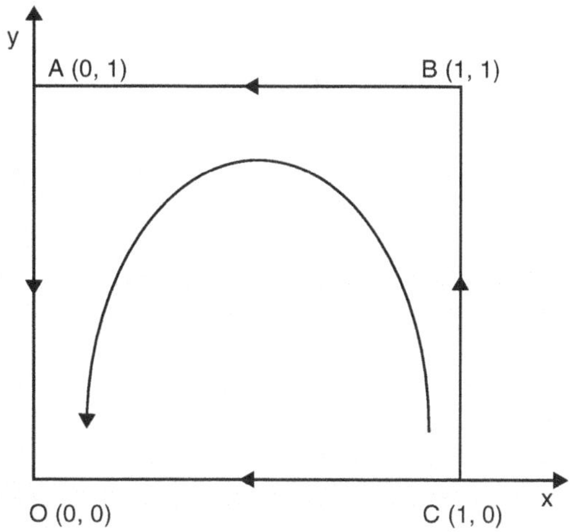

Fig. 17.4 Case 4 evolutionary game replication dynamic phase diagram

4. $(\vartheta_c+\mu_c)\gamma_v\beta_v\alpha_v+(\varphi-\rho_c)\gamma_c\beta_c\alpha_c-C_c < \pi_c\gamma_v\beta_v\alpha_v-\omega_c$ and $(\vartheta_v+\mu_v)\gamma_c\beta_c\alpha_c+(-\varphi-\rho_v)\gamma_v\beta_v\alpha_v-C_v > \pi_v\gamma_c\beta_c\alpha_c-\omega_v$, the results are shown in Table 17.5:

As shown in Fig. 17.4, the evolution strategy of the evolution game is $x^* = 0$, $y^* = 0$. In the course of the game, manufacturing companies found that chosen no outsourcing strategy is better than outsourcing strategies. At the same time TPL research and study found that chosen non-cooperative strategy is better than cooperative strategy. Long-term evolutionary game final result is manufacturing enterprises adopt no outsourcing strategy and TPL adopt non-cooperation strategy.

17.4 Manufacturing Enterprises and Tpl Outsourcing Strategy Analysis

In four cases, the stability strategy of case 1 is more realistic. The evolutionary equilibrium state is related to initial state and the critical point $D(x^*, y^*)$, however the size of the critical point depends on parameters of game model and its changes.

In the game model, α_i, β_i and γ_i are higher, the x^* and y^* values are smaller. And the area ABCD is smaller. The probability tend to point B is increase. In the case of other parameters unchanged, ϑ_i and μ_i are higher, point D is close to point O, the region ABCD area is larger. When π_i, ρ_i and C_i are decrease, the value of the D is smaller, the evolutionary game may end up at point B, so they cooperation probability is higher. When φ is greater, enterprises obtain more incentive. Then more and more enterprises incline to outsourcing, so effective incentive mechanism can promote occurrence of outsourcing. When the penalty ω_i is increase, the enterprise income of taking non-cooperation strategy is smaller, forcing companies to take cooperation.

17.5 Conclusion

Facing situation of China's manufacturing industry, this paper first describes that outsourcing is a new way for manufacturing enterprises to respond changing market; secondly, in the logistics outsourcing cooperation process, game behavior will inevitably occur. So this paper constructs outsourcing game model of the manufacturing enterprise and TPL, and system analyzes evolutionary game payoff matrix, replicates dynamic equation and obtains the evolution stability point. And analyzes optimal strategies that can be taken in the process of evolutionary game, so that both sides can get maximize benefits; Finally, according to the results of the model, this paper make recommendations on the problems that exist, provides effective incentive policies to stimulate enthusiasm of outsourcing cooperation and encourage common development of manufacturing enterprises and TPL.

Acknowledgements This work was financially supported by the National Natural Science Foundation project (71640022,71361011), the Jiangxi Province Social Science "Twelfth Five Year Plan" project(15TQ04),Science and technology research project of jiangxi province Ministry of education (no:14322). Humanities and social science research project of jiangxi provincecolleges and universities(No:gl1423).

References

1. Yang, B. (2010). Sublimation analysis information theory and knowledge transfer incentive mechanism of evolutionary game. *Information Science, 33*(7), 50–54.
2. Liu, J. (2015). Chinese manufacturing industry: current, challenges and trends. *Yuejiangzhong Science, 4*, 15–21.
3. Dong, G. (2015). Manufacturing companies choose TPL cooperation incentives game analysis. *Technology and Management, 17*, 70–74.
4. Zhao, S. (2015). Contract research fourth incentive-based outsourcing gradient effect. Hebei Province Yanshan University.
5. Zou, X. (2013). Chunlong Gu research under Asymmetric Information logistics outsourcing incentive model. *Zhongnan University of Technology, 5*, 130–135.
6. Geng, H. (2010). Manufacturing enterprises logistics outsourcing from the supply chain perspective, *5*, 52–53.
7. Zhuwen, Z. (2010). Duan Xiaoying quality service logistics outsourcing management game behavior. *Industrial Engineering, 13*, 45–48.

Chapter 18
Research on External Quality Inspection Technology of Tropical Fruits Based on Computer Vision

Kun Zhang, Xiaoyan Chen, and Haifeng Wang

Abstract With the computer vision technology in the image processing has been widely used, which for the automatic classification of fruit provides a research space. This paper mainly uses the method of computer vision, combined with the problem of grade quality detection of agricultural products in agricultural research hotspots. Taking tropical fruit of Hainan as the research object, taking mango as the experimental object, extracting the characteristics of fruit image, explore the differences in the external size and color of different types of fruits, and establish a visual quality inspection technology for tropical fruits based on computer vision.

K. Zhang
College of Ocean Information Engineering, Hainan Tropical Ocean University, Sanya, Hainan, China

State Key Laboratory of Marine Resources Utilization in South China Sea, Hainan University, Haikou, Hainan, China

College of Information Science and Technology, Hainan University, Haikou, Hainan, China

Sanya Key Laboratory of Computer Vision, Hainan Tropical Ocean University, Sanya, Hainan, China

X. Chen
College of Ocean Information Engineering, Hainan Tropical Ocean University, Sanya, Hainan, China

Sanya Key Laboratory of Computer Vision, Hainan Tropical Ocean University, Sanya, Hainan, China

H. Wang (✉)
College of Ocean Information Engineering, Hainan Tropical Ocean University, Sanya, Hainan, China

State Key Laboratory of Marine Resources Utilization in South China Sea, Hainan University, Haikou, Hainan, China

Sanya Key Laboratory of Computer Vision, Hainan Tropical Ocean University, Sanya, Hainan, China
e-mail: wyfxxz@163.com

M. Tavana, S. Patnaik (eds.), *Recent Developments in Data Science and Business Analytics*, Springer Proceedings in Business and Economics,
https://doi.org/10.1007/978-3-319-72745-5_18

Keywords Computer vision · Image processing · Detection technology · Analysis of algorithms

18.1 Introduction

In recent years, due to the fruit market demand and the rapid development of the logistics industry, Hainan tropical fruit market demand continues to rise, prices rose steadily [1, 2]. Hainan Province, tropical species are more abundant, such as mango, carambola, Sakya, guava, dragon fruit, eggs, etc., from the appearance of the characteristics of its shape characteristics are different, including regular and irregular shapes are [2]. However, the harvest and screening work of Hainan Tropical Fruits is still largely dependent on the manual or pipelined semi-artificial classification, not only the labor intensity, the high cost of picking and sorting, but also the high experience of the staff, the low efficiency of sorting, The lack of uniform pepper evaluation criteria [3]. Due to the low degree of mechanization and automation, resulting in a year in the fruit picking and sorting links, a large number of fruits have not yet entered the market has been rotten, the resulting economic losses are enormous [4]. Therefore, to solve the standardization of tropical fruit automatic sorting problem is imminent [5].

With the development of science and technology and progress, the continuous development of society, the modernization of large-scale production requirements continue to improve, the computer vision research has been in all aspects of society [6]. Agricultural product quality [7] inspection is an important part of the protection of product quality, how to detect high-quality products to consumers to become the primary purpose of quality testing system.

As we all know, Hainan tropical fruit with many varieties, not easy to store, the characteristics of the gap between the larger. Therefore, in the tropical fruit purchase links and storage links for the quality of fruit is particularly important to grasp. In order to standardize the classification criteria, improve the efficiency of sorting, at home and abroad have developed a number of vegetable automation grading system. In particular, some automatic classification systems based on machine vision are widely researched and applied because of their low cost, simple installation, wide monitoring range and stable grading performance, such as potato and tomato classification system based on machine vision [8]. However, after research, at home and abroad have not yet mature tropical fruit automatic grading system technology, on the tropical fruit of the external quality testing system is still blank. Therefore, the project aims to develop a computer vision based on the quality of tropical fruit detection technology, to promote the development of tropical fruit industry in Sanya and even the province, is of great significance [9].

This paper is organized as follows. In the next section, the authors make introduces the concept of computer vision technology, and describes the development of computer vision technology in Sect. 18.3. Describe the application of computer vision technology research in Sect. 18.4. The conclusion is given for the previous research in Sect. 18.5.

18.2 Overview of Computer Vision Technology

Computer vision technology is to use the image sensor to obtain the image of the detection object, and then the image into a data matrix, with a computer analysis, and replace the brain to complete the treatment and interpretation [10]. Typical machine vision system is generally computer-centric, mainly by the visual sensor, high-speed image acquisition system and special image processing system and other modules, as shown in Fig. 18.1.

Computer vision technology has a large amount of information, processing information comprehensive strong, high speed, multi-function, the target quantitative measurement [11]; at the same time to avoid the subjective factors on the detection results, easy to design and manufacture of automatic detection lines. In recent years, many researchers in the international use of the technology in the field of fruit quality automatic detection carried out a wide range of research. The computerized visual inspection of fruit quality is the use of computer vision technology to collect fruit images, application of image processing, pattern recognition and other technologies to measure the fruit size, shape, color, defects, sugar content, hardness and other indicators parameters, according to these parameters Fruit quality.

Appearance quality refers to the external shape of the fruit, mainly including size, color, shape, surface finish, surface defects, uniformity, etc., it plays a decisive role in fruit market competitiveness, a direct impact on fruit sales. The internal quality refers to the inside of the fruit, to meet the human sensory aspects and nutritional value of the characteristics, including sugar content, acidity, the amount of juice, crisp meat, thickness, vitamins, minerals, protein content and so on. Computer vision technology has been widely applied to the quality of these two aspects of testing [12].

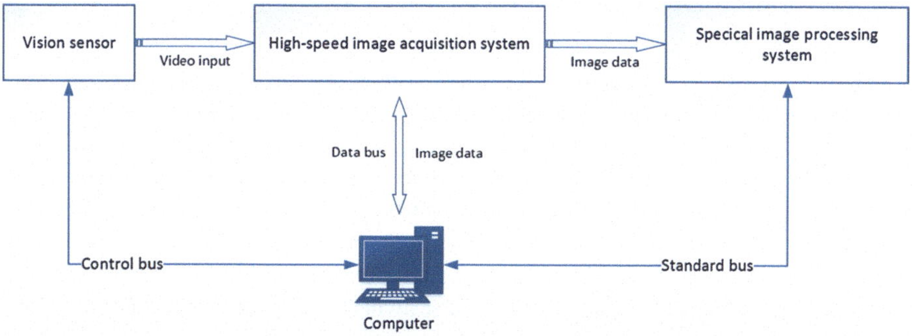

Fig. 18.1 Typical computer vision system

18.3 The Development of Computer Vision Detection Technology

18.3.1 Size Detection

Size is one of the important characteristics of fruit, according to the size of the fruit to detect grading, select the fruit size is basically the same, is conducive to packaging and processing and processing. Used to measure the size of the fruit size area, perimeter, length and length shaft size, volume and so on. The appearance of agricultural products is complex, irregular in shape, from different angles, showing different shapes and colors, thus leading to these features depend on the camera to obtain the image angle [13]. By combining the image information of different angles, the relative size of the target can be obtained. Throop et al. by the translation and rotation of the apple to obtain different angles of the image, according to these images to calculate the apple's equatorial radius and area; then, the apple as an ellipsoid, calculate the apple's long axis and short axis, and according to This estimate the size of the apple. The volume is also a common indicator of the size of the fruit, but this method is more complex and computationally larger than the area-based, perimeter, and so on. Koc studied the volume calculation algorithm of watermelon. By processing the collected images, the contours of the images were obtained on different projection planes, and the contour slices were formed, and then the contour slices were rotated along the x-axis. The slices are combined to measure the volume of the watermelon. In the domestic aspect, Ying Yi Bin et al. used the machine vision technology to accurately detect the fruit size, established the quantitative relationship between the point in the image and the point on the measured object, and used the boundary information of the object to find the centroid coordinates of the object, And then proposed an area correction algorithm, detection accuracy of 96%. On the basis of this, Rao Xiuqin et al. analyzed the various errors and causes of the size detection based on the relationship between the fruit, the camera lens and the image, and gave the calibration The formula for calculating the error and the formula for calculating the radius.

18.3.2 Shape Detection

The shape of the fruit is one of the important appearance qualities of the fruit. Deformation of the shape of the strange shape of the fruit is very low, or even simply cannot be sold, which is fruit quality testing must be considered [14]. The shape of agricultural and other agricultural and sideline products is irregular, how to accurately describe the shape of agricultural products, is one of the important issues. Xu et al. studied the method of classification of strawberry shape. First, we define four kinds of strawberries, such as long cone, square, conical and round, and extract the linear sequence from the contours of each strawberry. The length of a contour line eliminates the effect of fruit size on shape discrimination and uses k-means clustering algorithm to determine the classification of strawberries. Unay et al. used circularity features to characterize the shape of the fruit. Gao Hua 10 proposed a method using Fu Liye radius descriptor to describe and classify agricultural products image contour, and using radius descriptor to compute the area of image and similarity using Euclidean distance to determine the boundary shape measurement.

18.3.3 Color Detection

Appearance color is one of the most important appearance parameters of the fruit, consumers often according to the color of the fruit to decide whether to buy. In a computer vision image system, the color of a pixel is usually represented by three coordinate values of the fruit image in the RGB color space, i.e., (R, G, B). Blasco et al. used the pixel RGB averages and the simple thresholds for R / G to distinguish four different pomegranate varieties in real time, with a success rate of more than 90%. In the classification of apple color, Hou Wenjun proposed using HIS model to describe its color signs, and according to the characteristics of chroma histogram, using four chroma mean to replace the apple color value. On this basis, the BP neural network fruit grading system is established, and the classification accuracy is high, which can meet the grading requirements. Kang Qingqing on the apple surface color characteristics were analyzed, get three characteristic parameters: the surface red area ratio, the average surface color and surface color uniformity. The first two eigenvalues are obtained by calculating the cumulative frequency corresponding to the appropriate chromaticity value and the average of the chromaticity of each pixel in the apple image. The third eigenvalue is obtained by calculating the distance between the centroid of the red area in the apple image and the centroid of the whole apple image, and the difference in the average chromaticity of the same apple. The test results show that the classification accuracy of 88.9% or more.

18.4 Research on Application of Computer Vision Detection Technology

Based on the computer vision of Hainan tropical fruit detection, this study used Hainan Tropical fruit in the representative species – mango as the experimental object, the feature extraction and identification. Because the RGB color space is extremely sensitive to light, and HSV color space can overcome the influence of light, and more in line with human visual effects, therefore, in order to better extract the fruit surface color characteristics, the need to RGB color space into HSV color space The RGB color space to HSV color space conversion formula is:

$$H = \begin{cases} 60^{\circ} \times \frac{\text{G-B}}{\max(R,G,B) - \min(R,G,B)} + 0^{\circ}, \mathit{if} \max(R,G,B) = R \ and \ G \geq B \\ 60^{\circ} \times \frac{\text{G-B}}{\max(R,G,B) - \min(R,G,B)} + 360^{\circ}, \mathit{if} \max(R,G,B) = R \ and \ G < B \\ 60^{\circ} \times \frac{\text{R-G}}{\max(R,G,B) - \min(R,G,B)} + 240^{\circ}, \mathit{if} \max(R,G,B) = B \\ 60^{\circ} \times \frac{\text{B-R}}{\max(R,G,B) - \min(R,G,B)} + 120^{\circ}, \mathit{if} \max(R,G,B) = G \end{cases}$$

$$S = (\max(R,G,B) - \min(R,G,B))/\max(R,G,B)$$

$$V = \max(R,G,B)$$

In this model:

$V = \max(R,G,B) \in (0,1)$, $H \in (0^{\circ}, 360^{\circ})$, $V \in (0,1)$. Conversion color space experiment results shown in Fig. 18.2.

In order to extract the surface color characteristics of the fruit, a large number of experiments were done on the sample images. Here, the selected two types of mango

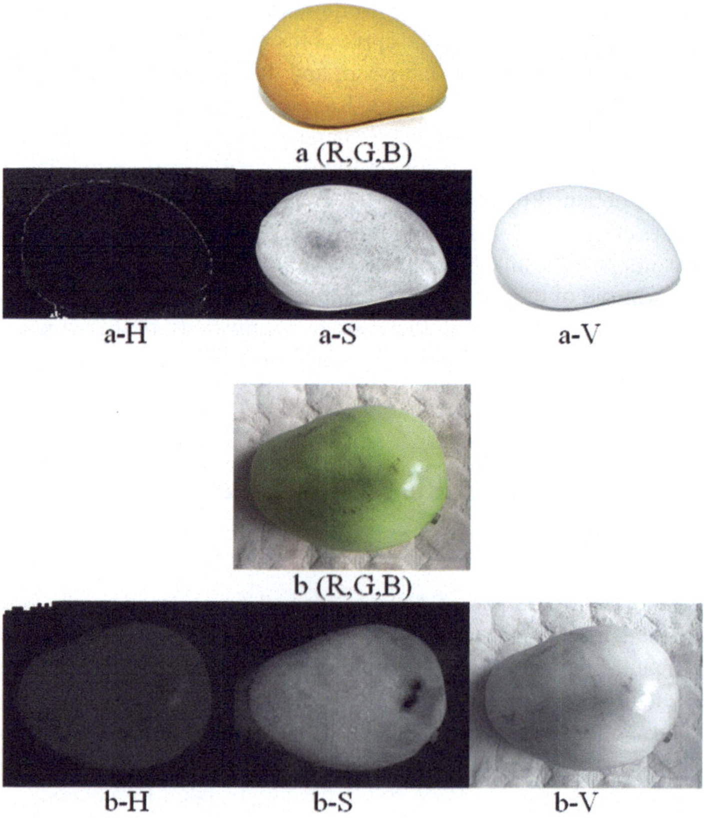

Fig. 18.2 R,G,B chart and HSV chart

Table 18.1 Color values, saturation values, and lightness values for each frame of a Class a image

	1	2	3	4	5	6	7	8	9	10
H	0.8358	0.8543	0.8616	0.8408	0.8527	0.8601	0.8524	0.8635	0.8533	0.8612
S	0.5023	0.5132	0.5116	0.5076	0.5112	0.5106	0.5076	0.5056	0.5012	0.5197
V	0.8016	0.7836	0.8026	0.8011	0.7964	0.8060	0.7955	0.8015	0.8073	0.8082

Remarks: H is the chromaticity value, S is the saturation value, and V is the lightness value

Table 18.2 Color values, saturation values, and lightness values for each frame of a Class b image

	1	2	3	4	5	6	7	8	9	10
H	0.8397	0.8368	0.8457	0.8520	0.8370	0.8385	0.8579	0.8621	0.8608	0.8704
S	0.4987	0.4960	0.5008	0.4919	0.5011	0.5032	0.4926	0.4996	0.5002	0.4908
V	0.7106	0.7986	0.8013	0.7715	0.7833	0.7717	0.7536	0.8008	0.7239	0.7926

Remarks: H is the chromaticity value, S is the saturation value, and V is the lightness value sample images are numbered: Type 1 image a Number: Serial number from 1 to 10; Type 2 image b Number: Serial number from 1 to 10. The chromaticity value, saturation value and brightness value of each frame are calculated in each type of image. Table 18.1 gives a large number of experimental results for HSV color space in this paper (Table 18.2).

By comparing the chromaticity, saturation, and brightness of each class, we find that the chromaticity H has a significant difference, and the corresponding saturation S for the chromaticity difference is also different, and the lightness V is almost indistinguishable. So the color value is selected as the surface color of the fruit.

18.5 Conclusions

In this study, tropical fruit was used as the object of study, and mango was used as the experimental object to extract the characteristics of fruit. In the study of fruit surface quality of computer vision, the color was an important feature. Color space is very sensitive to light, and HSV color space is more in line with human visual effects, HSV space in the color H can effectively distinguish the color, corresponding to different colors have different color values, it is the color And the gradient of the chromaticity space calculated by the Laplace operator can be used to detect the surface with the surface of the surface. Defective fruit, so use the H-space gradient as a fruit surface defect characteristics. In addition, the quality of the grading study is for a single fruit map, if the fruit is more than a combination of the situation will be complicated, you need to separate each fruit before the classification.

Acknowledgments This work is partially supported by the Agricultural Science and Technology Innovation Project of Sanya (No. 2015KJ15; No. 2015KJ16; 2016NK17); the Key Laboratory of Sanya Project (No. L1410).

References

1. Liu, Z. W., & Zhang, M. J. (1999). Comprehensive use of color and texture features of image retrieval. *Journal of China Institute Communications, 20*(05), 5.
2. Wang, G. Q., Yang, Z. B., Zheng, L. M., et al. (2003). Application of computer vision technology in detection of agricultural products. *Review of China Agricultural Science and Technology, 5*(03), 52–56.
3. Throop, J. A., Aneshansley, D. J., Anger, W. C., et al. (2005). Quality evaluation of Apples based on surface defects: Development of an automated inspection system. *Postharvest Biology and Technology, 36*(03), 281–290.
4. Koc, A. B. (2007). Determination of Watermelon volume using ellipsoid approximation and image processing. *Postharvest Biology and Technology, 38*(03), 366–371.
5. Xu, L. M., & Zhao, Y. C. (2010). Automated Strawberry grading system based on image processing. *Computers & Electronics in Agriculture, 71*(S1), S32–S39.

6. Unay, D., & Gosselin, B. (2007). Stem and Calyx recognition on 'Jonagold'Apples by pattern recognition. *Journal of Food Engineering, 78*(02), 597–605.
7. Gao, H., & Wang, Y. Q. (2010). Study on the shape classification of farm produce based on computer vision. *Computer Engineering and Applications, 40*(14), 227–229.
8. Blasco, J., Aleixos, N., & Moltó, E. (2003). Machine vision system for automatic quality grading of fruit. *Biosystems Engineering, 85*(04), 415–423.
9. Hou, W. J. (2006). *Method of Apple automatic grading based on machine vision*. Nanjing: Nanjing Forestry University, Master Dissertation.
10. Peng, L. G. (2014). *Study on Fruit quality detection based on computer vision*. Guiyang: Guizhou Minzu University, Master Dissertation.
11. Kang, Q. Q. (2007). *Study on the detecting and grading method of Apples with machine vision*. Beijing: China Agricultural University, Master Dissertation.
12. Zhang, K., Wang, H. F., & Dai, S. P. (2013). Research on tropical agricultural products quality recognition and detection system based on electronic nose technology. *Advanced Materials Research, 659*(05), 75–78.
13. Zhang, K., & Dai, S. P. (2012). Based on the electronic nose technology of tropical agricultural product. *Software, 33*(08), 36–38.
14. Wang, Y. F., Zhang, W. Z., & Zhang, K. (2016). Experimental study on the detection of tropical fruit quality based on gas identification technology, advances in engineering research. *Advances in Eengineering Research, 68*, 623–627.

Chapter 19
Wholesale Price Contract and Quantity Discount Contract Under Competition with Various Games

Qin Jiabao, Lei Quansheng, and He Yezi

Abstract This article analyzes supply chain contracts in a two-tier supply chain system consisting of two competing manufacturers and one retailer. The two manufacturers compete for the order of the only retailer in the downstream. While the manufacturer 1 and the retailer adopted the quantity discount contract and therefore playing the Stackelberg game between them, the other manufacturer – manufacturer 2 adopted a wholesale price contract with the retailer, and playing a bargaining game between them. We find that in the competitive scenario with various games, the quantity discount contract can further expand its superiority and effectively coordinate the supply chain with the presence of the Stackelberg-leader advantage. Quantity discount contract is more advantageous than wholesale price contract.

Keywords Bargaining game · Stackelberg game · Quantity discount · Wholesale price contract

19.1 Introduction

In recent years, in the various kinds of literature of supply chain, the study of supply chain contract has been a focus research. As a tool to coordinate the supply chain, supply chain contract can improve the overall performance of the supply chain in a large extent. Supply chain contracts possese various forms, such as wholesale price contract, revenue sharing contract, quantity discount contract and so on. These contracts are different not only in form, but also in terms of the impact of the parties to the contract. In addition, both sides of the supply chain are also a pair of game

Q. Jiabao (✉) · L. Quansheng
School of Automation, Beijing University of Posts and Telecommunications, Beijing, China
e-mail: buptqjb@163.com

H. Yezi
School of International Studies, Beijing University of Posts and Telecommunications, Beijing, China

M. Tavana, S. Patnaik (eds.), *Recent Developments in Data Science and Business Analytics*, Springer Proceedings in Business and Economics,
https://doi.org/10.1007/978-3-319-72745-5_19

participants, the impact of the form of the game on both sides cannot be ignored. The game can be in multiple forms, among which the Stackelberg game and bargaining game are more famous. This article mainly focuses on these two forms.

Previous articles on the supply chain contract research has already been in-depth. Cachon [1] pointed out that the newsvendor model is simple, but it is of great significance to the research of supply chain contract. Raju and Zhang [2] proved that if there is a dominant retailer, manufacturers preferred a single contract. Katok et al. [3] further studied the impact of wholesale price contracts on supply chain performance while fairness as the private information of supply chain members. Shang et al. [4] studied the impact of sharing retailers' demand information on competing manufacturers in the supply chain. Song et al. [5] examine the equilibrium price and quantity decisions for a price-setting news vendor.

Our research is similar to Feng and Lu [6]. Feng and Lu [6] studied the supply chain contract of the two-tier supply chain system under competition. The difference is that the competitive scenario in Feng and Lu [6] is carried out under the same game, Stackelberg game or bargaining game, and there is competition between the manufacturers and the retailers, with the focus comparison between two kinds of games. Our research has made some improvements based on his research: competition only between manufacturers, not retailers. In addition, in the two-tier supply chain system, there are two kinds of games simultaneously, Stackelberg game and bargaining game. Our research focuses on the comparison between supply chain contracts. For the form of contract, we focus on wholesale price contract and quantity discount contract. We realized that this competition is more complex, and it is of significance on supply chain contract.

19.2 Model

We consider a two-tier supply chain system, consisting of two manufacturers and one retailer, with manufacturers being represented by 1 and 2. Each manufacturer sells product to a single retailer, who sells the goods to customers in the market. The two manufacturers sell the same kind products, but there are also differences between them, and we use $\gamma \epsilon [0, 1)$ to represent the substitutability of the product. Manufacturer 1 adopts quantity discount contract with the retailer in the Stackelberg game. Manufacturer 2 adopts wholesale price contract with the retailer in the bargaining game.

The parameters involved in this article are summarized as the following:

p_i The price of the merchandise ordered by the retailer from the manufacturer i
q_i The order quantity of the retailer at the manufacturer i
C The unit cost of the product
β Price sensitivity of the merchandise
w The wholesale price of the product
π The profit of the retailer

π^{-i} The profit of the retailer without trading with manufacturer i
π_i The profit of manufacturer i

The market price of the goods ordered by the retailer from the manufacturer i is $p_i = \alpha - \beta(q_i + \gamma q_j)$, i, j $\in \{1,2\}$.

For a quantity discount contract, the wholesale price $W(q)$ is a function of the order quantity, which needs to satisfy the following relationship:

$W(q)$ in monotonically decreasing when $q \geq 0$, which means $W'(q) \leq 0$, however, $W(q) > c$ and $W''(q) \geq 0$.

First we calculate the optimal order quantity without considering game theory.

1. Retailer only trades with manufacturer 2 instead of manufacturer 1.

   Optimal order quantity $q_{\tilde{2}} = \dfrac{\alpha - w}{2\beta}$

   Maximum profit of the retailer $\pi^{-1} = \frac{(\alpha-w)^2}{4\beta}$

2. Retailer only trades with manufacturer 1 instead of manufacturer 2.

   Optimal order quantity $q_{\tilde{1}}$ needs to meet $\alpha - 2\beta q_{\tilde{1}} - W(q_{\tilde{1}}) - q_{\tilde{1}} W'(q_{\tilde{1}}) = 0$

   We introduce a new function $f(q_1) = q_1 W(q_1)$, then we can have $f'(q_{\tilde{1}}) = \alpha - 2\beta q_{\tilde{1}}$

   Maximum profit of the retailer $\pi^{-2} = \dfrac{\alpha^2 - [f'(q_{\tilde{1}})]^2}{4\beta} - f(q_{\tilde{1}})$

3. Retailer trades with both manufacturer 1 and manufacturer 2.

   In this case, profit of the retailer equals:

$$\pi = [\alpha - \beta(q_1 + \gamma q_2) - W(q_1)]q_1 + (\alpha - \beta(q_2 + \gamma q_1) - w)q_2 \tag{19.1}$$

$$H = \begin{vmatrix} \dfrac{\partial^2 \pi}{\partial {q_1}^2} & \dfrac{\partial^2 \pi}{\partial q_1 \partial q_2} \\ \dfrac{\partial^2 \pi}{\partial q_2 \partial q_1} & \dfrac{\partial^2 \pi}{\partial {q_2}^2} \end{vmatrix} \tag{19.2}$$

Let $\frac{\partial \pi}{\partial q_1} = 0$ and $\frac{\partial \pi}{\partial q_2} = 0$. If $H > 0$ and $\frac{\partial^2 \pi}{\partial {q_1}^2} < 0$, then q_1 and q_2 calculated are optimal order quantity.

We use the quantity discount contact in the form as following:

$$W(q_1) = \frac{a}{q_1} + b, b > c, a > 0$$

The function form of the wholesale price meets the requirement for $W(q_1)$.

In the situation where the retailer only trades with manufacturer 1:

$$q_{\tilde{1}} = \frac{\alpha - b}{2\beta} \tag{19.3}$$

$$\pi^{-2} = \frac{(\alpha - b)^2}{4\beta} - a \tag{19.4}$$

In the situation where the retailer trades with both manufacturer 1 and manufacturer 2:

$$q_1^* = \frac{\alpha(1-\gamma) + w\gamma - b}{2\beta(1-\gamma^2)} \tag{19.5}$$

$$q_2^* = \frac{\alpha(1-\gamma) - w + b\gamma}{2\beta(1-\gamma^2)} \tag{19.6}$$

$$H = 4\beta^2(1-\gamma^2) > 0 \tag{19.7}$$

$$\frac{\partial^2 \pi}{\partial {q_1}^2} = -2\beta < 0 \tag{19.8}$$

$$\pi = \frac{\alpha^2 - \alpha(w+b) + bw}{2\beta(1+\gamma)} + \frac{(w-b)^2}{4\beta(1-\gamma^2)} - a \tag{19.9}$$

$$\pi_1 = (b-c)\frac{\alpha(1-\gamma) + w\gamma - b}{2\beta(1-\gamma^2)} + a \tag{19.10}$$

$$\pi_2 = (w-c)\frac{\alpha(1-\gamma) - w + b\gamma}{2\beta(1-\gamma^2)} \tag{19.11}$$

Now we introduce the situation of game theory.

We define the bargaining power of the retailer, which equals $\theta \in (0, 1)$ and the bargaining power of manufacturer equals $(1-\theta)$ correspondingly.

We establish a new function:

$$\varphi = \left(\pi - \pi^{-2}\right)^{\theta} {\pi_2}^{(1-\theta)} \tag{19.12}$$

According to $\frac{\partial \varphi}{\partial w} = 0$ and $\frac{\partial \pi_1}{\partial b} = 0$, the optimal wholesale price w and the optimal parameter b satisfy the relation as following:

$$\begin{cases} (\theta - 1)\left[2(\alpha - w)(\alpha - b)(1-\gamma) - (\alpha - b)^2(1-\gamma^2) + (w-b)^2\right] \\ \quad * [2w - c - br - \alpha(1-\gamma)] = \\ \quad 2\theta(w-c)[b\gamma - w + \alpha(1-\gamma)]^2 \\ \quad w\gamma + c - 2b = \alpha(\gamma - 1) \end{cases} \tag{19.13}$$

Though those equations can be solved, the specific solution is not given here since there are too many parameters and the form of the solution is too complicated. The conclusions about the solution will be discussed in the next section.

19.3 Numerical Analysis

We put emphasis on discussion of γ and θ and their impacts on solution. Using the control variable method, we first assume that the unit cost of the product $c = 50$ and $\alpha = 100$, $\beta = 0.1$ in the retail price function.

Assume that $\gamma = 0.3$ $\gamma = 0.5$ $\gamma = 0.7$, in these three cases, we observe the effect of θ and γ on the results (Figs. 19.1 and 19.2).

We can see that whatever the value of γ, the wholesale price w and parameter b on the trend of θ is decreasing, and the wholesale price w on the decline rate of θ is significantly higher than the parameter b. Although the manufacturer 1 is not a participant in the bargaining game, however, the quantity discount contract is also affected by the bargaining power of the bargaining game, but the effect of bargaining power compared to the wholesale price contract in bargaining game is smaller. At the same time, we can see that, when θ is approaching to 1, that is, the retailer in the negotiation has an absolute advantage, the wholesale price w of the wholesale price contract is approaching to the cost of the product, but the parameter b of quantity discount contract is still greater than the cost price, so the quantity discount contract pricing has an advantage.

Through the vertical observation, we can see that the wholesale price w and parameter b on the trend of γ is decreasing. That means, with the increase in product substitution, the balanced wholesale price w and parameter b are both reduced, and the impact of γ on the wholesale price w is obviously greater than b. So in our hypothetical situation, the quantity discount contract has a bigger stability than the wholesale price contract.

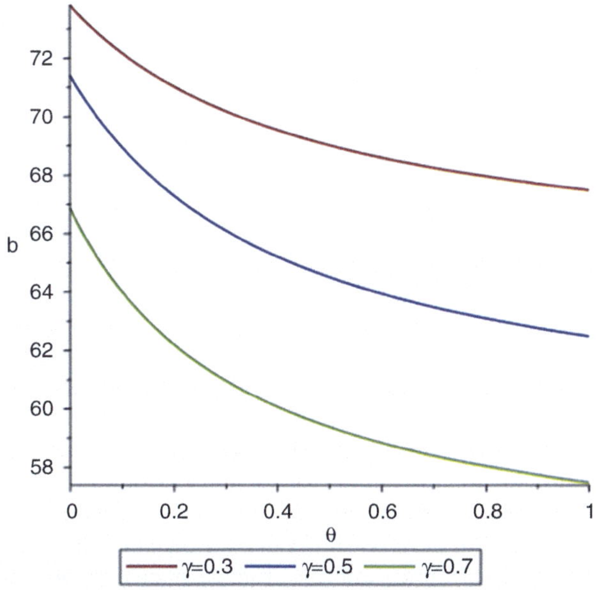

Fig. 19.1 Parameter b changing about θ when $\gamma = 0.3, 0.5, 0.7$

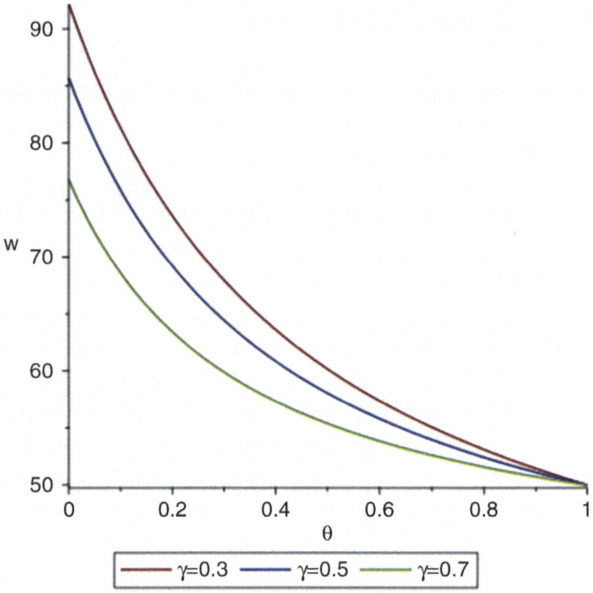

Fig. 19.2 Wholesale price w changing about θ when $\gamma = 0.3, 0.5, 0.7$

We choose one of the typical cases to carry out the following analysis, $\gamma = 0.5$ $\theta = 0.5$, in which the competition and the game are relatively balanced (Figs. 19.3 and 19.4).

When $\gamma = 0.5$ and $\theta = 0.5$, we can get $b = 64.5$ and $w = 58$.

Then we further get:

$$\begin{aligned} &q_1^* = 96.7 \ \ q_2^* = 161.7 \\ &p_1^* = 82.2 \ \ \ p_2^* = 74.1 \\ &\pi_1 = 1402.2 + a \quad \pi_2 = 1293.6 \\ &\pi = 4314.9 - a \end{aligned}$$

We can see that $\pi_1 > \pi_2$

19.4 Conclusion

In a competitive scenario consisting of two competing manufacturers and one retailer, we compare the wholesale contact with the quantity discount contact. Our model contains two kind of games, Stackelberg game and bargaining game. We don't make two manufacturers to adopt the same game, because we think it would be interesting to have two scenarios. In the meanwhile, the supply chain contacts which manufacturers adopt are also different, which increasing the asymmetry between channels. Our analysis shows that the quantity discount contract is more stable than the wholesale price contact. The parameter of quantity discount contract is not

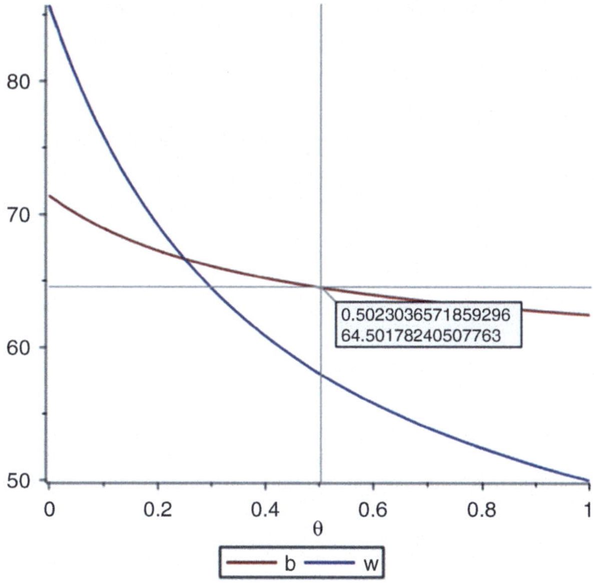

Fig. 19.3 Parameter b when $\gamma = 0.5$ $\theta = 0.5$

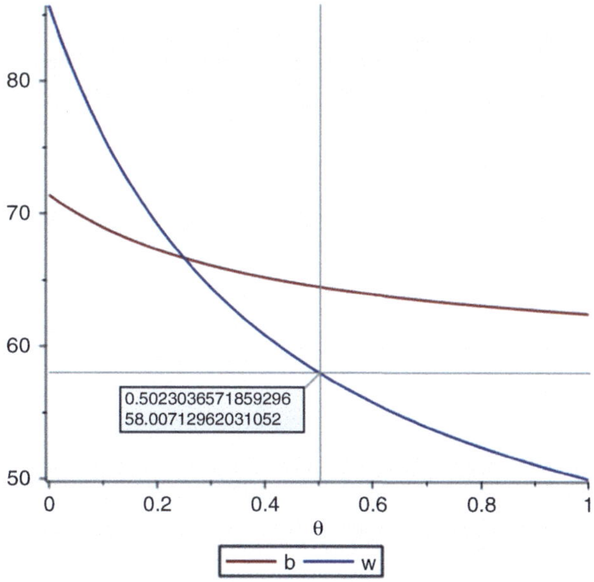

Fig. 19.4 Wholesale price w when $\gamma = 0.5$ $\theta = 0.5$

unique, but the wholesale price w is the only variable parameter of wholesale price contract, which limits the flexibility of wholesale price contract. In addition, participants' skills in the negotiation play an important role in bargaining game. If the bargaining power of manufacturer is lower, the wholesale price is easy to be suppressed, or even close to the cost price, which seriously restrain the manufacturer's profits. In practice, quantity discount contract is more advantageous than wholesale price contract. The form of quantity discount contract has great attraction for retailers. Even if the actual wholesale price of quantity discount contract is higher, retailers tend to order products under the contract. Quantity discount contract can bring considerable benefits to manufacturers The influence of bargaining power on quantity discount contract is low, and the manufacturer's pricing is not easy to be suppressed. The flexibility of quantity discount contract is higher, which is benefit for manufacturers, they can improve their earnings by changing parameters.

References

1. Cachon, G. P. (2003). Supply chain coordination with contracts. *Handbooks in Operations Research & Management Science, 11*(03), 227–339.
2. Raju, J., & Zhang, Z. J. (2005). Channel coordination in the presence of a dominant retailer. *Marketing Science, 24*(2), 254–262.
3. Katok, E., Olsen, T., & Pavlov, V. (2012). Wholesale pricing under mild and privately known concerns for fairness. *Production and Operations Management Society, 23*(2), 285–302.
4. Shang, W., Ha, A. Y., & Tong, S. (2015). Information sharing in a supply chain with a common retailer. *Management Science, 62*(1), 245–263.
5. Song, Y., Ray, S., & Li, S. (2008). Structural properties of buyback contracts for price-setting newsvendors. *Manufacturing & Service Operations Management, 10*(1), 1–18.
6. Feng, Q., & Lu, L. X. (2013). Supply chain contracting under competition: Bilateral Bargaining vs Stackelberg. *Production and Operations Management, 22*(3), 661–675.

Chapter 20
Evaluation of Science and Technology Service Industry in Shandong Province

Yanming Yang, Xi Wang, and Keliang Jia

Abstract In order to evaluate the development level of the science and technology service industry in Shandong Province, the paper built a two-level index system, and then designed an evaluation model of the development level of science and technology service industry based on factor analysis which could effectively reduce the error of subjective factors and the weight of experts to determine, finally calculated the comprehensive evaluation score of the 17 cities and analyzed the influence factor.

Keywords Shandong province · Science and technology service industry · Factor analysis · Evaluation model

20.1 Introduction

In October 2014, the "Opinions of the State Council on Accelerating the Development of Science and Technology Service Industry" proposed to improve the science and technology service industry statistical system. So, it is significance to establish the index system of science and technology service industry and evaluate it.

Muller and Zenker discussed the interaction of small and medium-sized enterprises and knowledge services in innovation [1]. Zhou Hong [2], Chen Chunming [3, 4] established four aspects to evaluate and compare the development level of the science and technology service industry in Tianjin and Heilongjiang province. Li and Kong selected 11 specific indexes and analyzed the development level of science

Y. Yang
Shandong Research Center of Science and Technology Statistics and Analysis, Shandong, China

X. Wang · K. Jia (✉)
School of Management Science and Engineering, Shandong University of Finance and Economics, Shandong, China
e-mail: jiakeliang@sina.com

M. Tavana, S. Patnaik (eds.), *Recent Developments in Data Science and Business Analytics*, Springer Proceedings in Business and Economics,
https://doi.org/10.1007/978-3-319-72745-5_20

and technology service industry in Guangdong Province by principal component analysis [5]. Zhou Huamei did the research of the competition level of science and technology service industry in 13 cities in Jiangsu Province by using principal component analysis and cluster analysis method [6]. Cui and Wei established the evaluation index system of the service capability of science and technology service industry and carried out it in 10 cities of Shanxi Province [7]. Yang et al. assessed the status of Guangdong's science and technology services in China and then did comparative analysis of competitiveness to other advanced regions of the country [8].

The multi-index comprehensive evaluation is used to establish the evaluation system with two levels, so as to evaluate the level of science and technology service industry. There are three kinds of comprehensive evaluation of multiple indicators, which are the subjective empowerment method, the objective weighting method and combination of subjective and objective method [9]. There are two main ways to implement objective weighting, which are principal component analysis and factor analysis. Both methods use the idea of dimensionality reduction, but the factor analysis method which used rotation method is more conducive to the interpretation of the factors. Therefore, the paper chooses the factor analysis method as the main method to evaluate the development level of science and technology service.

20.2 Establishment of a Comprehensive Evaluation System

The paper drew on the previous research results and combined with the actual investigation and research, referred to the policy and the demand for science and technology services statistical index information in "modern service industry science and technology development 'second five' special plan" and "the State Council on accelerating science and technology Service industry development of a number of opinions". In accordance with the principle of hierarchy, the paper established a two-level science and technology service industry statistical system, which ensured that the statistical index system could be a comprehensive measure of technology services. The paper established the index system from three dimensions, science and technology service industry, science and technology service industry investment and output. The basis of science and technology service industry refers to the basic environment of science and technology development and economic development, which composed of the proportion of the urban population to the total population (X_1), expenditure on financial education (X_2), public library collection (X_3), number of research and experimental development staff (X_4), number of research and experimental development staff (X_5). The investment in science and technology service industry included human investment and capital investment which composed of indicators such as R&D internal expenses (X_6), R&D staff equivalent to full time equivalent (X_7), the average wage of workers in the service industry (X_8), local financial science and technology funding, expenditure (X_9). The output of science and technology service industry was included of High-tech industry output value

(X_{10}), number of domestic patent applications (X_{11}), number of three domestic patent applications accepted (X_{12}),High-tech Zone Tax Revenue (X_{13}), the added value of the tertiary industry(X_{14}).

20.3 Data Collation and Empirical Analysis

20.3.1 Data Collation

Considering the availability of data, according to the classification of national economy (GB/T4754–2002), the paper chooses the industry statistics of research and experiment development, professional technical service, science and technology exchange and extension service and geological prospecting industry as technology service Industry statistics. The data mainly come from "Shandong Province Statistical Yearbook 2016", "Shandong Province Science and Technology Statistics Handbook 2016" etc. Then combined with the previous established evaluation index system, we collected the relevant data of 17 cities in Shandong Province, including Jinan, Tai'an, Weifang, Dezhou, Binzhou, Laiwu, Qingdao, Yantai, Rizhao, Dongying, Jining, Liaocheng, Linyi, Zaozhuang, Zibo, Weihai.

20.3.2 Empirical Analysis

Data Standardization Since the dimensions of the 14 indicators in this paper are quite different, the original data should be standardized. 14 indicators X_i data selected from 2015 in Shandong Province 17 cities were standardized as YX_i.

Correlation Coefficient Matrix Used the data and the SPSS 22.0 software for analysis, the variable coefficient correlation matrix was obtained. From the matrix, we could see that the correlation coefficients between the most of variables were large, which reflect the correlations between the original variables were strong, and the selected indicators were suitable for factor analysis.

KMO and Bartlett's Tests KMO and Bartlett's tests were performed on the standardized variable YX. The KMO value is between 0 and 1. The greater the KMO value means the better the factor analysis. Generally believed that KMO in 0.5 or more is suitable for factor analysis. The test results showed the KMO value $0.732 > 0.5$, so the sample data in this paper was suitable for factor analysis. The value of sig of Bartlett's sphere test was 0.000, or $P = 0.000 < 0.05$, which proved that the standardized indicators were suitable for factor analysis.

Extract the Common Factor Variance Interpretation The cumulative variance of the three components extracted from the Eigen values and variance contribution tables was $92.649\% > 90\%$, indicating that the explanations were good, so three factors were extracted to represent the original 14 variables.

Table 20.1 Rotated component matrix[a]

	Component		
	1	2	3
YX_1	0.275	0.210	0.909
YX_2	0.760	0.618	0.005
YX_3	0.608	0.705	0.227
YX_4	0.801	0.326	0.444
YX_5	0.843	0.408	0.306
YX_6	0.584	0.709	0.371
YX_7	0.782	0.346	0.469
YX_8	0.470	0.417	0.674
YX_9	0.552	0.719	0.286
YX_{10}	0.202	0.909	0.289
YX_{11}	0.847	0.356	0.347
YX_{12}	0.749	0.443	0.318
YX_{13}	0.743	0.208	0.443
YX_{14}	0.761	0.535	0.339

[a]Rotation converged in 25 iterations

Variance Maximum Rotation Because of the initial factor load matrix was not obvious for naming the factors' and interpreting, it was difficult to name the first three factors that we got. So we used the maximum variance method for rotation, after the rotation of the factor load matrix as shown in Table 20.1.

It could be seen from the Table 20.1, the first common factor in X_2, X_4, X_5, X_7, X_{11}, X_{12}, X_{13} and X_{14} had large load. The load values were 0.76, 0.801, 0.843, 0.782, 0.847, 0.749, 0.743 and 0.761, which indicated that the eight indicators had a strong correlation and could be classified as the first factor together. These factors reflected the expenditure on financial expenses, the growth of R & D personnel and the number of units, and the increase or decrease in the output of the technology services. So we named the first factor "growth factor of science and technology service industry."

The second common factor in X_3, X_6, X_9, X_{10} had large load. The load values were 0.705, 0.709, 0.719 and 0.990. These four indicators reflected the investment in technology services. So the second factor could be named as "the development of science and technology service industry input factor."

The third common factor in the X_1, X_8 had larger load. The load values are 0.909 and 0.674. These two indicators mainly represented the basis of the development of science and technology service industry. So the third common factor could be named "the basis of the development of science and technology services."

Factor Scores Calculation In order to better monitor the level of science and technology service in 17 prefecture-level cities in Shandong Province, we used the regression method to obtain the factor score coefficient matrix. Through the factor score coefficient matrix, the three common factors could be expressed as 15 statistical indicators of the linear function.

$$\begin{aligned} F_1 = \; & -0.295YX_1 + 0.228YX_2 - 0.02YX_3 + 0.220YX_4 \\ & +0.268YX_5 - 0.1YX_6 + 0.187YX_7 - 0.161YX_8 \\ & -0.093YX_9 - 0.443YX_{10} + 0.280YX_{11} + 0.176YX_{12} \\ & +0.241YX_{13} + 0.13YX_{14} \end{aligned} \tag{20.1}$$

$$\begin{aligned} F_2 = \; & -0.089YX_1 + 0.152YX_2 + 0.275YX_3 - 0.189YX_4 \\ & -0.119YX_5 + 0.269YX_6 - 0.166YX_7 - 0.027YX_8 \\ & +0.307YX_9 + 0.653YX_{10} - 0.172YX_{11} - 0.041YX_{12} \\ & -0.263YX_{13} + 0.030YX_{14} \end{aligned} \tag{20.2}$$

$$\begin{aligned} F_3 = \; & 0.773YX_1 - 0.407YX_2 - 0.130YX_3 + 0.064YX_4 \\ & -0.107YX_5 + 0.25YX_6 + 0.094YX_7 + 0.422YX_8 \\ & -0.49YX_9 + 0.075YX_{10} - 0.059YX_{11} - 0.059YX_{12} \\ & +0.104YX_{13} - 0.057YX_{14} \end{aligned} \tag{20.3}$$

Calculate the Comprehensive Score According to the formula $F = 0.876^{*} F1 + 0.07^{*} F2 + 0.054^{*} F3$, we got the scores of the three common factors of 17 prefecture-level cities in Shandong Province and the comprehensive score. The specific results were shown in Table 20.2.

Table 20.2 Scores of 17 cities in Shandong province

Region	F1	F2	F3	Overall score F	Rank
Jinan	2.17257	−1.36157	1.48466	1.89	1
Qingdao	1.87747	2.021	0.79132	1.83	2
Zibo	0.07953	−0.39011	1.36291	0.12	6
Zaozhuang	−0.29084	−0.82507	−0.25162	−0.33	9
Dongying	−1.78815	0.56942	1.68811	−1.44	17
Yantai	−0.71756	2.38466	0.21223	−0.45	11
Weifang	1.108	0.63141	−0.92049	0.97	3
Jining	0.66632	−0.48609	−0.70614	0.51	4
Tai'an	−0.03118	−0.58732	0.03612	−0.07	7
Weihai	−0.63961	0.12299	0.61575	−0.52	13
Rizhao	−0.58947	−0.87224	−0.04224	−0.58	15
Laiwu	−0.71438	−1.16919	0.35486	−0.69	16
Linyi	0.44634	0.00543	−1.01218	0.34	5
Dezhou	−0.52668	0.20022	−0.85949	−0.49	12
Liaocheng	−0.43855	0.02866	−0.98833	−0.44	10
Binzhou	−0.58524	−0.22849	0.16641	−0.52	13
Heze	−0.02855	−0.04368	−1.93189	−0.13	8

20.4 Conclusion Analysis

From Table 20.2 we could know the F value which was the score of science and technology service industry development level of 17 cities in Shandong Province, Jinan ranked first. Followed by Qingdao, Weifang, Jining, Linyi and Zibo. Dongying, Laiwu, Rizhao and other regions owned low comprehensive score, the development of science and technology service industry was relatively backward.

1. Sorting the 17 prefecture-level cities in accordance with the first public factor "science and technology service industry development growth factor", Jinan was of the highest score of 2.17257. Followed by Qingdao, Weifang, Jining, Linyi, which scores were 1.87747, 1.108, 0.66632 and 0.44634. The growth factor of the development of science and technology service industry represented a series of indicators, such as X_2, X_4, X_5, X_7, X_{11}, X_{12}, X_{13} and X_{14}. Jinan owned the highest score on the social factors of the development of science and technology services, mainly because of Jinan invested huge in science and technology services and scientific and technical personnel sufficient. For instance, in the case of R&D it was higher at least 8 times compare with the lowest value of prefecture-level city.
2. In the second public factor "the input factor of development of science and technology service industry", Yantai, owned the highest score, 2.38466, followed by Qingdao, Weifang, Dongying, respectively, 0.63141, 0.56942. The second public factor mainly showed the investment situation of the science and technology service industry which contained the construction of the society for the library, the financial science and technology funding and so on. Through the analysis of the original data, in Shandong Province, in the case of the city's science and technology services' innovation and R&D expenditure and local financial and technological funding, Qingdao City owned the first place, followed by Yantai. Yantai, Qingdao, Weifang and Dongying which were located in the top four prefecture-level cities won the advantage of economic development by close to the ocean, the government for the four coastal cities with higher technology services investment.
3. In the third main factor "the basis of the development of science and technology service industry," Dongying got the highest score, followed by Jinan, Qingdao, Weihai; scores were 1.68811, 1.48466, 0.79132, 0.61575. Dongying and Jinan in the third main factor had a small gap, but there was a large gap between Qingdao and Jinan. The main reason was that the proportion of urban population to the total population and the average wage of workers in the service industry determines the basis of the development of science and technology service industry.

Acknowledgments This work is supported by Humanity and Social Science foundation of Education Ministry of China (No. 14YJC860011), Shandong province key R & D projects (No. 2015GSF122002).

References

1. Muller, E., & Zenker, A. (2001). Business services as actors of knowledge transformation: the role of KIBSin regional and national innovation systems. *Research Policy, 30*(9), 1501–1516.
2. Hong, Z., & Chen, S. (2013). Empirical research on evaluation of the development level to promote science and technology service industry: Take Tianjin as an Example. *Transaction of Guizhou University of Finance and Economics*, (04), 92–97.
3. Chunming, C., & Xue, F. (2014). Science and technology service industry development status and countermeasures. *Study & Exploration*, (04), 100–104.
4. Chunming, C. (2014). Heilongjiang Province science and technology service industry development research. *Social Science in Heilongjiang*, (03), 56–59.
5. Li, S., & Kong, Y. (2013). Present situation' evaluation and tactics of science and technology service industry in Guangdong province. *Science and Technology Management Research*, (10), 115–119.
6. Zhou, M., Xu, J., & Wang, X. (2010). Empirical study on the comprehensive evaluation of region technology service industry competitive capability Level. *Science & Technology Progress and Policy, 27*(8), 137–140.
7. Cui, Y., & Wei, G. (2013). Study on the evaluation of technology service industry service capability. *Journal of Xi'an Polytechnic University, 27*(6), 811–815.
8. Yang, B., Zhenkun, Y., & Yi, X. (2013). Evaluating the competitiveness of S&T service industries in China: Toward a Statistical Analysis. *Journal of Guangdong University of Foreign Studies, 24*(2), 32–49.
9. Yang, Y. (2006). Multi-index comprehensive evaluation of the method of empowerment. *Journal of Statistics and Decision*, (13), 17–19.

Part III
Financial Analytics

Chapter 21
CEO's Background Characteristics, Financing Preference and Firm Performance – Empirical Evidence From China's A-Share Listed Companies

Yongzhuang Li, Xuan Wang, and Dehuan Zhang

Abstract We carry out regression analysis using data of 300 companies listed in China's A-share market in 2015, and introduce the shareholdings of top ten shareholders and corporate current ratio in regression models as control variables. We reach the follow conclusions: in China, (1) gender and education background of a CEO and equity incentives have no significant impact on firm performance; (2) among China's current A-share listed companies, excessive self-confidence does exist in CEOs, and has a negative impact on firm performance; (3) Corporate financing preference plays a partial mediator role in the relationship between CEO's self-confidence and firm performance.

Keywords CEO's background characteristics · Financing preference · Firm performance

21.1 Introduction

With the rapid development of China's economy, the number of listed companies in China is increasing, and the size of their assets is also growing. As of December 2016, the number of A-share listed companies in China has exceeded 3000, and the total market value has reached RMB 50 trillion. China's companies have been developing rapidly because China has a good economic environment, political

Y. Li (✉)
Business School, Central University of Finance and Economics (CUFE), Beijing, China

Academy of Innovation Management, CUFE, Beijing, China
e-mail: liyongzhuang507@163.com

X. Wang
Business School, Central University of Finance and Economics (CUFE), Beijing, China

D. Zhang
Academy of Innovation Management, CUFE, Beijing, China

M. Tavana, S. Patnaik (eds.), *Recent Developments in Data Science and Business Analytics*, Springer Proceedings in Business and Economics,
https://doi.org/10.1007/978-3-319-72745-5_21

stability, and relatively low labor costs. In addition, there is an important factor that they have learned management experience from foreign companies, establishing an improved corporate governance mechanism. Listed companies generally have a complete management system, upon which the owners (i.e. shareholders) and managers (i.e. the management) are able to implement mutual cooperation. The shareholders own the shares of a company, while the managers are mainly responsible for the operation of a company. As a senior officer of a company, CEO is a very important position in the corporate management. The development of a company largely depends on the CEO's strategic insight and operational thinking. A person's panoramic strategic thinking does not come into being in one day but forms through continuous accumulation and learning, and will be affected by his/her own conditions. A CEO's own conditions include gender, age, degree of education and other factors, which will all affect the CEO's strategic decision. Financing decision is one of the important decisions in the development of a company. Already listed companies are mainly financed from two sources. One is equity financing, that is, to raise funds at the expense of the company's equity through the issuance of new shares or the transfer of existing shares. The company adopting this financing method generally does not need to pay interest but dividends only. The other way of financing is debt financing, that is, to borrow money from banks and other financial institutions in mortgage, pledge and other forms, which is usually interest-bearing. Whether it is equity financing or debt financing, the purpose is to solve the problem of insufficient funds of the company, to provide sufficient funds for the company to invest. The method of financing adopted will affect not only a company's capital structure, but also its final performance.

21.2 Purpose

Most of the previous researches on corporate governance, with respect to the CEO, focused on one aspect, such as gender, age, education, etc., without giving an all-round consideration of the CEO's background characteristics. As a senior officer of the company, the CEO is a complex person, who will be influenced by other factors in addition to his/her gender and education in the decision-making for the company, such as the existence of equity incentive policy in the company, the existence of excessive self-confidence of the manager, etc. These factors are intertwined with each other, which will have an impact on the CEO's final strategic decision-making. Moreover, due to China's unique cultural background and economic development process, China has different economic environment and economic system than those of foreign countries, which will also have an impact on the final conclusions. Some listed companies in A-share market were selected as samples to study the relationship between the background characteristics of the CEO, the tendency of corporate financing and the firm performance in the context of China. It is intended to find out, through a comprehensive analysis, how the background characteristics of the CEO influence the financing preference of the company and whether such financing preference can make a positive effect on the performance of the company, to provide suggestions on the relevant decision-making for companies.

21.3 Literature Review and Hypotheses

According to the gender theory, the gender system is a kind of cultural structure, and is the reappearance of the existence of an individual in the society. As the society has higher expectations for men, men are more likely to realize a higher value through adventures. When facing risks, women often show aversion, because women are more conservative and cautious. Bernasek and Shwiff [10] believed that, compared with men, women are more conservative, and will be more cautious when making investment. Barber and Odean [8] also argued that women are more hesitant and cautious in the face of uncertainty and more conservative in making financial decisions. Huang and Kisgen [14] found in the research on the data of American listed companies that female executives have a certain influence on the capital structure and financing preference of the companies, and they thought the influence comes from the higher confidence that men have relative to women when faced with the risk. As men are more confident about their investments, they tend to be more risky than women. Of course, this behavior does not exist only in economics, but psychology also indicates that men and women will have different attitudes towards risks. Arch believed [7] it is because men and women have different cognition of risks. Men are born with the spirit of adventure, daring to face the risks, while women are focused on avoiding risk and seeking security.

With regards to CEO gender and performance, Lin Jianbin [2] concluded that female executives will bring down the performance of companies through a regression analysis of the relationship between executive gender and firm performance with all the listed companies in Shanghai and Shenzhen stock markets as samples. Through further analysis, it was found that the performance of female executives was lower than that of male executives because of the difference in educational level rather than in gender. In fact, after removing the educational level, the impact of gender on firm performance is negligible. Actually, gender is not a key factor in determining firm performance, and other variables such as experience and social capital play the primary role. We therefore derive the following hypotheses:

Hypothesis 1 CEO gender is irrelevant to firm performance;

Hypothesis 2 Male CEOs are more inclined to debt financing, while female CEOs prefer equity financing;

High-level management theory points out that people are rational within certain limits in economics, and cannot take the most scientific choice in making strategic decisions, as they will be affected by personal psychological cognition. The high-level management theory mainly focuses on the relationship among the background of the executives, strategic decision-making and firm performance. The so-called corporate strategy decision-making is the decision-making under the CEO's cognition. Different cognitive backgrounds will have different influence on the final strategic decision. In the empirical study, Jiang [5] found that the chairman's educational background also has a profound impact on his/her financing tendencies. A chairman with economic management background is more inclined to debt financing, resulting in relatively higher debt to asset ratio, so that his/her company will have higher financial leverage. By analyzing the data of technology-intensive listed companies during 2006–2009, Tang Yingmei et al. [4] found that the educational level of a CEO was related to the company's risk preference. The CEOs with management background would prefer debt financing, resulting in a higher debt to asset ratio.

In general, the higher educational background a CEO has, he/she will have further strategic vision, and he/she will have a more profound and comprehensive perspective to look at and think about things. As the highly educated people will have a more global vision, they will be able to carry out analysis based on their knowledge learned in the past in the development of strategic decisions, which will be more rational. At the same time, highly educated CEOs tend to be more innovative, and will break the shackles of traditional thinking, injecting new ideas in the research and development. In case of lack of corporate funds, they will be more inclined to get loans to invest in further research and development. However, although the companies in fact tend to innovation, the innovation is also accompanied by a large number of investments. Whether these investments can be recovered is also an unknown. Gottesman and Morey [12], using Tobin's q as a measure of firm performance, found that there was no significant correlation between CEO's educational background and firm performance. The CEOs who have MBA education background did not contribute more to business performance than the CEOs who do not. They believed that this was because the education that CEOs have received may not give them practical help in their career. Lu and Zhang [3] studied the data of all the listed companies from 2003 to 2012, and found that the CEO's education level had no significant effect on the performance of the company. However, the higher level of education the CEO has, the higher the value of the company will be. And they believed higher educational level has positive effect on the market, but will not bring plentiful profits in the short term. Thus:

Hypothesis 3 There is no correlation between the CEO's education and the performance of the company;

Hypothesis 4 There is a positive correlation between CEO's educational background and corporate debt level;

A CEO's overconfidence comes from his/her cognitive behavior, which is expressed as over optimism about the future. In the view of an overconfident CEO, the future of the company is fully promising, the company will continue to profit, and the company's value will continue to enhance. Therefore, in making financing decisions, they will give priority to debt financing, because in their view, the cost of equity financing is high. As the future business situation is considered promising, the repayment of debt and interest can be guaranteed. So the more confident the CEO is, the higher the debt to asset ratio may be. Heaton's research confirmed this fact. If a manager is overly confident, he/she tends to use debt financing rather than equity financing [13].

However, as this cognitive behavior will easily lead to blind optimism, the overestimation of the future prospects of the company will result in excessive investment and even high frequency of mergers and acquisitions, especially when the internal funds are sufficient. Billett and Qian [11] found that self-confident managers tend to initiate more acquisitions, but subsequent acquisitions often prove to be of little value. Often, too much investment may not necessarily bring benefits, and overestimation of the company is often difficult to realize. Barber and Odean [9] found that the more transactions investors make, the lower the net return on the unit investment is, which provided an indirect proof that excessive self-confidence reduces individual performance. Thus:

Hypothesis 5 There is a negative correlation between CEO's self-confidence and firm performance;

Hypothesis 6 There is a positive correlation between CEO's self-confidence and firm debt level;

Equity incentive has been received attention gradually in recent years. There are more than 100 A-share listed companies currently that have implemented equity incentive, and many others are in planning. The managerial entrenchment theory points out that due to the uncertainty of debt financing, executives in the term of office will generally avoid excessive debt which may lead to the performance instability of the company, so general debt shortage is the result. Hu and Gai [1] found that when a company implements equity incentive, the common interests between the CEO and the shareholders become closer. The company is more inclined to choose the best way of financing, which is conducive to the formation of the best asset structure. Yang et al. [6] found that equity incentive has a certain impact on the financing tendency of executives. In the consideration of the long-term performance of the company, the equity incentive suppresses the defensive behavior in financing decision-making, and if it is a mixed arrangement of equity, the suppressing effect will be significantly enhanced.

According to the agency theory, the company will be entrusted by the owner to the CEO-based management. The agency costs are formed as the shareholders are concerned about the long-term development of the company, while the executives focus on short-term corporate value changes during their term of office. They pursue different values. The behavior of executives in the term of office is often short-sighted, which is not necessarily the best for the future development of the company, and they may even damage the future development of the company in pursuit of short-term interests. Equity incentive means, by giving the executives a certain amount of the company's shares, turning the executives to be one of the owners of the company. The interests of the executives and the company are linked together by the shares. Thus:

Hypothesis 7 There is a positive correlation between equity incentive and firm performance.

Hypothesis 8 There is a positive correlation between equity incentive and firm debt level;

The gender, degree of education, and degree of self-confidence of the executive and equity incentive will all affect the final management approach, while the change of management will affect the strategic decision of the company. The background characteristics of the executive will affect the company's choice of strategy, particularly financing strategy, and the changes in financing strategy will affect the company's capital structure. Financing decision-making, as a very important part of corporate management, will inevitably be affected. Equity financing is relatively simple, which does not require repaying the principal and interest like debt financing, and the risk is low. But debt financing is one of the most effective ways to obtain a lot of cash in short term in case of limited cash flow. Loans obtained through bank lending or other means allow a company to invest in projects which will have positive net present value, and have a significant impact on the company's performance. Thus:

Hypothesis 9 Financing preference plays a mediating role between managerial background and firm performance.

21.4 Data Sources and Research Design

21.4.1 Data Sources

The data used in this paper were sourced from Wind database and Zdatabase. The relevant financial data disclosed by A-share listed companies in the database were used as analysis samples to obtain relevant data. In the research, the ST / ST * class data were removed, while enterprises with high debt to asset ratio and those with incomplete data were excluded. Due to late implementation of equity incentive in China, this approach has not been generalized. The data could be obtained from 190 companies only. And then the data of the companies without equity incentive were randomly added. In the end, the data disclosed by 300 A-share listed companies in 2015 were selected as samples, and E-views software was used for regression analysis to obtain the relevant results.

21.4.2 Variables and Measures

1. Gender of CEO

We use dummy variables in measuring gender, with 1 representing male and 0 representing female.

2. Educational level of CEO

We find that CEO's educational level can be categorized into two groups: MBA and non-MBA. And most other scholars also classified CEOs according to whether they have received MBA education. So variables 0 and 1 were used, with 0 representing non-MBA, and 1 representing MBA.

3. Equity incentive

As there are multiple forms of equity incentive, and the specific way of implementation is different in each company, a comprehensive classification cannot be realized. Therefore, two types of companies were selected to conduct comparative analysis. One type of the companies has implemented equity incentive plans, while the other type has not. No further subdivision of equity incentive was conducted.

4. Self-confidence level of CEO

In the past researches, as the factors leading to the self-confidence of executives are complicated, some scholars simply used age, gender and educational level as the substitute variables, while some adopted CEO's shareholding changes, evaluation by mainstream media, CEO's pay ratio or profit forecast to measure. In this paper, CEO's profitability forecast is used. If a CEO is confident in himself/herself or the company, then he/she will be more optimistic about the future of the company, that is, he/she tends to overestimate the future business income and profitability of the company. This method can be used to verify whether the CEO is confident.

5. Financing preference

Financing preference refers to the degree of preference of a company for equity financing or debt financing. With reference to the previous literatures, the debt to asset ratio was taken as an alternative to a CEO's financing preference. The higher debt to asset ratio means the CEO is more inclined to use debt financing to obtain funds.

6. Firm performance

There are multiple ways to measure firm performance. In this paper, the performance of a company was considered mainly from the income. A higher the income indicates a higher ability to obtain profits. Therefore, the return on net assets was used to measure firm performance. The higher this value is the better performance it indicates.

21.4.3 Control Variables

An enterprise is a complex, and each aspect of it is interrelated. Thus it is necessary to control other extraneous variables to reduce their impact on the final result during data analysis. The control variables adopted in this paper include current ratio,

Table 21.1 Definition of variables

Variable type	Variable name	Variable symbol	Variable definition
Dependent variable	Firm performance	ROE	Return On Equity
Independent variable	Gender of CEO	G	Gender of CEO, 1 for male, 0 for female
	Educational level of CEO	EDU	Educational level of CEO, 1 for MBA, 0 for non-MBA
	Self-confidence level of CEO	CON	Self-confidence level of CEO
	Equity incentive	EI	Whether the listed company implements equity incentive measures for the executives
Mediating variable	Financing preference	DR	The debt to asset ratio of the company
Control variable	Ownership concentration	CR10	The shareholdings of top 10 shareholders
	Current ratio	CR	Current assets / current liabilities

shareholdings of top ten shareholders, etc., and the impact of these variables on firm financing preference and performance was eliminated. The current ratio is the ratio of a company's current assets to its current liabilities, and can be used measure its ability to pay its debts in the short term. The shareholdings of top ten shareholders represent the degree of dispersion of the company's share. Higher shareholdings of top ten shareholders indicate a more centralized ownership structure of the company. Centralized ownership structure may have an impact on the management right of the CEO. The higher the concentration of ownership is, the stronger the control of the shareholders to the company will be, which may weaken the CEO's ability to control the company. The impact of these factors should also be taken into account in the study of the background characteristics and financing preferences of CEOs and firm performance. Therefore, these two tables were selected as the main control variables to exclude their impact on the final result. For the definition of specific variables, see the following (Table 21.1):

21.5 Model Specification and Empirical Analysis

21.5.1 The Model

In studying the impact of CEO's background characteristics on firm performance, we set up individual models in order to separate the interaction between variables so that the individual relationship between firm performance and all the four background characteristics can be clearly identified. Moreover, as financing preference is supposed to play a mediator role in the relationship between CEO's background characteristics and firm performance, we also use regression model to analyze its mediator role. Hence, we first specify models to analyze the impacts of CEO's four background characteristics on firm performance.

In order to test the impact of CEO's gender on firm performance, we specify the following model:

$$ROE = \beta_0 + \beta_1 G + \beta_2 CR + \beta_3 CR10 + \varepsilon \tag{21.1}$$

We set up another model to test the impact of CEO's gender on firm's financing preference:

$$DR = \beta_0 + \beta_1 G + \beta_2 CR + \beta_3 CR10 + \varepsilon \tag{21.2}$$

We use the following model to investigate the impact of CEO's education on firm performance:

$$ROE = \beta_0 + \beta_1 EDU + \beta_2 CR + \beta_3 CR10 + \varepsilon \tag{21.3}$$

To test the impact of CEO's education on firm's financing preference, we specify the following model:

$$DR = \beta_0 + \beta_1 EDU + \beta_2 CR + \beta_3 CR10 + \varepsilon \tag{21.4}$$

The model for testing the impact of CEO's self-confidence on firm performance is as follows:

$$ROE = \beta_0 + \beta_1 CON + \beta_2 CR + \beta_3 CR10 + \varepsilon \tag{21.5}$$

The following model is used to test the impact of CEO's self-confidence on firm's financing preference:

$$DR = \beta_0 + \beta_1 CON + \beta_2 CR + \beta_3 CR10 + \varepsilon \tag{21.6}$$

The impact of CEO's equity incentive on firm performance is to be tested by the following model:

$$ROE = \beta_0 + \beta_1 EI + \beta_2 CR + \beta_3 CR10 + \varepsilon \tag{21.7}$$

To test the impact of CEO's equity incentive on firm's financing preference, we use the following model:

$$DR = \beta_0 + \beta_1 EI + \beta_2 CR + \beta_3 CR10 + \varepsilon \tag{21.8}$$

Finally, a three-step procedure is used to test the mediator role that financing preference plays in the relationship between CEO's background characteristics and firm performance. The first step is to test the relationships between the independent variables and the dependent variable, and the second step is to test the relationship between the independent variables and the mediator variables. Tests in these two steps will be achieved by the above models. The final step is to test the impact of both the independent variables and the mediator variables on the dependent variable. Hence, we specify the following models to test the mediator role that financing preference plays in the relationship between CEO's background characteristics and firm performance.

$$ROE = \beta_0 + \beta_1 G + \beta_2 DR + \beta_3 CR + \beta_4 CR10 + \varepsilon \quad (21.9)$$

$$ROE = \beta_0 + \beta_1 EDU + \beta_2 DR + \beta_3 CR + \beta_4 CR10 + \varepsilon \quad (21.10)$$

$$ROE = \beta_0 + \beta_1 CON + \beta_2 DR + \beta_3 CR + \beta_4 CR10 + \varepsilon \quad (21.11)$$

$$ROE = \beta_0 + \beta_1 EI + \beta_2 DR + \beta_3 CR + \beta_4 CR10 + \varepsilon \quad (21.12)$$

21.5.2 Descriptive Statistics

The descriptive statistics on the above variables can be summarized in Table 21.2:

The statistic summary in Table 21.2 demonstrates the current status of CEO's background characteristics, financing preference and firm performance of China's listed companies. First, for firm performance, we can see that the overall firm performance of China's A-share listed companies is in a good condition, and their mean of ROE reaches 7.21%, with relatively low standard deviation, maintaining a steadily fast growth. The maximum ROE reaches 43.2459%, while the minimum is −35.2728% (excluding ST companies). As to CEO's background characteristics, the statistics on gender show that around 93% of CEO posts of China's listed companies are still dominated by men. For education, the number of CEOs with MBA degrees slightly exceeds the number of CEOs without master's degrees in business administration, and 55% of CEOs have received MBA degrees. Meanwhile, most CEOs who are without MBA degrees are of older age, while younger CEOs tend to more likely have MBA degrees. As to CEO's self-confidence, on the current A-share market, CEOs are self-confident in expecting a better profitability for their companies, and believing that their companies will have a better growth path. For CEO's equity incentive, as among the nearly 3000 A-share listed companies, only about

Table 21.2 Descriptive statistics summary

Variables	Min.	Max.	Mean	SD
ROE (%)	−35.2728	43.2459	7.2171	8.376048
G	0	1	0.9333	0.249861
EDU	0	1	0.5500	0.498325
CON	0	1	0.4533	0.498649
EI	0	1	0.2100	0.407989
DR (%)	4.2585	85.2307	35.0264	20.41085
CR10 (%)	17.5600	95.2500	55.6365	15.60883
CR (%)	0.2889	23.3669	2.9606	2.950083

190 companies adopt equity incentive scheme, the equity incentive policy is not widely adopted. The average debt to asset ratio for current A-share listed companies is about 35%, which is within a reasonable range. However, we have to notice that some companies have a much higher debt to asset ratio, closing to 85%, while some other companies have a debt to asset ratio as low as around 4%, which indicate insufficient liabilities. As to other control variables, the average shareholdings of top 10 shareholders are 55%, with the highest reaching 95%, and this indicate that China's A-share market currently have a high degree of firm concentration.

21.5.3 Regression Results (Table 21.3)

The regression results of Model 1 show that the gender of CEO does not have a significant effect on firm performance (CEO gender has a negative coefficient for firm performance, but the effect is not significant, as *p*-value is greater than 0.1). Meanwhile, female CEOs do not necessarily improve firm performance, thus Hypothesis 1 is proved to be true.

The regression results of Model 3 indicate that there is no significant relationship between the education level of the CEO and the firm's performance (the education level of the CEO has a negative coefficient and the p-value is greater than 0.1, indicating insignificance). This implies that higher education level of CEOs do not actually improve firm performance. The result, excluding other factors, overthrows Hypothesis 3.

According to the results of Model 5, there is a significantly negative correlation between CEO's self-confidence and firm performance (the coefficient of CEO's self-confidence is negative and the p-value is less than 0.01), which means that CEO's higher self-confidence tends to bring about a decline in firm performance, while firms under the leadership of more conservative CEOs will have better business performance. Without considering the impact of other factors, we can conclude that Hypothesis 5 is proved to be true.

Table 21.3 Results of CEO's background characteristics and firm performance

Variable type	Variables	Model 1	Model 3	Model 5	Model 7
Control variables	CR10	0.1347***	0.1368***	0.1226***	0.1343***
	CR	0.4092***	0.4131***	0.3621**	0.3890**
Independent variables	G	−1.5982			
	EDU		−0.3483		
	CON			−2.5349***	
	EI				−1.5658
Adjusted R^2		0.0970	0.0952	0.1163	0.1004

* indicates P < 0.1; ** indicates P < 0.05; *** indicates P < 0.01

Table 21.4 Results of CEO's background characteristics and firm's financing preference

Variable type	Variables	Model 2	Model 4	Model 6	Model 8
Control variables	CR10	−0.1647***	−0.1586***	−0.1258**	−0.1457**
	CR	−3.9690***	−3.9190***	−3.8250***	−3.7905***
Independent variables	G	−0.6284			
	EDU		6.0560***		
	CON			6.6671***	
	EI				10.2900***
Adjusted R^2		0.3598	0.3815	0.3849	0.4011

* indicates P < 0.1; ** indicates P < 0.05; *** indicates P < 0.01

Results of Model 7 show that there is no correlation between equity incentive and firm performance (the equity incentive coefficient is negative and the p-value is much greater). This means that equity incentive policies do not substantially improve firm performance, thus the Hypothesis 7 has been proved wrong (Table 21.4).

The regression results of Model 2 indicate that the gender of CEO does not have significant impact on a firm's debt to asset ratio (the negative gender coefficient and greater p-value indicate insignificance), and this is to say that the CEO's gender does not have effect on firm's financing preference, thus proving the Hypothesis 2.

According to the results of Model 4, CEO's education level is positively correlated with the firm's debt to asset ratio (positive coefficient with p-value less than 0.01). If a firm's CEO has received a higher level of education, the firm tends to prefer debt financing and to have a higher debt to asset ratio, thus proving the Hypothesis 4.

The regression results of Model 6 confirm positive correlation between CEO's self-confidence and firm's debt to asset ratio (the positive coefficient between these two variables with a p-value less than 0.01). CEO's stronger self-confidence will lead to more debt financing in the firm, and the debt to asset ratio of the firm will be higher, thus proving the Hypothesis 6 to be true.

Results of Model 8 indicate a positive correlation between equity incentive and firm's debt to asset ratio (the positive coefficient between these two variables with a p-value less than 0.01), showing that firms which adopt equity incentive schemes will have higher debt to asset ratio. This proves Hypothesis 8.

21.5.4 Testing the Mediator Role of Financing Preference

Because there may be correlation among the background characteristics of CEOs, we should verify the mediator role of financing preference separately. As we have shown that CEO's gender, education level, and equity incentive have no significant effect on firm performance, and CEO's gender does not have significant effect on firm's financing preference, we can prove that financing preference is not the

Table 21.5 Testing the mediator role of financing preference

Variable type	Variables	Model 11
Control variables	CR10	0.1038***
	CR	−0.2121
Independent variables	CON	−1.5341*
Mediator variables	DR	−0.1501***
Adjusted R^2		0.1986

* indicates P < 0.1; ** indicates P < 0.05; *** indicates P < 0.01

mediator variable for CEO's gender, education level, and equity incentive. Hence, we only need to verify Model 12 (Table 21.5).

We can see, from the results of Model 11, that the absolute value of the coefficient of the independent variable CEO's self-confidence is reduced when adding the mediator variable debt to asset ratio to the model (reducing from −2.5349 in Model 5 to −1.5341 in Model 11; even though p-value is raised from less than 0.01 to less than 0.1, it is still relatively significant), reducing the negative effect of CEO's self-confidence on firm performance. Therefore, it can be shown that the debt to asset ratio plays an adjusting role in the relationship between CEO's self-confidence and firm performance, proving part of Hypothesis 9, which is that financing preference only plays part of the mediator role between CEO's self-confidence and firm performance.

21.6 Robustness Test

To deliver stable and reliable test results, we use proxy variables to carry out robustness test. In the robustness test, as no suitable indicators can be used as proxies for the independent variables and adjustment variables, we choose to change the dependent variable. The dependent variable is firm performance, which has a number of proxies. We select return on equity (ROE) as the proxy for its comprehensive meaning. To ensure that the previous conclusions reached before the dependent variable is replaced can be verified, we replace ROE with ROA (return on assets) for analysis, with testing procedures unchanged. The results remain consistent with the previous analysis, and the analysis passes the robustness test.

21.7 Conclusions

21.7.1 CEO's Gender, Education Level, and Equity Incentive Have No Significant Relationship with Firm Performance

The analysis shows that, in selecting CEO for improving firm performance, gender factor has no effect. Although men are generally considered to be more risk-loving and decisive, CEO's gender is proved to have little impact on firm performance. In fact, there is no clear relationship between CEO's education and firm performance, and we suppose the reasons for this may lies in: first, the education a CEO has received becomes outdated, or the knowledge obtained may be forgotten; second, the theoretical knowledge has little value in business practice.

Most surprisingly, equity incentive does not have substantial effect on firm performance. Firms which adopt equity incentive schemes do not improve their business performance. We speculate that, on one hand, in China's immature capital market, equity incentive policy cannot fully play its role in motivating CEOs; on the other hand, China's equity incentive practice is still in its early development stage. Firms with equity incentive plans have not seen improvement in performance, resulting in China's currently underdeveloped equity incentive schemes. However, we believe that as more equity incentive plans are adopted by China's listed companies, their effect on firm performance will be increasingly significant. After all, most related studies have found the positive correlation between equity incentive and firm performance.

21.7.2 CEO's Self-Confidence Is Negatively Correlated with Firm Performance

CEO's self-confidence is often translated into over optimism about business future and subjectivism in decision-making, resulting in not taking actual conditions into consideration and mistakes. Studies show that CEO's self-confidence comes from self-awareness, and it cannot be eliminated by learning. This awareness is not helpful in correcting mistakes, and, as data show, it can have negative effect on firm performance.

21.7.3 Firm's Financing Preference Plays Partial Role in the Relationship Between CEO's Self-Confidence and Firm Performance

Overly confident CEOs tends to expect the stock price of their firms to keep rising, and they are likely to choose debt financing. They are usually confident in future steady profitability of their firms, expecting sufficient cash supply and lower risk in solvency. In their opinion, their rising stock price makes equity financing costly. Hence, self-confident CEOs will have positive effect on firm's debt financing, and they will not choose equity financing. As higher debt to asset ratio brings down firm performance, self-confident CEOs will worsen firm performance, and financing preference plays a partial mediator role in this relationship.

21.8 Implications

21.8.1 Implications for CEO Selection

Firms should avoid sex discrimination in selecting their CEOs, because gender difference does not have an impact on firm performance. Neither female CEOs nor male CEOs can make big difference to the firms. The characteristics and ability of a CEO is of importance in improving a firm's performance.

For ability, education is not the only criterion, and we have proved that CEO's higher education level does not necessarily related to better firm performance. Therefore, education is only part of the picture, and other important factors such as qualifications and experience may also contribute to CEO's ability.

Self-confidence will never be the primary criterion in selecting CEO because overly confident CEOs are subjective, and they will find it hard to recognize their own mistakes. Some of them may even transform their optimism into adventurism. For this reason, in CEO selection we should consider the factor of a CEO candidate's self-confidence.

As for equity incentive, since China is in its early stage in equity incentive development, we do not have rich samples to reach definitive conclusions. We believe that firms should take its constraints into account when considering adopting equity incentive policy, and equity incentive plans should be developed in accordance with the characteristics of the firm. We should be cautious in learning international experience and make adjustment when adopting international equity incentive schemes.

21.8.2 Implications for Corporate Governance

Some have compared a firm's cash to its blood to demonstrate the importance of cash to a firm. The development of a firm requires a lot of cash, the lack of which will result in bankruptcy of the firm. Firms have two ways to obtain external finance for their cash, equity financing and debt financing, both of which can supply the cash much needed by the firms and can help their business operations. However, in developing financing strategies, firms should take actual business conditions into consideration. For firms with higher debt and debt to asset ratio, equity financing will be a better option to avoid financial difficulties.

References

1. Hu, J. B., & Gai, D. (2014). Executive equity incentive and bank credit decision – Based on empirical evidence of private listed companies in China. *Accounting Research, 4*, 58–96.
2. Lin, J. B. (2014). An empirical study on executive gender and corporate financial performance of listed companies. *Development Research, 09*, 54–59.
3. Lu, Y., & Zhang, M. Y. (2015). CEO's education and firm performance. *Journal of Tsinghua University (Science and Technology), 55*(04), 428–442.
4. Tang, Y. M., Wang, H. M., & Bai, Y. F. (2011). CEO characteristics, risk preference and corporate R & D expenditure – Based on Technology Intensive Industry. *Forum on Science and Technology in China, 10*, 89–95.
5. Jiang, W. (2011). Chairman's personal characteristics, overconfidence and capital structure. *Economic Management Journal, 2*, 78–86.
6. Yang, Z. Q., Shi, S. P., Shi, B. R., & Cao, X. Y. (2016). Mixed ownership, equity incentive and defensive behavior in financing decision-making – evidence based on dynamic weighing theory. *Journal of Finance and Economics, 8*, 108–120.
7. Arch, E. C. (1993). Risk-taking: A motivational basis for sex differences. *Psychological Reports, 73*(1), 3–11.
8. Barber, B. M., & Odean, T. (2001). Boys will be boys: Gender, overconfidence, and common stock investment. *Quarterly Journal of Economics, 116*(1), 261–292.
9. Barber, B. M., & Odean, T. (1999). The courage of misguided convictions. *Financial Analysts Journal, 55*(6), 41–55.
10. Bernasek, A., & Shwiff, S. (2001). Gender, risk, and retirement. *Journal of Economic Issues, 35* (2), 345–356.
11. Billett, M. T., & Qian, Y. (2008). Are overconfident CEOs born or made? Evidence of self-attribution bias from frequent acquirers. *Management Science, 54*(6), 1037–1051.
12. Gottesman, A. A., & Morey, M. R. (2010). CEO educational background and firm financial performance. *Journal of Applied Finance, 2*, 70–82.
13. Heaton, J. B. (2002). Managerial optimism and corporate finance. *Financial Management, 31* (2), 33–45.
14. Huang, J., & Kisgen, D. J. (2013). Gender and corporate finance: Are male executives overconfident relative to female executives? *Journal of Financial Economics, 108*(3), 822–839.

Chapter 22
Security Risk Management Approach for Improving Information Security Return of Investment

Xichun Li, Mahmoud Al-Shawabkeh, and Zhiyuan Li

Abstract In particular, the reason when organisation competitors win business, is the gain of organisation sensitive and important data. This happen as results of having a local employee active as an agent inside the organisation or because of a persistent threat attack. The data sharing, data protection, data retention, data risk management, and personnel physical security are the responsibilities of the organisation Chief Security Officer (CSO), who solves the intellectual property theft problems before and when occurred, by combining approached of Information Risk Management (IRM) and Information Security Governance (ISG). In this paper, we propose a new novel security management approach that improve information security return of investment.

Keywords Information security · Risk management · Return of investment

22.1 Overview

The organisation objectives, security requirements, organisation size, processes, and structure are the highly effective reasons that influence the organisation seeking a strategic decision of implementing information security management system. Implementing information security management system preserves the organisation data integrity, confidentiality and availability, which are the three major security needs of the organisation [1, 2]. In general, risk management is simply can be defined as "the process of defining, understanding, and responding to actions that

X. Li (✉)
Guangxi Normal University for Nationalities, Chongzuo, China
e-mail: 291495244@qq.com

M. Al-Shawabkeh
National Defense University of Malaysia, Kuala Lumpur, Malaysia

Z. Li
ChongQing Fivision Technology, ChongQing, China

M. Tavana, S. Patnaik (eds.), *Recent Developments in Data Science and Business Analytics*, Springer Proceedings in Business and Economics,
https://doi.org/10.1007/978-3-319-72745-5_22

may lead to a failure in any of the main three information security factors, the confidentiality, integrity or availability of an information system" [3]. The proposed novel solution in this paper helps the organisation chief security officer to set directions, procedures and a visions of solving organisation security and intellectual properties by solving a continuous problem of developing and implementing an information security approach to protect sensitive organisation data. The research problem statement explained by the research problem statement "Does implementing the information security standards improve information security system return of investment?".

22.2 The Review of Literature

To protect organisation data and sensitive information from the attackers gaols, the security approaches, methods, and the characteristics of information systems are designed based on the organisation motivations and need to protect the data. This section describes the various approaches that are proposed and used in the organizations information security programs.

22.2.1 The Information Security Management System (ISMS)

Standard for providing overall requirements and needs of create, implement, maintain, and enhance the information security system of the organisation is the information security management system ISO/IEC-27001:2013. As defined in ISO/IEC 27002:2013, the overall objectives of information security management system (ISMS) are the risk reduction and the impact on operations by increasing opportunities of the business and the gaunt of its continuity [4]. The main ISO/IEC 27001:2013 standard requirement is the development and maintaining of security in the organisation, which should be planned carful and be able to define all controls and risks that may occur in the organisation information system. Steps of implementing the ISO/IEC 27001:2013 standard are shown in Fig. 22.1. The information security standard ISO/IEC 27001:2013 is one of the best information security method when the organisation need to monitor, review, maintain and improve the security of the information system. The reason is because by

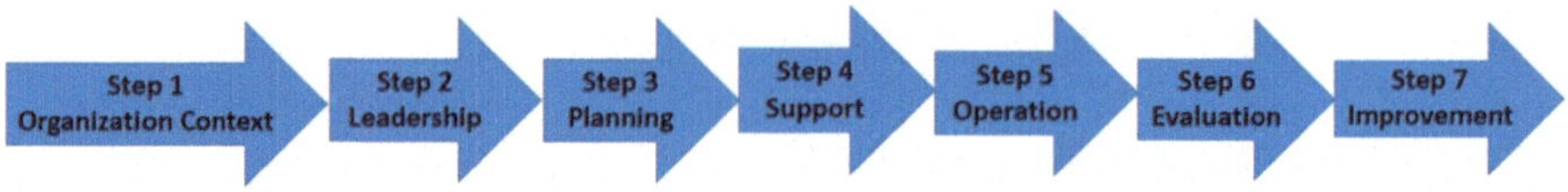

Fig. 22.1 The ISO/IEC 27001:2013 standard implementation steps

implementing this standard, the organisation information system must be audited and certified. Furthermore, the ISO/IEC 27001:2013 standard focus on the process of the risk make it the better than other approaches as USA national institute standards and technology (NIST), or Control Objectives for IT and Related Technology (COBIT). The ISO/IEC 27001:2013 standard has 134 classic controls, this is one of the disadvantages of this method. A reason must be stated if any of this controls not implemented, this require more work to be done by the team of information security in the organisation.

22.2.2 Standards of US National Institute of Standards & Technology (NIST)

The Federal Information Processing Standards Publications (FIPS PUBS) has issued from National Institute of Standards and Technology (NIST) in United States of America [5]. In this method, the Federal Information Security Management Act (FISMA) proposed to emphasize public sector organisations in US for developing, documenting, and implementing a program that is secure and support the operations of the organisation or systems provided by other sources. Directions of FISMA are based on the objectives of having a certain level of information security system to protect from specific risk; and based on the lowest security requirements of systems and information. The FISMA slandered specifies 17 of information and information system security categories and the lowest security requirements for each of them. With the NIST special publication 800–53, the organisations need to apply the lowest information security requirements [6, 7].

22.2.3 Business Framework (COBIT)

The Control Objectives for IT and Related Technology (COBIT), is a framework developed by Information Systems Audit and Control Association (ISACA) for both public and private sector organisations for the management of information technology. COBIT is a global association aim to provide accepted security practices and tools to improve trust of information system. When the organisation have a technology and a business process responsibility, then must use COBIT as a best choice for providing reliability, quality, and controlling security of information system [8].

22.2.4 Operationally Critical Threat, Asset, and Vulnerability Evaluation (OCTAVE^{SM})

To have a systematic, comprehensive, and context driven evaluation of information risk in the organisation, the OCTAVESM is the best choice for implementing security standard. This method is to understand Integrity, confidentiality, and availability of information security to make decision on protection mechanisms for the information system. The three phases of OCTAVE are:

Phase I: To build the security requirements for the organisation.
Phase II: To identify vulnerability for the information system structure.
Phase III: To define the strategy for the security risk management need to be implemented.

22.3 The Proposed Solution

Adapting from NIST 800–30, the systematic risk management method can be used by the organisation as follows:

I. Involvement of employees: by involving both management and technical employees as the managers, administrators of networks, Application support specialists, Senior IT technicians, and chief information security officers.
II. Provide system categorisation: By applying FIPS to categorise the information system security based on confidentiality, integrity, and availability.
III. Threads Identification: Table 22.1 shows some possible information system threat.

From Table 22.1. The following is the categorization of threats that harm the information system, which are:

Table 22.1 Some information system threats

The threat	Threat type
Accidental Disclosure	Accidentally release of sensitive information to an unauthorised person
Usage of Communication Bandwidth	Usage of communication bandwidth intentionally or non-intentionally for not planed reasons
Interruption of Communication	The information transferred on communication link can be interrupted
Nature Act	Information availability can be occurring by natural occurrences as tornadoes and earthquakes
Disruption of Electrical Interference	Denial of service to an authorised user or the modification of information may happen by the failure of commercial power
Software and Programs Alteration	The deletion or modification of programs or operating systems that affect confidentiality, integrity, and availability of information
Configuration Error of System	When installation or updating software or hardware this configuration error may occur
Data Alteration	The deletion or modification of data by users that affect confidentiality, integrity, and availability of information

(a) Threats occurs by humans:

- Disclosure of Sensitive data
- Employees Disgruntled
- Expose of social media expose

(b) Environment and Physical Threats

- Threats of hazardous materials as the server room beside the fire exit or smokers room
- Threats of Physical Security as having an open plan structure
- Threats of Theft or Vandalism as having rarely check of employee ID card

(c) Technical Threats

- Malicious code, or malicious malware, or malicious spyware
- Access by an authorised user
- Lake of logging as not secure passwords
- Use of known unsecure software as for example the use of Widows XP that is now is not supported by Microsoft

IV. Vulnerability Identification

When asking the question: "what may go wrong?" to discover weakness of information system then categorise the security breach and the violation of the implemented security policy [7]. Following are some listed information security vulnerabilities:

- Procedures and directives are not defined clearly. Furthermore, the weak define and test of planes.
- No recovery procedures planed specially for the networks
- No security training provided and weak backup of information and programs
- Not having alternate communication service, storage databases, and processing sites

V. Controls Identification

The following Table 22.2. list some controls to be identified.

I. Likelihood Exploited (Table 22.3)

II. Severity Impact (Table 22.4)

III. Level of Risk (Table 22.5)

Table 22.2 List of known controls

Controls of organisation	Controls of humans	Controls of technical	Controls of process
Structure design	employees	Architecture	Technical processes
Controls of compliance	Culture of users	OS and applications	Man-Machine interface
Culture of organisation	Factors of users	Infrastructure	Infrastructure life cycle
	Emergence	Technical infrastructure	

Table 22.3 List of likelihood exploited

Type	Level	Likelihood
A	25% and below	Low
B	Between 26–75%	Moderate
B	76% and above	High

Table 22.4 Severity impacts

Level	Likelihood	The confidentiality	The integrity	The availability
A	Low	Confidentiality loss means less on organisation effect	Integrity loss means limited on organisation effect	Availability loss means limited on organisation effect
B	Moderate	Confidentiality loss means serious on organisation effect	Integrity loss means serious on organisation effect	Availability loss means serious on the organisation effect
C	high	Confidentiality loss means severe on organisation effect	Integrity loss means severe on organisation effect	Availability loss means severe on the organisation effect

22.4 Conclusion

This paper proposes a novel information security method that focus on most important variables of information security including data and information to be protected, controls to be applied, theft types and sources to be identified, and the objectives of protecting the information system. Information security return of investment increased by implementing this method that have an effect of reducing information loss and improving its confidentiality, integrity, and availability of security of the system. Finding of this research help organisations management to implement a successful information security policy. Nevertheless, full successful implementation of secure information system require commitment from organisation employees and management.

Table 22.5 Level of risk

Type	Access	Attacker	Reasons		Effect			
			Accidental	Intentional	Disclose	Modify	Loss	Interrupt
Threat of people	Networking	Inside & Outside	Y	Y	Y	Y	Y	Y
Environmental /physical	Physical	Inside	Y	Y	Y	No	Y	Y
Technical	Network/physical	Inside & Outside	Y	Y	Y	Y	Y	Y

References

1. Mahmoud, A., & Xichun, L. (2017). A new counterfeiting approach: Computer security evaluation of fuel rationing system, in 2nd Joint International Mechanical. *Electronic and Information Technology Conference (JIMET).*
2. Al-Shawabkeh M., Xichun Li, Mohamed S. (2016). Leading Change: Adaption of Information Security in University Announcement System, Proceedings of the 2016 Joint International Information Technology, Mechanical and Electronic Engineering, http://production.atlantis-press.com/proceedings/jimec-16/25861516, September 2016, ISBN 978-94-6252-234-3, ISSN 2352-5401, doi:10.2991/jimec-16.2016.16
3. Brotby K. (2009). *Information security governance: A practical development and implementation approach.* Apr 2009, ISBN: 978-0-470-13118-3. John Wiley & Sons, Inc., Hoboken, New Jersey.
4. ISO/IEC27002:2013, Information technology – Security techniques – Code of practice for information security controls, code of practice for information security controls. International Organization for Standardization (ISO), Switzerland, (2013). p. 80.
5. FIPS-200, FIPS publication 200 minimum security requirements for federal information and information systems. (2006).
6. NIST-800-53, Security and privacy controls for Federal Information Systems and Organizations. National Institute of Standards and Technology. (2013).
7. N. I. of S. and T. S. P. (2002). NIST-SP-800-30, SP 800–30, Risk management guide for information technology systems.
8. De Haes, S. (2009). The risk IT practitioner guide. ISACA. www.isaca.org

Chapter 23
Comparative Analysis on Investment and Financing Models of Urban Rail Transportation

Xiaodi Wang

Abstract Urban rail transportation is trend of urban traffic development in the world. The utility model has advantages of high volume, high-speed, high security, and significant environmental protection. Government should give priority to establish an effective investment and financing mechanism to solve construction funding issues. For social benefits purpose, authors analyze and compare characteristics of London, Tokyo and Beijing in rail transit, investment and financing mode, summarize similarities and differences to establish investment and financing mode to meet Beijing city railway transit development, provides necessary background knowledge. This paper discusses policy direction and development trend of rail transportation construction investment and financing system transformation process in recent years. The authors propose future rail transit construction references in Beijing.

Keywords Urban rail transportation · Investment and financing mode · World city · Comparative analysis

In recent years, with rapid development of urban scale and economic construction in China, urban population has increased rapidly. To achieve purpose of controlling urban growth, increasing urban capacity and relieving traffic congestion, railway transportation such as subway, light railway, Inter City Express (ICE) has become an inevitable choice to alleviate problem of relieving traffic congestion in large city.

The world cities of Tokyo and London have already built a perfect urban rail transit network, and actively explored innovative ways of rail transit investment and financing methods to promote further development. Urban rail transit has characteristics of large amount of investment, long payback cycle length, operating risk and public welfare. Initial stage of rail transit construction, the government must give

X. Wang (✉)
International Information Research Department, Beijing Institute of Science and Technology, Beijing, China
e-mail: wangxd@bjstinfo.com.cn

M. Tavana, S. Patnaik (eds.), *Recent Developments in Data Science and Business Analytics*, Springer Proceedings in Business and Economics,
https://doi.org/10.1007/978-3-319-72745-5_23

priority to establish an effective investment and financing mechanism to solve construction funding issues [1].

23.1 Urban Rail Transportation Industry Development Advantage

In the tenth 5-year plan, China's governments has included subway traffic development into the outline of the National Economic and Social Development 5-Year Plan, and token it as a major strategy for stimulating rapid, steady and sustainable development of national economy. Many cities have invested a large number of manpower and material resources, formulated comprehensive subway development general plan, and carried out in-depth feasibility research [2]. In recent years, China's government has continuously increased in investment in subway construction (Fig. 23.1).

China has made great efforts to develop urban rail transit, mainly because it has several advantages:

Strong transport capacity Rail transportation is an important approach to solve traffic issues in large cities. Such as subway transport, it has high-volume passenger capacity and powerful urban rail transit system transportation capacity. A carriage capacity is about 150–200 people, and one-way transmission capacity is up to 80 thousand people per hour.

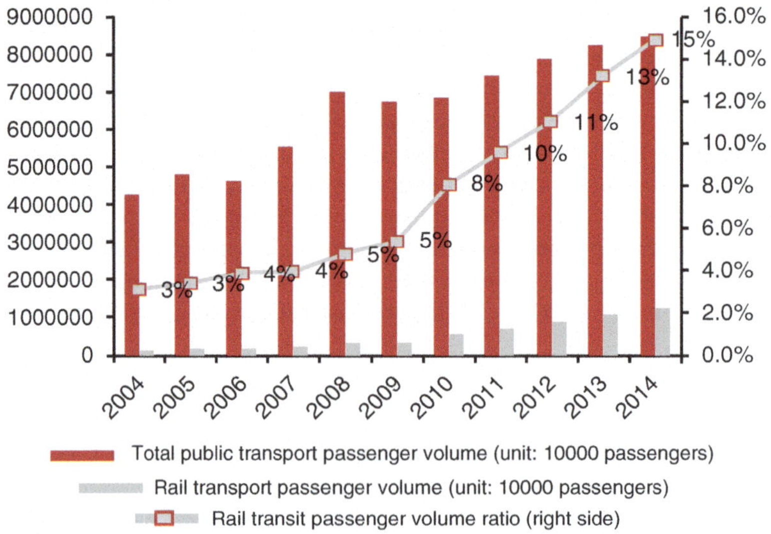

Fig. 23.1 Proportion of China's rail transportation in public transport [6]

Fast running speed General operation of city bus running speed is only 20 km/h, and road will be more crowded during rush-hour traffic. Underground trains are not subject to ground restrictions with the highest speed of 80–100 km fast forward.

Small occupation of land Rail transportation construction can save land resources. Especially subway transportation facilities, it will reduce land resources waste, economize ground space and conserve quality resources.

Energy consumption According to statistics, rail transportation energy consumption per unit can save 30 KJ more than public transport, and save 2132 KJ more than automobile. With Beijing metro as an example, most of subway trains/track catenary DC power transmission system, traction system is VVVF, which greatly use energy sparingly.

23.2 London Rail Transportation Investment and Financing Models

The London Underground Ltd is a government company, responsible for entire subway system which is including metro vehicles, stations operation, and decision on formulation and adjustment of metro services and fares. London Underground is specifically responsible for metro system operation; maintenance and upgrading of subway tunnels, metro vehicles, stations and signal systems are carried out by several different network lines.

23.2.1 London Rail Transportation Investment and Financing in PPP Model

London rail transportation introduced private equity investors and established Item Company, and was in progress by PPP model. PPP (Public–Private-Partnerships) model is a long-term cooperative partnership established between the public and private institution to provide public services. In March 1998, UK government launched PPP model to address issues, private infrastructure companies were in charge of maintenance and facility, while public sectors were responsible for operation. UK has become an advocate of infrastructure projects and public services financing through the PPP model.

UK government has adopted a railway transit construction investment policy, and has taken a series of preferential measures to private institution investors and financial group, such as investment insurance, tax incentive. Thus, investors have the lowest rate of return guaranteed; UK government would grant government subsidies in failure situation to promote society investment in rail transportation construction.

23.2.2 London Rail Transportation Investment and Financing in Bot Model

BOT (Build-Operate-Transfer) is a form of project financing, wherein a private entity receives a concession from the private or public sector to finance, design, construct, and operate a facility stated in the concession contract.[1]

Part of Light rail transit project in London adopted by BOT model. Since 90s, there were three private institutions in UK have signed with government investment, construction and operation track lines in light rail transit projects. They were CroydonTram in London, Manchester Metroiink (Manchester medium volume rail transit system), and Midiandiink in Bermingham. Their common feature: partial route was to reform former national railway or disused railway; combined with light railway form, flexible route planning, and rapid transit formation in urban center; lead real estate redevelopment along abandoned railway.

23.2.3 London Rail Transportation Investment and Financing in PFI Model

PFI (Private finance initiative) is a way of creating PPPs by funding public infrastructure projects with private capital. PFI and its variants have now been adopted in many countries as part of the wider programme of privatisation and financialisation, and presented as a means for increasing accountability and efficiency for public spending.[2]

In 1992, UK government had introduced PFI concept and gradually matured to consider private capital introducing public construction under pressures of controlling fiscal expenditure and improving infrastructure [5]. UK HM Treasury had introduced PFI official definition: public institution had bought high quality services in a long-term agreement from private sector, which included delivery achievement of mutual agreement, corresponding maintenance and repairmen construction of necessary infrastructure. London public rail transportation routes were government planned and private sector operated. In most instances, all bus services had commercial practices by competitive bidding for a 3 year agreement. UK rail transit privatization has drawn on the experience of bus service industry, implemented bidding system, and introduced actual strength, good credit and quality service private companies [4].

[1]BOT model: https://en.wikipedia.org/wiki/Build%E2%80%93operate%E2%80%93transfer.

[2]PFI model: https://en.wikipedia.org/wiki/Private_finance_initiative.

23.3 Tokyo Rail Transportation Investment and Financing Systems

Because urban rail transportation investment has the attributes of public welfare, most of urban rail transit constructions in big cities were dominated by government investment.

23.3.1 Tokyo Rail Transportation Investment and Financing Model

There are five main ways to raise urban rail transit construction funding in Japan, mainly there are mainly including government grants, users' burden, beneficiaries' burden, bonds issuance and loan floatation.

In April 1982, Tokyo government has encouraged investment in urban rail construction, increased urban railway transport capacity, and established "Special Urban Railway Construction Accumulation Funding System". The system has stipulated in view of railway as a public facility, railway sector should implement corporation tax reduction and fixed assets tax in preferential tax policy. Various subsidies and tax incentives for rail transit investors by the government has promoted investment in urban rail transit construction by the public society. There were set up a semi-public and semi-private rail transportation enterprise. The Japanese private railway length in Tokyo has all within the scope of rail transit lines, ratios were as high as 34% and 46%, there is no precedent in any other cities.

23.3.2 Tokyo Rail Transport Construction Tod Model

In urban planning, a transit-oriented development (TOD) is a type of urban development that maximizes the amount of residential, business and leisure space within walking distance of public transport. A TOD model includes a central transit stop, such as train stations, light railway and bus stations, which are surrounded by a high-density mixed-use area with lower-density areas spreading out from this center.[3]

In Japan, TOD model began in the early 1920s, was first adopted by the Hankyu railway company in Osaka and achieved great success. In order to enhance public transport attraction, Japanese government formulate "The Third National Comprehensive Development Plan", and proposed road transportation facilities supply in urbanized area should maintain adequate coordination with urban land use development. Many private rail companies in Tokyo operate industry comprehensive management along the railway route like Retail, real estate, buses and hotels. Government has unified land use, railway construction planning and infrastructure facilities.

[3]TOD model:https://en.wikipedia.org/wiki/Transit-oriented_development.

23.3.3 Tokyo Metro Projects Investment and Financing Mode

Tokyo Metropolitan Government Bureau of Transportation is responsible for Tokyo Metro track line construction and maintenance. Toei transportation and Battalion subway are operating units for Tokyo metro transportation, which are managed by Tokyo Metropolitan Government Bureau of Transportation.

Public rail transportation
Government subsidiaries distribution condition: the high speed railway construction underground allowance system in 1962 implemented the maximum amount was 6.5% of interest difference; in 1978, construction allowance was 10.5–70% of construction costs. Payment period was 5–10 years; railway construction funding in 1991 provided a one-time interest free loan with 40% project cost of Railway Construction Company and Battalion subway construction project.

Privately-operated rail transportation
Government subsidiaries distribution condition: P line system in 1972, in order to promote comprehensive development of rail transportation and residential buildings along track line, financial subsiding targets construct multiple track and satellite unban subways for three city circle to optimize city construction and industrial structure. The railway construction company was responsible for project construction and providing cost of project implementation, private developers reimbursement could pay 50 times in 25 years, interest rate was less than 5%; urban rail accumulation funding system in 1996, part of metro equipment modification project cost was added to existing operating fares; station comprehensive reform allowance system in 2001 implemented barrier free, earthquake prevention, fire protection and other renovation project; underground high-speed railway construction allowance system in 2002 applied to private lines.

23.4 Transformation Process of Beijing Rail Transportation Investment and Financing Models

In 2016, Beijing has issued the "Beijing 13th 5-Year Railway Transportation Construction Plan". It proposed to rail transit and high-capacity rapid transportation and ground transportation based integrated public transport system as development priorities. In order to give full play to urban public railway transport, people were encouraged to give priority to rail transit travelling. Average daily passenger volume will also increase from 11 million passengers to more than 18 million and 500 thousand passengers; rail transportation will become urban public transport priority selection, proportion will be accounting for more than 58%.

23.4.1 *Transformation Process of Investment and Financing Models*

Beijing earliest subway Line 1 and Line 2 have entirely led by government programs with funds allocated by Beijing municipal government. It reflected government arrangement investing and proposed integrated subway project model.

Beijing Subway BATONG Line construction management mode has reflected characteristics of financing Item Company. Financing rising has mainly composed by three parts: Beijing municipal government investment, National Development Bank loans and foreign particular loans, state-controlled enterprises investment. Government investment, government loan guarantee and stock system operation were main features [3].

The Beijing Metro Line 4 and Line 14 was adopted traditional PPP model, and introduce social investors through franchising. Social investors were responsible for projects investment and regular operation for a certain period. Subway construction has introduced social capital by "equity financing + franchising" compound PPP model, established a bridge between rail transit construction and insurance equity.

23.4.2 *Investment and Financing Models of Urban Rail Transportation Comparison Between London, Tokyo and Beijing*

Tokyo, London and Beijing rail transit project financing (Table 23.1) have also taken government leading mode, adopt preferential policies to attract private sector

Table. 23.1 Investment and financing models of urban rail transportation comparison

City	Government responsibility	Social institution responsibility	Impact on the public	Suitable application
Tokyo	100% govt. construction investment; social investment increase, government grants.	Enterprises are responsible for operation investment; private high-speed rail connect with bus industry function.	Government formulates pricing 4 years once.	Municipal Road assets include tunnels, railways and stations.
London	100% govt. construction investment; preferential policies (investment insurance, tax incentives).	3 infrastructure companies maintenance and construct subway system within 30 years franchising.	Govt. pricing welfare operations; profit communities provide funding.	Municipal Road assets include tunnels, railways and stations.
Beijing	100% govt. investment before 2004; 70% total investment, financing and construction.	MTR CORP. joints cooperation mode in investment, construction and operation by 2004.	Low fare policy; business circle along subway line.	Better govt. finance situation; perfect fiscal system.

investment, and cultivated market mainstays and innovation types. It has advantages of maximum capital, long term and stability.

23.5 Conclusion

Beijing should draw on the experience of Tokyo, London and other world city in rail transit investment and mature financing experiences, actively formulate preferential policies to effectively mobilize social resources in rail transportation investment construction, and implement investment diversification strategy. The government should actively guide large-scale suburban land development and rail transportation mode. Through implementing diversified investment and financing strategy to attract various economic elements participate in rail transit construction, so as to promote rapid and healthy development of suburban county economy.

References

1. Yang, Q. (2011). *Transport business China*. Tokyo: Rail Transportation. (In Chinese), (5), 95.
2. Xu, X. (2009). The method of strategic analysis of metropolitan railway traffic development and application. *China Science and Technology Information*, *16*, 2009–2016.
3. Xiao, X., Zhu, X., & Xiao, X. (2013). Optimization research of fare system for rail transit in Beijing economic & trade update. *Economic and Trade Update*, *24*, 135–136. (In Chinese).
4. Phang, SY. (2006). *Public private partnership in urban rail transit: Recent international experience* (pp. 57–72). USA: Taylor and Francis, 47(1).
5. Chen, J. X., & Fang, J. (2011). On the public private partnership financing strategies of urban rail transit. *Advanced Materials Research, 255–260*, 4095–4099.
6. *Forecast and industry trends of Urban Rail Transit Investment in China*. China Industry Information. http://www.chyxx.com/industry/201611/469469.html. 2016.11.21.

Chapter 24
Measuring Systemic Risk in the Chinese Financial System Based on Asymmetric Exponential Power Distribution

Helong Li, Tianqi Luo, Liuling Li, and Tiancheng Liu

Abstract We propose an extension of CoVaR approach by employing the Asymmetric Exponential Power Distribution (AEPD) to capture the properties of financial data series such as fat-tailedness and skewness. We prove the new model with AEPD has better goodness-of-fit than traditional model with Gaussian distribution, which means a higher precision. Basing on the Chinese stock market data and the new model, we measure the contribution of 29 financial institutions in bank, security, insurance and other industries.

Keywords Asymmetric Exponential Power Distribution (AEPD) · Systemic Risk · Conditional Value-at-Risk (CoVaR)

24.1 Introduction

The global financial crisis of 2008 has alerted the public to the importance of systemic risk. In this context, Adrian and Brunnermeier [11] proposed Conditional Value-at-Risk (CoVaR) to measure systemic risk contributions from individual institutions to the financial system. Then CoVaR method is improved from different perspectives e.g. [1, 2] and practiced in many countries e.g. [3–6]. AR-GARCH was applied to the CoVaR estimation by Gao and Pan [7] to capture the time-varying systemic risk exposure of an individual institution, based on the Student-t

H. Li (✉) · T. Luo
School of Economics and Commerce, South China University of Technology, Guangzhou, Guangdong, People's Republic of China
e-mail: hlongli@scut.edu.cn; luotianqi816@163.com

L. Li
Institute of Statistics and Econometrics, Economics School, Nankai University, Tianjing, People's Republic of China

T. Liu
School of Computer, South China University of Technology, Guangzhou, People's Republic of China

M. Tavana, S. Patnaik (eds.), *Recent Developments in Data Science and Business Analytics*, Springer Proceedings in Business and Economics,
https://doi.org/10.1007/978-3-319-72745-5_24

distribution. Liu and Gu [8] and Lin et al. [9] estimated the CoVaR of the real estate industry and the insurance institutions by AR-GARCH model, assuming the innovation follows the Gaussian and the Student-t distribution respectively.

However, the Gaussian and the Student-t distribution cannot accommodate properties such as asymmetric fat-tailedness and skewness of financial series. Zhu and Zinde-Walsh [10] proposed AEPD which contained the skewness parameter and the decay rates of left and right tails. They proved that AEPD had good performance in error fitting and VaR forecasting.

Hence, AEPD is used as the innovations of the AR-GARCH and CoVaR model. The result of goodness-of-fit confirms that our approach is an adequate method in improving systemic risk measurement. By the new method, we estimate Chinese financial systemic risk from 2008 to 2016.

24.2 Model and Methodology

24.2.1 *VaR*

Given the daily returns of a particular institution i (R_t^i) and the confidence level q, VaR is defined as the q-quantile of the return distribution of institution i.

$$\mathrm{Prob}\left(R_t^i \leq VaR_{q,t}^i\right) = q \tag{24.1}$$

VaR of each institution i is computed by estimating the following model

$$R_t^i = \beta_1^i + \beta_2^i R_{t-1}^i + \beta_3^i R_t^m + u_t^i \tag{24.2}$$

$$\sigma_t^i 2 = \theta_1^i + \theta_2^i {u_{t-1}^i}^2 + \theta_3^i {\sigma_{t-1}^i}^2 \tag{24.3}$$

$$u_t^i = \sigma_t^i \varepsilon_t^i \tag{24.4}$$

$$\varepsilon_t^i \sim AEPD\left(\alpha^i, p_1^i, p_2^i\right) \tag{24.5}$$

where $\xi^i = \left(\alpha^i, p_1^i, p_2^i, \beta_1^i, \beta_2^i, \beta_3^i, \theta_1^i, \theta_2^i, \theta_3^i\right)$ is parameter vector to be estimated by Maximum Likelihood Estimation method. R_t^m is the return of stock market. ε_t^i is the innovation with zero mean and unit variance. σ_t^i is the conditional standard deviation, i.e., volatility. With the innovation series ε_t^i sorted, the q-quantile value z_q^i is obtained. The VaR of institution i can be estimated as follow

$$VaR_{q,t}^i = \widehat{R_t^i} + z_q^i \sigma_t^i \tag{24.6}$$

The same process is repeated for financial system by substituting R_t^{sys} with R_t^i to obtain the parameter vector $\xi^{sys} = \left(\alpha^{sys}, p_1^{sys}, p_2^{sys}, \beta_1^{sys}, \beta_2^{sys}, \beta_3^{sys}, \theta_1^{sys}, \theta_2^{sys}, \theta_3^{sys}\right)$ and the unconditional $VaR_{q,t}^{sys}$.

24.2.2 CoVaR and ΔCoVaR

Adrian and Brunnermeier [11] defined $CoVaR_{q,t}^{sys|i}$ as financial system's VaR conditional on institution i being in financial distress in time t, i.e., its return being at its VaR.

$$Prob\left(R_t^{sys} \le CoVaR_{q,t}^{sys|i} | R_t^i = VaR_{q,t}^i\right) = q \tag{24.7}$$

For each i, we estimate the following model

$$R_t^{sys} = \beta_1^{sys|i} + \beta_2^{sys|i} R_{t-1}^{sys} + \beta_3^{sys|i} R_t^i + u_t^{sys|i} \tag{24.8}$$

$$\sigma_t^{sys|i2} = \theta_1^{sys|i} + \theta_2^{sys|i} u_{t-1}^{sys|i2} + \theta_3^{sys|i} \sigma_{t-1}^{sys|i2} \tag{24.9}$$

$$u_t^{sys|i} = \sigma_t^{sys|i} \times \varepsilon_t^{sys|i} \tag{24.10}$$

$$\varepsilon_t^{sys|i} \sim AEPD\left(\alpha^{sys|i}, p_1^{sys|i}, p_2^{sys|i}\right) \tag{24.11}$$

where $\xi^{sys|i} = \left(\alpha^{sys|i}, p_1^{sys|i}, p_2^{sys|i}, \beta_1^{sys|i}, \beta_2^{sys|i}, \beta_3^{sys|i}, \theta_1^{sys|i}, \theta_2^{sys|i}, \theta_3^{sys|i}\right)$ is parameter vector to be estimated for each institution. Then we generate the predicted values from these regressions to obtain $CoVaR_{q,t}^{sys|i}$

$$CoVaR_{q,t}^{sys|i} = \widehat{\beta_1^{sys|i}} + \widehat{\beta_2^{sys|i}} R_{t-1}^{sys} + \widehat{\beta_3^{sys|i}} VaR_{q,t}^i + z_q^i \times \sigma_t^{sys|i} \tag{24.12}$$

We define the systemic risk contribution of a particular financial institution i by $\Delta CoVaR_{q,t}^{sys|i}$, which means the change degree on financial system's VaR conditional on the financial distress of institution i for unit change on institution i's VaR.

$$\Delta CoVaR_{q,t}^{sys|i} = \frac{CoVaR_{q,t}^{sys|i} - VaR_{q,t}^{sys}}{VaR_{q,t}^i} \tag{24.13}$$

24.2.3 AEPD

The recent distribution class, Asymmetric Exponential Power Distribution, is introduced to fit all the innovations above (ε_t^i, ε_t^{sys} and $\varepsilon_t^{sys|i}$). To embed this distribution in the ARX-GARCH model, AEPD is standardized by the same way of Zhu and Galbraith [12] and Li et al. [13]. The probability density function is given by

$$f(z)=\begin{cases}\delta\left(\frac{\alpha}{\alpha*}\right)K(p_1)\exp\left(-\frac{1}{p_1}\left|\frac{\omega+z\delta}{2\alpha*}\right|^{p_1}\right), z\le-\frac{\omega}{\delta}\\ \delta\left(\frac{1-\alpha}{1-\alpha*}\right)K(p_2)\exp\left(-\frac{1}{p_2}\left|\frac{\omega+z\delta}{2(1-\alpha*)}\right|^{p_2}\right), z>-\frac{\omega}{\delta}\end{cases} \tag{24.14}$$

$$\alpha^*=\frac{\alpha K(p_1)}{\alpha K(p_1)+(1-\alpha)K(p_2)} \tag{24.15}$$

$$K(p)=\frac{1}{2p^{1/p}\Gamma(1+1/p)} \tag{24.16}$$

The $\alpha\in(0,1)$ is the skewness parameter. $p_1>0$ and $p_2>0$ are the left and right tail parameters. When $\alpha=0.5$, $p_1=p_2=2$, AEPD will be reduced to Normal(0,1). The mean and variance of the Standard AEPD are given as follows

$$\omega=\frac{1}{B}\left[(1-\alpha)^2\frac{p_2\times\Gamma(2/p_2)}{\Gamma^2(1/p_2)}-\alpha^2\frac{p_1\times\Gamma(2/p_1)}{\Gamma^2(1/p_1)}\right] \tag{24.17}$$

$$\begin{aligned}\delta^2=&\frac{1}{B^2}\left\{(1-\alpha)^3\times\frac{p_2{}^2\times\Gamma(3/p_2)}{\Gamma^3(1/p_2)}+\alpha^3\times\frac{p_1{}^2\times\Gamma(3/p_1)}{\Gamma^3(1/p_1)}\right.\\&\left.-\left[(1-\alpha)^2\times\frac{p_2\times\Gamma(2/p_2)}{\Gamma^2(1/p_2)}+\alpha^2\times\frac{p_1\times\Gamma(2/p_1)}{\Gamma^2(1/p_1)}\right]^2\right\}\end{aligned} \tag{24.18}$$

$$B=\alpha K(p_1)+(1-\alpha)K(p_2) \tag{24.19}$$

24.3 Data and Results

24.3.1 Data

We consider 29 Chinese financial institutions listed before 2008. The codes and names are presented by industry group in Appendix Table. 24.1. The CSI 300 Financials Index and Shanghai Composite Index are used as proxies for the financial system and market respectively. Sample period is from 1/2/2008 to 30/12/2016. The VaR and CoVaR are computed at $q=5\%$ confidence level. The data is obtained from Wind database. Summary statistics for daily returns are presented in Appendix Table. 24.2.

24.3.2 Goodness-of-Fit

Assuming the innovation follows the AEPD distribution and the Gaussian distribution respectively, we obtain the estimated parameters and innovation series. Then, the kernel density estimation curves of innovation series and the distributional

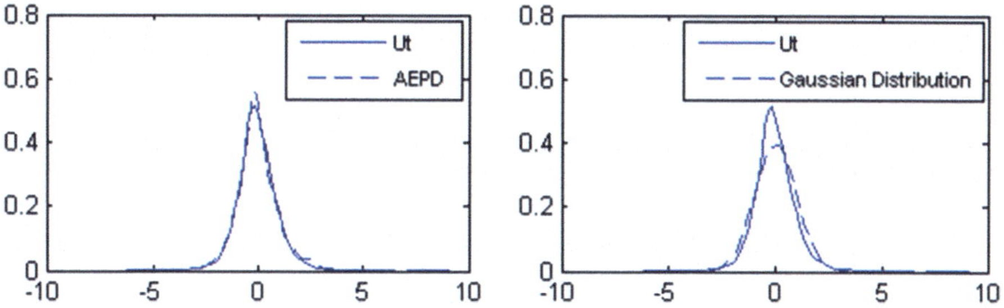

Fig. 24.1 Simulated density and distributional fit of innovation (Bank of China)

simulated curves are compared. We take innovation in Eq. (24.10) of the Bank of China as an example. Figure 24.1 compares the difference between innovation (expressed as Ut) and the AEPD, as well as the difference between innovation and the Gaussian distribution. The simulated curve of AEPD is closer to the innovation than the Gaussian distribution, which shows better goodness-of-fit. The results of goodness-of-fit for other equations and institutions are almost similar.

24.3.3 Chinese Systemic Risk

We obtain ΔCoVaR for each institution i and time period t, through the empirical calculation. The result shows that the state-owned banks (Bank of China, Industrial and Commercial Bank, Construction Bank and Bank of Communications) are the systemically important financial institutions. It is in line with the system-critical banks issued by the Financial Stability Board.

Average ΔCoVaR is computed for each group. The Monetary and Financial Services Group has the greatest contribution to systemic risk (0.68). The second systemically important financial industry is Others (0.64), the third is the Insurance (0.56), and the last is the Capital Market Services (0.51).

According to the time trend of ΔCoVaR presented in Appendix Fig. 24.2, Chinese systemic risk can be divided into three periods: (1) Between 2008 and 2009, the systemic risks of China were considerably high, which may be related to the remaining influence of international financial crisis. (2) From 2009 to 2013, the systemic risk showed a downward trend and the overall ΔCoVaR reduced by about 0.15. (3) Since 2014, the systemic risk witnessed a dramatical increase. Chinese systemic risk in 2015 was even higher than the level of financial crisis in 2008.

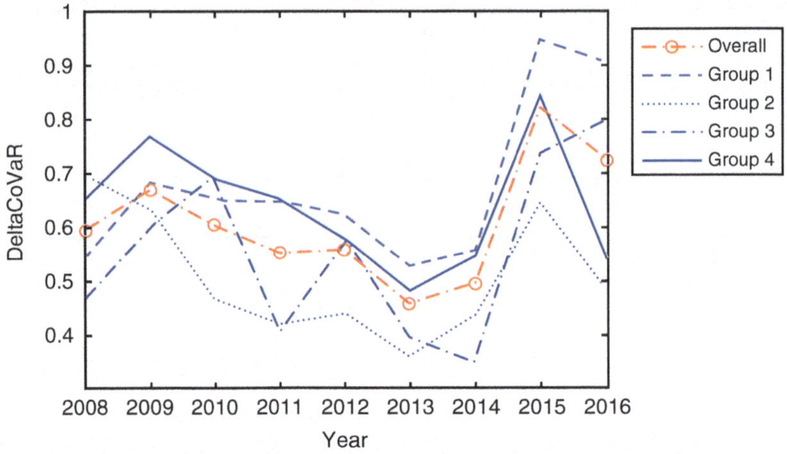

Fig. 24.2 The time trend of ΔCoVaR

24.4 Conclusion

To capture the skewed and fat-tailed property of financial data series, this paper generalizes the CoVaR model of Adrian and Brunnermeier [11] by introducing AEPD of Zhu and Zinde-Walsh [10]. By comparing the kernel density estimation curve of innovation series and the distributional simulated curve, we prove that the goodness-of-fit of AEPD is better than the Gaussian distribution.

Using the AEPD and the CoVaR model, we calculate the systemic risk contribution by four industry groups. The result shows that Monetary and Financial Services Group was the largest contributors to systemic risk, especially the four state-owned banks. Since 2014, Chinese systemic risk becomes considerable, thus the regulators should increase efforts to control the risk of infection.

Appendix

Table. 24.1 Stock codes, names and classifications of Chinese financial institutions

Group1: Monetary and financial services	Group2: Capital market services
000001 Ping An Bank	000686 Northeast Securities
002142 Bank of Ningbo	000712 Guangdong Golden Dragon
600000 Shanghai Pudong Development Bank	000728 Guoyuan Securities
600015 Hua Xia Bank	000783 Changjiang Securities
600016 China Minsheng Banking	600030 CITIC Securities
600036 China Merchants Bank	600109 Sinolink Securities
601009 Bank of Nanjing	600369 Southwest Securities Investment
601166 Industrial Bank	600837 HAITONG Securities
601169 Bank of Beijing	601099 The Pacific Securities
601328 Bank of Communications	
601398 Industrial & Commercial Bank of China	
601939 China Construction Bank	
601988 Bank of China	
601998 China CITIC Bank	
Group3: Insurance	**Group4: Others**
601318 Ping An Insurance	000563 Shaanxi International Trust
601601 China Pacific Insurance	600643 Shanghai AJ
601628 China Life Insurance	600816 Anxin Trust

Table. 24.2 Summary statistics for the daily returns

	Mean	Std	Skewness	Kurtosis	JB-stat	P-value
CSI 300 Financials Index	−0.0002	0.0209	−0.19	6.29	999.90	0.00
Shanghai Composite Index	−0.0002	0.0175	−0.53	7.09	1636.70	0.00
Ping An Bank	0.0000	0.0255	0.06	6.46	1098.04	0.00
China Minsheng Banking	0.0002	0.0225	0.09	6.92	1407.51	0.00
Bank of China	−0.0001	0.0173	0.32	10.83	5643.22	0.00
Northeast Securities	−0.0002	0.0343	−0.16	4.60	242.92	0.00
Changjiang Securities	−0.0001	0.0331	−0.15	4.93	349.67	0.00
HAITONG Securities	−0.0002	0.0320	−0.10	5.32	496.13	0.00
Ping An Insurance	−0.0001	0.0250	−0.14	6.07	870.66	0.00
China Pacific Insurance	−0.0002	0.0258	−0.01	5.21	446.45	0.00
SH China Life Insurance	−0.0003	0.0254	0.20	5.92	792.08	0.00
Shaanxi International Trust	0.0002	0.0338	−0.24	4.76	302.32	0.00
Shanghai AJ	0.0001	0.0318	−0.09	5.23	456.25	0.00
Anxin Trust	0.0003	0.0321	−0.11	5.31	492.33	0.00

Notes: The daily returns are calculated as $R_t = 100\,\% \times [\ln(P_t) - \ln(P_{t-1})]$, where P_t is the market price of a stock on the closing of day t. Given space limitations, the table shows the summary statistics for 12 institutions (3 for each groups).

References

1. Giulio, G., & Tolga Ergün, A. (2013). Systemic risk measurement: Multivariate GARCH estimation of CoVaR. *Journal of Banking & Finance, 37*, 3169–3180.
2. López-Espinosa, G., Moreno, A., Rubia, A., & Valderrama, L. (2015). Systemic risk and asymmetric responses in the financial industry. *Journal of Banking & Finance, 58*, 471–485.
3. Slijkerman, F. J., Schoenmaker, D., & de Vries, C. G. (2013). Systemic risk and diversification across European banks and insurers. *Journal of Banking & Finance, 37*, 773–785.
4. Straetmans, S., & Chaudhry, S. M. (2015). Tail risk and systemic risk of US and Eurozone financial institutions in the wake of the global financial crisis. *Journal of International Money and Finance, 58*, 191–223.
5. Yun, J., & Moonb, H. (2014). Measuring systemic risk in the Korean banking sector via dynamic conditional correlation models. *Pacific-Basin Finance Journal, 27*, 94–114.
6. Zhang, R., He, X., & Qi, Y. (2015). Measuring systemic risk of China's financial system under extreme condition. *Statistical Research, 32*(9), 30–38.
7. Gao, G., & Pan, Y. (2011). Banking systemic risk based on dynamic CoVaR estimation. *Journal of Shanghai Jiaotong University, 45*(12), 1753–1759.
8. Liu, X., & Gu, S. (2014). Research on risk spillovers from the real estate department to financial system based on AR-GARCH-CoVaR. *Systems Engineering - Theory & Practice, 34*(s1), 106–111.
9. Lin, H., Liu, T., & Zhang, P. (2012). An empirical study on systemic risk spillover effects of insurance institutions- based on AR-GARCH-CoVaR model, Beida CCISSR Forum.
10. Zhu, D., & Zinde-Walsh, V. (2009). Properties and estimation of asymmetric exponential power distribution. *Journal of Econometrics, 148*, 86–99.
11. Tobias, A., & Brunnermeier, M. K. (2011). *CoVaR*, Working paper, Federal Reserve Bank of New York.
12. Zhu, D., & Galbraith, J. W. (2011). Modeling and forecasting expected shortfall with the generalized asymmetric Student-t and asymmetric exponential power distributions. *Journal of Empirical Finance, 18*, 765–778.
13. Li, L., Gan, Q., Zhuo, Z., & Mizrach, B. (2014). Testing the CAPM theory based on a new model for Fama-French 25 portfolio returns. *Theoretical Economics Letters, 04*(8), 666–680.

Chapter 25
Research on Liquidity Preferences of Mutual Fund

Yinda Chen

Abstract This paper discusses the relationship between expected volatility and liquidity preferences of open-end mutual fund. I use actively managed stock open-end mutual funds in Chinese market for empirical analysis and found that expected volatility is expected to affect liquidity preferences of mutual fund; not only that, managers will be expected to hold more liquidity assets in their portfolio when market fluctuates.

Keywords Open-end Mutual Fund · Liquidity Preferences · Volatility

25.1 Introduction

The purpose of this paper is to test factors which influence the liquidity preferences of open-end mutual fund through the empirical research of open-end mutual funds in Chinese market, and try to bring different thinking about the liquidity management of open-end mutual fund.

25.2 Review of Relevant Theory

For liquidity issues, from "active management" perspective, people often explore how fund managers use a variety of liquidity management strategies to run open-end mutual fund to achieve higher returns and lower liquidity risk. However, the study of Cao (2013) tells us that this kind of active management behavior also increases the duration, fee, and turnover of those funds with timing ability [1]. On the basis of active strategy, Yawei Yao (2009), Geng Tian (2013) use empirical analysis to

Y. Chen (✉)
School of Economics Shanghai University, Shanghai, China
e-mail: dada_yj_1118@126.com

M. Tavana, S. Patnaik (eds.), *Recent Developments in Data Science and Business Analytics*, Springer Proceedings in Business and Economics,
https://doi.org/10.1007/978-3-319-72745-5_25

further explore the open-end mutual fund's liquidity, investor behavior characteristics, investors purchase and redemption flow and other liquidity variables on mutual fund marketing strategy, operational management, performance and so on [2, 3]. On the other hand, people also explored the causes of the liquidity risk of mutual fund and the means of management. Qing Zhang (2007) and Kun Dai (2008), respectively, do some systematic studies on mutual fund's liquidity risk. Those studies mainly relate to mutual fund's liquidity risk theory, measures and strategies on establishment of mutual fund's liquidity risk management system [4, 5]. They strive to inspect the root causes of liquidity risk characteristics of open-end mutual fund.

25.3 Hypothesis of Liquidity Preferences

As we all know, an important function of mutual fund is to provide appropriate liquidity management for investors. I assume that when market volatility is large, fund investors will increase the demand for liquidity. The reason is that the price of the assets in portfolio will be affected by market volatility. As the price of portfolio volatiles, the performance of the fund may be lower than that of benchmark, leading to that fund investors withdraw the invested funds because of anxious about market volatility. And besides, higher market volatility may further exacerbate liquidity risk and market panic [6]. If mutual fund does not have enough liquidity to respond to redemption of investors, fund manager may have to liquidate those illiquid assets at a lower price and mutual fund will bear liquidity costs which can damage mutual fund's performance [7].

It can be seen that, when the market fluctuates significantly, mutual fund is more likely to encounter large redemption than ever before because of the increase in liquidity demand of investors and the resulting short of liquidity in the portfolio. This is also the case in the Liquidity/Loss Spiral Effect Theory [8]. As a result, the risk of funds' withdrawal from mutual fund investors increases with market volatility, and fund managers will naturally hoard more liquid assets to deal with such crises. Increasing the overall liquidity of the portfolio in advance can alleviate the further expansion of the "spiral" to a certain extent, stopping spreading panic about liquidity in investors and reducing the probability and magnitude of capital outflows. Therefore, as compared with the case where the liquidity is not prepared in advance, there will be less pressure to deal with the liquidity demand of investors' outflow and lower costs to acquire liquidity if fund managers carry out liquidity prevention in anticipation of investors' outflows in advance.

The discussion above can be summarized as follows: Liquidity preferences of an open-end mutual fund should increase as market expected volatility increases.

25.4 Introduction of Data Sources and Variables

The data for empirical analysis of this paper are all collected from CSMAR. The sample of mutual funds is open-ended, mostly focus on stock assets, established between 2001 and 2016 and operated at least one quarter. After screening and processing the raw data of samples, I have got 226 stock open-end mutual funds and a total of 8649 quarterly frequency variables. The description of variables is as follows:

I use GARCH model to construct the variable EVOL_M to quantify market expected volatility, GARCH model is as follows [9]:

$$EVOL\_M_t = \mu + \alpha a_{t-1}^2 + \beta EVOL\_M_{t-1} \tag{25.1}$$

In addition to market expected volatility, I also take into account a market variable as a control variable to capture the liquidity impact of mutual fund: market illiquidity ratio. It is defined as an average value weighted by market value of illiquidity ratio at the end of each quarter for all stocks in Chinese A-share market, expressed in ILLQ_M. Among them, illiquidity ratio of individual stock is now widely used to measure liquidity which is proposed by Amihud first in 2002 [10]:

$$ILLQ\_M_{i,m} = \frac{1}{D_{i,m}} \sum\nolimits_{t=1}^{D_{i,m}} \frac{|r_{i,t}|}{dvol_{i,t}} \tag{25.2}$$

In order to do quantitative analysis on liquidity preferences of mutual fund, I construct following indicators as dependent variables for liquidity preferences:

(i) Ratio of cash holdings to TNA, expressed as CRATE_P;
(ii) Illiquidity ratio, similar to ILLQ_M, defined as an average value weighted by market value of illiquidity ratio at the end of each quarter for those stocks held by mutual fund, expressed as ILLQ_P.

Furthermore, I construct another two indicators as independent variables to control those sensitive factors of liquidity preferences that EVOL_M, ILLQ_M cannot cover:

(i) Ratio of Net flow of funds, expressed as NFLOW_P:

$$NFLOW\_P = \frac{TNA\_P_{i,q} - TNA\_P_{i,q-1}(1 + R_{i,q})}{TNA\_P_{i,q-1}} \tag{25.3}$$

where $TNA\_P_{i,q}$ is net asset value of mutual fund at the end of quarter q; $R_{i,q}$ is the return of mutual fund during quarter q;

(ii) Volatility of net flow of funds, expressed as NFLOW_VOL_P, a variance calculated from quarterly NFLOW_P for past 5 years.

25.5 Empirical Analysis

25.5.1 *Should Market Expected Volatility Affect Mutual Fund's Liquidity Preferences?*

In this part, I focus on mutual fund's net outflow of funds so that I can test whether there is correlation between market expected volatility and mutual fund's outflow (mutual fund investors' redemption behavior). For net outflow of funds, this paper constructs a variable FRACTION, defined as the fraction of the amount of mutual funds experiencing large outflow which lie in the bottom 25% tail of normal distribution of quarterly NFLOW_P. I do the descriptive statistics for those variables in previous section (Table. 25.1). It can be seen that NFLOW_P at bottom 25% is −12.27%, this explains why the average CRATE_P of mutual funds' sample is 13.98%. Part of the description of the statistics as follows:

According to FRACTION, EVOL_M and other market variables to construct time series regression model, the regression equation is as follows:

$$FRACTION_{q+1} = \mu + \alpha EVOL\_M_q + \beta ILLQ\_M_q + \sum\nolimits_{j=0}^{m} \gamma_j FRACTION_{q-j} + \varepsilon_q \tag{25.4}$$

where, the dependent variable is the fraction of the amount of mutual funds experiencing large outflow in quarter q + 1; $EVOL\_M_q$ is market expected volatility in the quarter q; $ILLQ\_M_q$ is market illiquidity ratio at the end of quarter q. I use the partial autocorrelation function to choose the number of lags of the dependent variable to be included in the regressions.

The regression results are shown in Table. 25.2. EVOL_M is positively correlated with the dependent variable, which indicates that high expected volatility is one of the important signals of mutual fund's liquidity demand. In terms of economic magnitude, it can also be seen that if market expected volatility increases by 10%, the number of mutual funds experiencing large outflow will increase 3.21%.

Table 25.1 Part of the descriptive statistics

	Mean	Standard deviation	Skewness	Kurtosis	p25	Median	p75
CRATE_P	13.978	8.623	1.722	4.793	8.520	11.460	17.610
NFLOW_P	4.555	73.271	14.682	304.310	−12.265	−3.200	5.045

Table 25.2 Regression results of EVOL_M, etc. and FRACTION

	Fraction
EVOL_M	0.321**
	(2.453)
ILLQ_M	−0.191***
	(3.09)
Dependentq	0.065**
	(2.519)
Dependentq-1	0.210*
	(1.779)
Dependentq-2	0.018**
	(2.15)
Constant	0.128**
	(2.413)
Observations	58
Adj-R2	0.285

*/**/*** respectively indicate significant levels at 10%, 5% and 1%

25.5.2 *Market Expected Volatility and Mutual Fund's Liquid Asset Positions*

Hypothesis of Liquidity preferences tells us that fund managers will show greater liquidity preferences when the market is expected to fluctuate significantly. In order to test the conclusion, we construct the panel regression equation as follows:

$$LIQPREF_{i,q} = \mu + v_i + \alpha EVOL\_M_q + \beta ILLQ\_M_q + \gamma X_{i,q} + \varepsilon_{i,q} \tag{25.5}$$

Where $LIQPREF_{i,\ q}$ represents one of dependent variables of mutual fund's liquidity preferences at the end of quarter q; v_i is the fixed factor of the mutual fund that cannot be directly observed; EVOL_Mq is market expected volatility in the quarter q; ILLQ_Mq is market illiquidity ratio at the end of quarter q; variable vector $X_{i,\ q}$ is a set of variables describing the characteristics of mutual fund at the end of quarter q, including NFLOW_P and NFLOW_VOL_P. The fixed factor in the model is to capture factors that have an impact on mutual fund's liquidity preferences but do not change over time and are independent to the characteristics of mutual fund.

The regression results are shown in Table. 25.3 below. There is a significant correlation between market expected volatility and mutual fund's liquidity preferences from statistical point of view. It is consistent with the hypothesis, that is, fund managers have stronger preferences to configure the liquid assets when market is expected to fluctuate. In terms of economic magnitude, we can also see that market expected volatility and mutual fund's liquidity preferences are significant. All stock open-end mutual funds in sample will average an increase of 1.25 percentage points of cash holdings, accounting for 9%, as compared to the mean cash holding of

Table 25.3 Regression results of EVOL_M, etc. and liquidity preferences

	CRATE_P	ILLQ_P
EVOL_M	0.125***	−0.017***
	−6.91	−4.89
ILLQ_M	0.155***	0.091***
	−5.761	−2.705
NFLOW_P	0.001**	−0.004***
	−2.543	−3.041
NFLOW_VOL_P	−0.001	0.002***
	−1.171	−5.95
Constant	0.325**	0.039
	−2.107	−0.072
Observations	8649	8649
Adj-R^2	0.326	0.087

*/**/*** respectively indicate significant levels at 10%, 5% and 1%

13.98% when EVOL_M increases by 10 percentage points; identically, the average mutual funds will reduce illiquidity ratio of 0.17 units by 15% relative to the mean illiquidity ratio of 1.13.

25.6 Conclusion

This paper explores the influencing factors of liquidity preferences of mutual fund. Through empirical analysis, we conclude that fund managers show strong liquidity preferences when market is expected to have higher volatility. It is not difficult to see from the analysis conclusion that liquidity preferences hypothesis is to help the mutual funds to minimize the operating costs of redemption by the investors. In general, this paper will help us to better understand mutual fund's liquidity management of some of the dynamic adjustment behavior and bring us a different perspective on mutual fund's active management.

Acknowledgments First of all, I want to express my deep respect and sincere thanks to Prof. Gui Yongping. This paper was completed under the guidance of Prof. Gui. Although he was always in busy, he still spared his time to give me academic guidance and help, especially to provide me with a good learning environment, so I benefit from benefits.

Moreover, to my wife, who gave me great support and encouragement in my life, the confidence and strength she gave me to study hard.

Finally, I'll also thanks to all my classmates, friends, teachers and loved ones who care about me, support me and help me.

References

1. Cao, C., Simin, T. T., & Wang, Y. (2013). Do mutual fund managers time market liquidity?[J]. *Journal of Financial Markets, 16*, 279–307.
2. Yao, Y. (2009). *Research on liquidity and stock portfolio investment management[D]*. Shanghai: Shanghai Jiao Tong University.
3. Tian, G. A. (2013). *An empirical study on factors affecting mutual fund flows[D]*. Harbin: Harbin Institute of Technology.
4. Zhang, Q. (2007). *Research on liquidity risk of mutual fund[D]*. Beijing: University of International Business and Economics.
5. Dai, K. (2008). *Liquidity risk analysis and management strategy of mutual fund[D]*. Chengdu: Southwestern University of Finance and Economics.
6. Liu, X., & Mello, A. S. (2011). The fragile capital structure of hedge funds and the limits to arbitrage[J]. *Journal of Financial Economics, 102*, 491–506.
7. Coval, J., & Stafford, E. (2007). Asset fire sales (and purchases) in equity markets[J]. *Journal of Financial Economics, 86*, 479–512.
8. Brunnermeier, M. K., & Pedersen, L. H. (2009). Market liquidity and funding liquidity[J]. *Review of Financial Studies, 22*, 2201–2238.
9. Zheng, Z., & Huang, Y. (2010). Volatility forecast: GARCH model and implied volatility[J]. *The Journal of Quantitative & Technical Economics, 1*, 140–150.
10. Amihud, Y. (2002). Illiquidity and stock returns: Cross-section and time-series effects[J]. *Journal of Financial Markets, 5*, 31–56.

Chapter 26
Relationship of the Financial Agglomeration and Fiscal Expenditure Scale of Yangtze River Delta

Liu Jiacheng, Chen Yujie, and Liu Nan

Abstract This paper taking Yangtze River Delta as an example, select Shanghai, Jiangsu province and Zhejiang province relevant data in 2004–2015; Use SPSS software run principal component analysis to research current situation of financial agglomeration in this three provinces (city); Through the establishment of panel data model, using least squares regression analysis, this paper studies the impact of fiscal expenditure on financial agglomeration in the Yangtze River Delta. The study shows that the fiscal expenditure of Yangtze River Delta has a significant positive correlation on financial Agglomeration. Then we put forward some suggestions for improving the status quo.

Keywords Fiscal expenditure · Financial agglomeration · Yangtze River Delta

26.1 Introduction

Financial agglomeration refers to the financial related enterprises gather in one area, with further cooperation and competition based on the enterprise division, using the similarity and complementary to form interdependent forms of industrial organization. Fiscal expenditure reflects the government's use of social resources at a given period. It is one of the main means that the government affects social and economic activities. So the relationship between fiscal expenditure and the development of economy has become widely concerned by public recently. According to Keynes's fiscal theory, positive fiscal policy promotes economic growth through the multiplier

L. Jiacheng
School of Economics and Management of Hainan University, Hubei Jianshi, China

C. Yujie (✉)
School of Economics and Management of Hainan University, Shandong Linyi, China
e-mail: cyjqpzm@163.com

L. Nan
School of Economics and Management of Hainan University, Shandong Rushan, China

M. Tavana, S. Patnaik (eds.), *Recent Developments in Data Science and Business Analytics*, Springer Proceedings in Business and Economics,
https://doi.org/10.1007/978-3-319-72745-5_26

effect. This effect is weakened after the increase of fiscal expenditure to a certain level because of the crowding out effect [1].

In this paper, by quantitatively analyzing the influence of fiscal expenditure on the agglomeration of financial industry, we reveal the relationship between financial agglomeration and fiscal expenditure in three provinces (cities) of the Yangtze River Delta region. Then we put forward some suggestions for improving the status quo.

26.2 Status Analysis of Financial Agglomeration and Fiscal Expenditure

26.2.1 Status Analysis of Financial Agglomeration

This paper uses location entropy to measure the level of financial agglomeration in three economic circles. Location entropy refers to the measurement of the distribution of specific factors in a region. It is the index that reflects the specialized level of a specific industry, which was first proposed and applied by P. Haggett. The formula is

$$LQ_{\text{ij}} = \frac{q_{ij}/q_j}{q_i/q}$$

q_{ij} is i industry's output value, quanty or other related indicators in j region, q_i is the related indicators of all industries in j region, q is the related indicators of all sectors of the country. The greater entropy shows the higher level of agglomeration. Considering the data representativeness and availability, this paper selected relevant financial data in the statistical yearbook of three provinces (cities) in 2004–2015. For the selection of explained variables, we consider the reality situation of the financial industry's development and the evaluation of industrial agglomeration model. Taking Ma Jun (2012) as a reference, we choose the location entropy of financial value (*LQv*), deposit balance(*LQs*), loan balance (*LQl*)and insurance premium income (*LQr*)to show the degree of financial agglomeration [2].

In order to compare the degree of financial agglomeration more intuitively, we use SPSS software to do principal component analysis. Principal component analysis (PCA) represents multiple variables of a description information with a sum variable, which reduces the dimensionality of the data to facilitate the description. In this paper, we take the location entropy of financial value, location entropy of financial value(*LQv*), deposit balance(*LQs*), loan balance (*LQl*)and insurance premium income(*LQr*) as indicators to show the four-dimensional spatial variables of the financial agglomeration degree, then reduce the four-dimensional variables to one-dimensional variable.

Describe the financial agglomeration index of the three provinces (cities) as a trend map, and get the Fig. 26.1. From Fig. 26.1, we can see that the financial agglomeration index of the Yangtze River Delta region is greater than 1, indicating that the three provinces (cities) have obvious agglomeration effect. It is not difficult to find from the trend that the financial agglomeration index decreased year by year

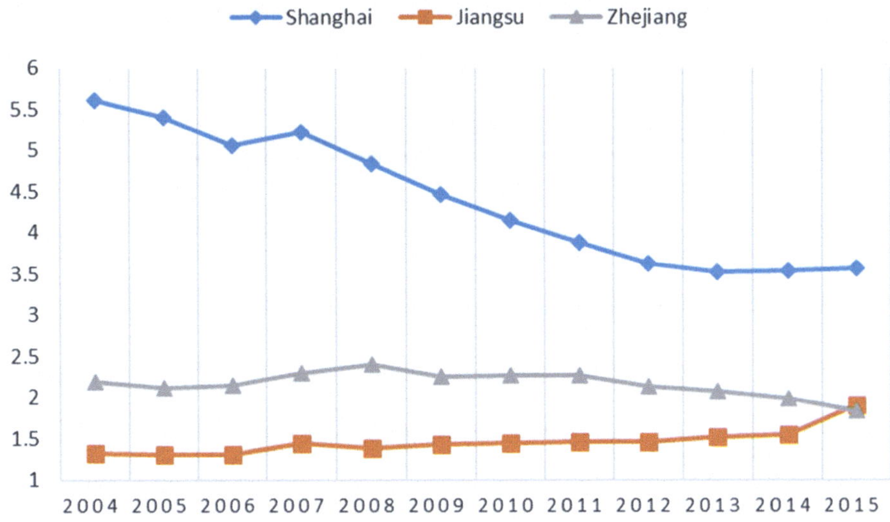

Fig. 26.1 Composite index of financial agglomeration in three provinces (cities)

in Shanghai city and Zhejiang Province, Jiangsu province's financial agglomeration index increased year by year, indicating homogenization trend in the Yangtze River Delta region. There is a good foundation of industry cooperation in this region. We can speculate by "Myrdal– Hirschman Prognosis" that the polarization effect of Shanghai is weakened and the diffusion effect is enhanced.

26.2.2 Current Situation of Financial Expenditure in Yangtze River Delta

By analyzing hundreds of countries' financial expenditure data, Romer [3] (1986) found that there is a reasonable range of the ratio of fiscal expenditure to GDP. The fiscal expenditure has a positive effect on the economy in this range, otherwise, if beyond this range, it has a negative effect. Based on Ram's evaluation index, in this paper eg stands for the ratio of national fiscal expenditure to national GDP. EGsh, EGjs and EGzj respectively stands for the ratio of local fiscal expenditure to local GDP divided by eg in Shanghai, Jiangsu and Zhejiang, which shows the scale of fiscal expenditure in the three provinces (cities), named as the scale of relative fiscal expenditure, and the trend chart is shown in Fig. 26.2.

In contrast to Figs. 26.1 and 26.2, it is easy to find that the fiscal expenditure of Shanghai, Jiangsu and Zhejiang provinces is positively related to their financial agglomeration, however, the relationship between fiscal expenditure and the scale of financial agglomeration can not be strictly proved by observation. We construct econometric models to empirically analyze the linear relationship between fiscal expenditure and financial agglomeration below.

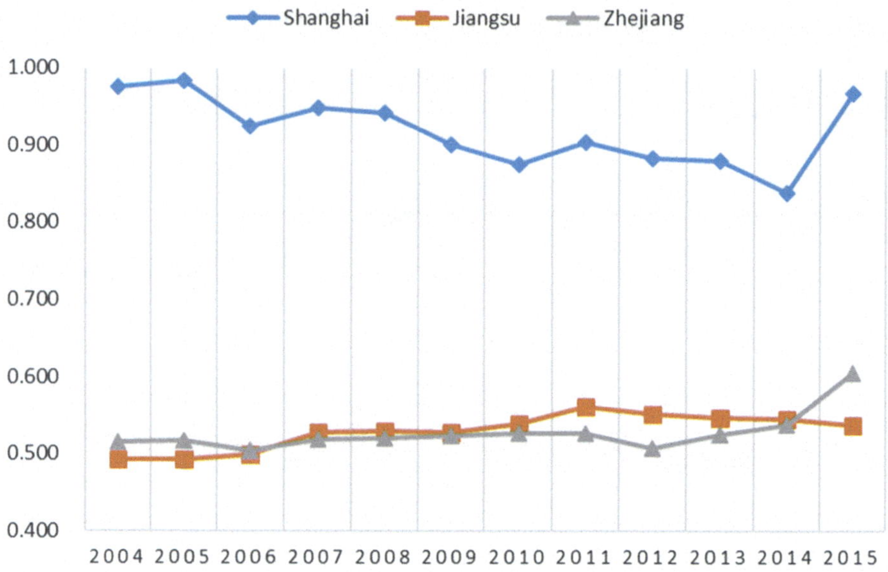

Fig. 26.2 Scale of relative fiscal expenditure in three provinces (cities)

26.3 Empirical Study on the Influence of Fiscal Expenditure on Financial Agglomeration in the Yangtze River Delta

The regression model of fiscal expenditure and financial agglomeration is set as follows:

$$LQ = C1_i + aEG_{it} + \varepsilon_{it};\ \ LQit^v = C2_i + \beta EG_{it} + \varepsilon_{it};\ \ LQit^s = C3_i + \lambda EG_{it} + \varepsilon_{it}; LQit^l = C4_i + \gamma EG_{it} + \varepsilon_{it}; LQit^r = C5_i + \sigma EG_{it} + \varepsilon_{it};$$

In order to reflect the influence of fiscal expenditure on the financial agglomeration in Shanghai, Jiangsu and Zhejiang, we take the relative fiscal expenditure scale of three provinces (cities) as independent variable, and respectively use financial agglomeration index (*LQ*), financial value location entropy (LQ^V), deposit balance location entropy (LQ^S), loan balance location entropy (LQ^L) and insurance premium income location entropy (LQ^R) as dependent variables to regress.

In the above equation, the subscript i is the identification of three provinces (cities) in the Yangtze River Delta (i = sh, js, zj), the subscript t is the identifier of each year (t = 2004, 2005...2015), a, β, λ, γ and σ are the influence coefficients, C_i is the intercept term, and ε_{it} is the random error term.

From Table 26.1, we find that the coefficients of each variable are positive, indicating that the relative fiscal expenditure of the Yangtze River Delta has a significant positive impact on the financial agglomeration. Separately using the location entropy of financial value, deposit balance, loan balance and insurance

Table 26.1 Regression estimation results of panel model

Variable	Effects on LQ	Effects on LQ^V	Effects on LQS	Effects on LQ^I	Effects on LQ^r
Constant term	1.910916** (0.897318)	−2.186616*** (−1.87533)	−2.288745*** (−2.956217)	−1.664927 (−1.289458)	−0.619388*** (4.0066653)
EG	6.962665*** (1.287589)	7.624969*** (4.558700)	7.383556*** (6.707199)	6.703891*** (3.627303)	2.562422*** (14.13651)
R^2	0.865059	0.836547	0.916999	0.732634	0.844438
Model selection	stochastic model	stochastic model	stochastic model	stochastic model	fixed model

Note: * * *, * * are respectively statistical tests of significance levels 1% and 5%

premium income as explanatory variables to estimate, results also show that the relationship between fiscal expenditure and financial agglomeration in Yangtze River Delta region is significant positive correlation. It shows that fiscal expenditure controls the flow of resources under the market conditions, and improves the overall competitive advantage of the financial sector in the Yangtze River Delta region.

26.4 Suggestions

Firstly, the government should improve the efficiency of fiscal expenditure. The scale of fiscal expenditure should be coordinated with economic development, and the proportion of fiscal expenditure and GDP should be kept within a reasonable range [4]. we should limit the fiscal expenditure of Shanghai especially the administrative expenses and other non productive expenditure, reduce staff, streamline structure and improve the efficiency. The government should optimize the expenditure structure and increase the science and education expenditure. We should establish a fiscal and financial coordination mechanism [5], fiscal and financial policy should be coordinated. We can establish fiscal and financial policy coordination mechanism and framework. All departments should carry out consultation and communication, and implement a multi-level policy coordination mechanism.

Secondly, we should strengthen the trickle down effect of fiscal expenditure in Shanghai. The Yangtze River Delta region should apply transfer payment, public investment, government procurement, financial resources spillover compensation, production transfer compensation to carry out financial leverage, promote financial resources flows across the region, and strengthen the financial radiation ability of Shanghai. Then, we can increase the financial coordination institutions, set up cross regional financial coordination institutions, and establish a joint regular meeting system of governors in the Pearl River Delta region, so as to make up for the shortage of the horizontal financial coordination institutions in the Yangtze River Delta region.

References

1. Vedder, R. K., & Gallaway, L. E. (1982). Productivity and wages in the American economy: A tale of two centuries. *Business and Economic History, 11*(4), 162–170.
2. Ma, J., & Guan, Y. F. (2015) The three major economic circles of financial industry agglomeration and influencing factors of, Journal of Commercial Economics, *16*, 65–66.
3. Romer, P. M. (1986). Increasing returns and long-run growth(1986). *Journal of Political Economy, 94*(5), 1002–1037.
4. Yang, Y. C. (2009). *Economic growth model and empirical research using institutional factors.* [D]. Shandong University.
5. Liu, H., Tian, G. X., & Zheng, J. C. (2011). Empirical analysis of the scale of fiscal expenditure in China. *Inquiry Into Economic issues, 4*, 77–79.

Chapter 27
Applying Data Processing Method for Relationship Discovery in the Stock Market

Mouataz Zreika, Jie Hua, and Guohua Wang

Abstract Decision making in the stock market is often made based on current events and the historical data analysis. In addition, related stock trends may affect investors' future decisions. To extract such relationship between stocks, a proposed methodology applies data processing techniques on raw data collected from the Australian Stock Market, to provide investors another angle of view, comes with initiative potential connections analysis between listed corporations, which is based on pure mathematics computing.

Keywords Data processing · The stock market · Relationship analysis · Time-series chart

27.1 Introduction

There is a large amount of raw data involved that varies over time in the stock market [3]. And stock investment decisions require many factors such as current events of the economy nationally, regionally and globally etc. especially, the analytics on relevant historical data plays an important role in future investments. Although the stock market system is extremely difficult to model with reasonable accuracy due to its data complexity features [1]. Hence, predicting the return of the stock investment plays an important role in the financial sector [2, 8].

Data analytics and data visualization techniques have been applied in the stock market analysis. Data processing contains the capability of pre/post processing, it

M. Zreika
Western Sydney University, Sydney, NSW, Australia

J. Hua (✉)
University of Technology Sydney, Sydney, NSW, Australia
e-mail: Jie.hua@alumni.uts.edu.au

G. Wang
South China University of Technology, Guangzhou, China

M. Tavana, S. Patnaik (eds.), *Recent Developments in Data Science and Business Analytics*, Springer Proceedings in Business and Economics,
https://doi.org/10.1007/978-3-319-72745-5_27

involves transforming raw data, which is essential for the predictive analytics [4]. Some of its functionalities include the discovery of data attribute descriptions, relationship analysis, and cleansing etc. On the other hand, data visualization techniques provide visual representations of complex data, which is one of the most efficient ways to assist investors to have a clear overview of movements of the stock market.

In this study, an approach is proposed which applies data processing method to assist investors digging deeper relationships between stocks. The early outcome of our approach represents potential connections between selected stocks and match the traditional time-series chart's results.

27.2 Existing Work

Various techniques were discussed by Sekar et al. which offered the capabilities of predicting future stock value (close value), based on current events analytics in related areas. In Sekar's research, visual representation was involved as well. Five methods were combined to analyze stocks, stock value trends (going up/down) were computed automatically. Comparing to the chance with a high level of significance, the experimental results of that algorithm was more accurate [6].

Tsang et al. introduced a bank system that implemented trading alert features, and the backpropagation neural network (NN5) applied in the study was tested in a practical case for decision making purpose. The empirical results showed reasonable accuracy in predicting short term price changing, and its application is based on 'small' amount of data gathered as well [7].

The CRISP-DM methodology was introduced by Al-Radaideh et al. based on raw data collected from the Amman Stock Exchange (ASE), which applied a decision tree to analyze/classify the historical data, furthermore to assist investors on future decision making in the stock market. But due to the stock data analytics complexity, such as current events in related fields etc. the methodology was not ready for the practical adoption [5].

In regards to the data visualization's adoption in stock market, treemap concepts were designed to facilitate financial decision-making, use both hierarchy and similarity information, provide an overview of large amount of hierarchical financial data and allow users to alter aspects of visual display dynamically, such as FolioMap (Jungmeister & Turo 1992), Self-Organizing Map (Kohonen, 1997; Joseph & Indratmo, 2003), FundExplorer (Csallner et al. 2003) and ordered treemap (Shneiderman & Wattenberg 2001) etc. [3, 9–14]. Wyer and Eades proposed a system that adopted visualization techniques to present data movements based on the UK Stock Market data [15]. Simunic conducted a methodology that shows chart clusters according to their similarities [16].

27.3 Proposed Approach

The proposed approach addresses relationship analytics in the entire stock market, not only on the particular chosen stocks (which can be done via existing methods such as time-series chart etc.), hence potential connections can be determined initiatively to improve stakeholder's decision making.

The approach proposed in this study is represented as:

1. Data Collection. Stock raw data were collected from the Australian Securities Exchange (ASX); Relevant data attributes include Open Value, Close Value, and Volume etc. In our experiment, Close Value is applied for testing due to its importance.
2. Data preparation. This step involves data filtering/cleansing/formatting etc.
3. Data processing. Crossing-comparisons on cleaned data;
4. Stocks relationship analytics based on results from step (3) (Fig. 27.1).

All raw data were collected from the ASX, including data attributes such as Close Value etc. In experiments, we only applied raw data of stocks which still exist nowadays (from 02/01/1997 to 30/06/2017) and the total absence days of each stock were less than 40% of the entire stock open days in last 20 years.

Detailed methodology of cross-comparison on data processing are as following:

1. Let $S = \{R, D\}$, $R_m = \{r_1, r_2, \ldots, r_m\}$ indicates every value change rate between two continuous business days on the single selected stock; $D_m = \{d_1, d_2, \ldots, d_m\}$ presents any two continuous business days; $m = |R_m| = |D_m|$ expresses the number of the datasets of changing rates and dates;
2. $A = \{S\} = \{S_1, S_2, \ldots, S_n\}$, $n = |S|$ represents the number of all stocks involved;

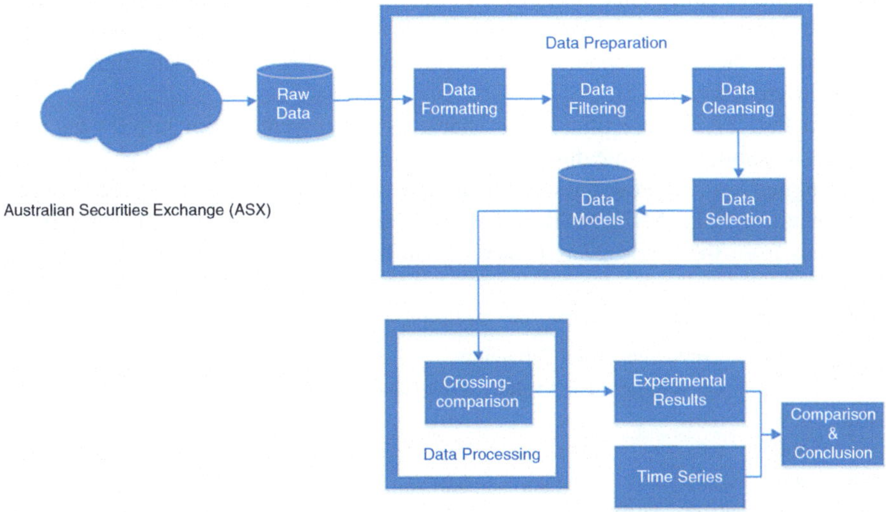

Fig. 27.1 Proposed approach

3. $C_s = \{c_{s1}, c_{s2}, \ldots, c_{si}\}$, $i = |C_s|$ shows the number of detailed cross-comparison results on one specific stock S; Suppose there are two set of changing rates on stocks S_1 and S_2, then
4. $c_{s1\text{-}s2} = \left(\sum_{K=0}^{n} \binom{n}{k} \parallel (S_1 - S_2)\right)/n$, ($S_1 = \{R_1, D_1\}$; $S_2 = \{R_2, D_2\}$; $|(S_1 - S_2)|$ presents the cross-comparisons between every $\{R_1\}$ and $\{R_2\}$; and n presents the number of matched two business days for computing, which means only the changing rates happens on the same time period of two stocks are taken into account.

27.4 Case Study

In experiment evaluation, raw data collected include 5088 stocks in last twenty years in Australia, around 6.4 million data entries (before formatted/filtered); after cross-comparison on related rates between every two stocks, nearly 194 million raw data were finalized; 352 connections that involve AGL have been finalized (which have 'strong' connections between each other, 'weak' connections have been removed); for description purpose, only five connections are given as an example in Table 27.1 below, selected stocks connected to AGL have been presented for analysis, including AGL: AGL Energy Limited; NAB: National Australia Bank Ltd.; WOW: Woolworths Limited; SAR: Saracen Mineral Holdings Limited; ALK: Alkane Resources Limited; ANZ: Australia and New Zealand Banking Group.

As shown in Table 27.1 an example of finalized crossing-comparison results is given, which includes data attributes such as Open Value, Close Value, and Volume etc. in our case study, relationship analytics were based on Close Value comparison only. The fewer difference between two stocks, the stronger connection between them then. Hence, AGL and NAB, ANZ, WOW have similar 'stronger' connections; and there are 'weaker' connections between AGL and SAR/ALK etc. due to their larger difference values.

From Fig. 27.2. three time-series charts were generated from http://www.asx.com.au, in detail, those time durations were ranged from 1 month to 10 years, and in all changing trend representations, ALK and SAR come with bigger differences from others' (AGL, ANZ, NAB and WOW), and those results match comparison outcomes from the Table 27.1. Hence, it can be seen that the outcomes from the

Table 27.1 Example of cross-comparisons results on AGL

Stock	Stock	Cross-comparison Difference					
		Open value	High value	Low value	Close value	Volume	Date match
AGL	ANZ	0.01	0.01	0.01	0.01	10.30	0.077
AGL	NAB	0.01	0.01	0.01	0.01	0.68	0.155
AGL	WOW	0.01	0.01	0.01	0.01	2.53	0.116
AGL	ALK	0.03	0.03	0.03	0.03	3.13	7.608
AGL	SAR	0.04	0.04	0.04	0.04	1.88	64.775

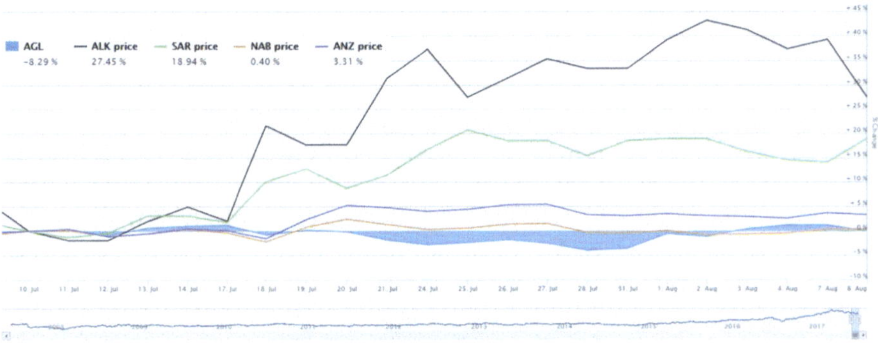

(a) one month's comparison

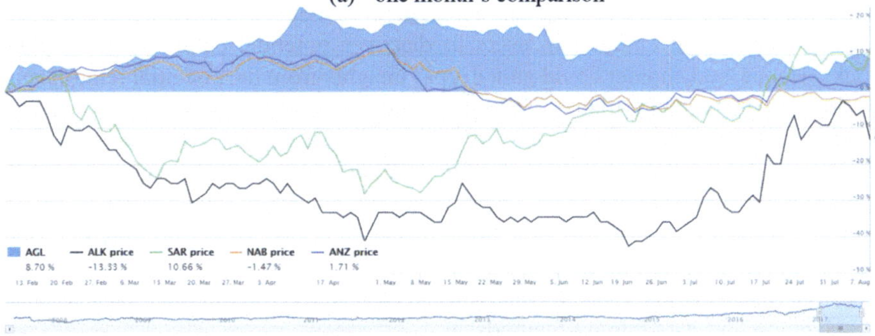

(b) six months' comparison

(c) ten years' comparison

Fig. 27.2 Time series chart on selected stocks (price changing rate trends/comparisons from http://www.asx.com.au) (**a**) one month's comparison, (**b**) six months' comparison (**c**) ten years' comparison

proposed approach provide similar relationship representation as time series chart's in this case.

Unlike the charts' feature – which can only compare preselected stocks, our approach can offer investors an entire analytics on all stocks (115 stocks in this case), and then potential relationships could be figured out easily.

27.5 Conclusions

The stock market is one domain comes with thousands of companies where complex data arise, and the performance is normally presented and measured by chart diagrams etc. showing price changing in a time period, as well as applying treemap for representing volume and relationship etc. Yet, those methods rarely offer the capability of potential relationship analytics, stakeholders can normally analyze connections between preselected stocks, which may ignore some related stocks, then leads to an incomplete decision on future investments.

Here we propose a method that adopts data processing methodology, to offer stakeholder a new angle of view on relationship analytics between stocks, comes with connection weight measurement, based on pure math calculation. Comparing to existing methods, it could help users to discover potential connections between stocks. Hence, stakeholder could gain deeper insights from the potential relationship of stocks in the market initiatively and may adjust investments accordingly based on related stocks' trends.

This study cannot replace the existing methods in the stock market analytics though, but offering stakeholders a new angle of views on the stock relationship, to complete all aspects related to their investments, hence to assist on future decision making in the stock market.

In our future work, more techniques/algorithms will be applied, such as data visualization and data mining methods.

References

1. Wang, Y. F. (2003). Mining stock price using fuzzy rough set system. *Expert Systems with Applications, 24*, 13–23.
2. Matías, J. M., & Reboredo, J. C. (2012). Forecasting performance of nonlinear models for intraday stock returns. *Journal of Forecasting, 31*, 172–188. https://doi.org/10.1002/for.1218.
3. Joseph, J., & Indratmo, I. (2013) *Visualizing stock market data with self-organizing map*. North America: Florida Artificial Intelligence Research Society Conference.
4. Asadi, S., Hadavandi, E., Mehmanpazir, F., & Nakhostin, M. M. (2012). Hybridization of evolutionary Levenberg-Marquardt neural networks and data pre-processing for stock market prediction. *Knowledge-Based Systems, 35*, 245–258.
5. Al-Radaideh, Q., Assaf, A., & Alnagi, E. (2013). *Predicting stock prices using data mining techniques. International Arab Conference on Information Technology (ACIT'2013).*
6. Sekar, P. S., Kannan, K. S., Sathik, M. M., & Arumugam, P. (2010). Financial stock market forecast using data mining techniques. *Proceedings of The International MultiConference of Engineers and Computer Sciece, I*, 5.
7. Tsang, P. M., et al. (2007). Design and implementation of NN5 for Hong Kong stock price forecasting. *Engineering Applications of Artificial Intelligence, 20*(4), 453–461.
8. Huang, C.-Y., & Lin, P. K. P. (2014). Application of integrated data mining techniques in stock market forecasting. *Cogent Economics & Finance, 2*(1), 1–18.
9. Jungmeister, W. A., & Turo, D. (1992). *Adapting treemaps to stock portfolio visualization*. Tech. Rep. CS-TR-2996, Computer Science Department, University of Maryland, College Park, MD.

10. Csallner, C., Handte, M., Lehmann, O., & Stasko, J. (2003). *Fundexplorer: Supporting the diversification of mutual fund portfolios using context treemaps*. In Information Visualization, 2003. IEEE Symposium on (pp. 203–208). IEEE, INFOVIS.
11. Kohonen, T. (1998). The self-organizing map. *Neurocomputing, 21*(1), 1–6.
12. Wattenberg, M. (1999). Visualizing the stock market. In *CHI'99 extended abstracts on Human factors in computing systems* (pp. 188–189). New York, NY: ACM.
13. Shneiderman, B., & Wattenberg, M. (2001). Ordered treemap layouts. *IEEE Symposium on Information Visualization : Proceedings, INFOVIS, 2001*, 2–7.
14. Bederson, B. B., Shneiderman, B., & Wattenberg, M. (2002). Ordered and quantum treemaps: Making effective use of 2D space to display hierarchies. *ACM Transactions on Graphics, 21*(4), 833–854.
15. Dwyer, T., & Eades, P. (2002). *Visualising a fund manager flow graph with columns and worms. In Information Visualisation*, Proceedings. sixth international conference on (pp. 147–152). IEEE.
16. Šimunić, K. (2003). *Visualization of stock market charts*. In Proceedings International Conference in Central Europe on Computer Graphics, Visualization and Computer Vision (WSCG), 2003.

Chapter 28
A Study on Assets Categorizations and Optimal Allocation via an Improved Algorithm

Guang Liu

Abstract The efficiency of assets allocation is essential to investment performance. To deal with the dynamic optimization of industry asset allocation, an improved algorithm (the Index Hierarchical Structure Algorithm, IHSA) is applied to 20 industry indexes according to different sample period, and proved to deduce stabilized industry categorizations. Then, a series of different size industry portfolios are constructed by the algorithm, and their monthly rate of returns are compared with 369 open-ended funds (including 263 growth funds and 106 index funds). The results indicate that the performance of each industry portfolio is better than most of the sample funds in the early sample period, especially than index funds. The algorithm provides a useful guidance for industry allocation and active asset management.

Keywords Index hierarchical structure algorithm · Industry categorization · Asset allocation · Topological space

28.1 Introduction

Related researches have pointed out that industry selection is as important as individual share selection in asset allocation. Either from top to bottom or from bottom to top, optimizing the industry proportions is an important step for asset allocation, and has a crucial impact on the final performance. When conducting a cross-industry portfolio, industry is one of the most important factors in cross-section returns. Industry selection is more important than regional selection [1]. Kotter and Lel [2] taken sovereign wealth funds (SWFs) as samples, finding that different industry selection will cause different returns. Their research laid theoretical foundation for cross-industry allocation. Chen et al. [3] compared the

G. Liu (✉)
Guangzhou University, Guangzhou, China
e-mail: mrliuguang@126.com

M. Tavana, S. Patnaik (eds.), *Recent Developments in Data Science and Business Analytics*, Springer Proceedings in Business and Economics,
https://doi.org/10.1007/978-3-319-72745-5_28

performance of seven kinds of international portfolios, finding the "industry diversification" asset allocation strategy is relatively better. Industry selection has already become an important part for evaluating fund managers' investment ability.

Industry selection being able to improve the portfolio performance, the reason is that industry selection can reduce the portfolio dependence on the industry's basic information, and win better diversification effect. The more the industries are diversified, the better the portfolio is performed. However, the industry diversification, as well as individual share diversification, has its boundary. It is constrained by the tradeoff between cost and benefit. Foreign studies have proved that fund managers from Australia, the United States and other countries have industry selection ability [4, 5]. Unfortunately, domestic researched have not found significant evidence for industry selection ability of China's fund managers [6]. Considering the importance of industry selection and the shortage of fund managers' industry selection ability in China, it is necessary to use appropriate methods to balance industry characteristics with individual share properties for dynamic portfolio.

Benzécri [7] firstly studied on the relationship between ultrametric space (UC) and the index hierarchical structure (IHS). Song and Liu [8] improved the construction of IHS basing on the M-V rule, and named it as index hierarchical structure algorithm(IHSA). They further empirically proved that IHSA has strong asset selection adaptability and practical guidance. We still refer to the very method here, and test its stability and availability in cross-industry selection.

For its application, Mantegna [9] first pointed out that the IHST of financial assets is associated with their economic properties. Therefore, the IHST can be used to assets selection. Miccich et al. [10] also proved the topological stability of MinST. Using multivariable information, Brida and Risso [11] constructed MinST and IHST, and illustrated by Monte Carlo simulation that asset correlations are stable rather than random. Furthermore, several scholars applied the algorithm to stock markets of European Union, Brazil and so on, obtaining stable categorizations too [12, 13].

The assets categorizations stability of IHSA is demonstrated by both mathematical derivation and empirical test, providing theoretical basis for the intrinsic link between assets. However, the IHSA has been hardly used to investment until now. The main reason is that the categorizations are based on maximum correlations, which is contrary to asset allocation criterion basing on minimum correlations. Inspired by MPT, we attempt to construct the maximum spanning tree (MaxST) of the sample industries here, test the stability of industry categorizations and availability of industry selection, and provide effective guidance for dynamic portfolio.

28.2 Principles and Methodology

We refer to the method proposed by [7–9]. Remarkably, we command some appropriate improvements on the original algorithm when constructing IHST. The steps of modified IHSA can be summarized as follows.

S1 For any n given financial assets and their return time series $r_i\{i = 1, 2, \ldots, n\}$, their correlation coefficients ρ_{ij} during a period Tcan be calculated by $\rho_{ij} = \left[\sum_{t=1}^{T}(r_i - \bar{r}_i)(r_j - \bar{r}_j)\right] / \left[\sqrt{\sum_{t=1}^{T}(r_i - \bar{r}_i)^2}\sqrt{\sum_{t=1}^{T}(r_j - \bar{r}_j)^2}\right]$.

S2 For all components $\tilde{r}_{ik}$ of an n-dimensional vector $\tilde{r}_i$, the ED d_{ij} between them can be defined by Eq. (28.1).

$$d_{ij}^2 = \left\|\tilde{r}_i - \tilde{r}_j\right\|^2 = \sum_{k=1}^{n}\left(\tilde{r}_{ik} - \tilde{r}_{jk}\right)^2 \tag{28.1}$$

Where $\tilde{r}_i = (r_i - \bar{r}_i) / \left[\sqrt{\sum_{t=1}^{T} r_i^2 - \left(\sum_{t=1}^{N} r_i\right)^2}\right]$. For $\sum_{k=1}^{n}\tilde{r}_{ik}^2 = 1$ and $\sum_{k=1}^{n}\tilde{r}_{ik}\tilde{r}_{jk} = \rho_{ij}$, then

$$d_{ij} = \sqrt{2(1 - \rho_{ij})} \tag{28.2}$$

Mantegna [9] pointed out that the ED defined by Eq. (28.2) meets the three properties of distance metrics, i.e. it can be treated as distance metrics.

S3 In order to satisfy the above ED's to a certain topological structure, the ultrametric distances(UD) between assets will be further defined by Eq.(28.3).

$$\widehat{d}_{ij} \leq \max\left\{\widehat{d}_{ik}, \widehat{d}_{kj}\right\} \tag{28.3}$$

Then an $n \times n$ UC can be derived. We choose the bigger instead of the smaller between $\widehat{d}_{ik}$ and $\widehat{d}_{kj}$ here, which is the very difference from [7] or [9].

S4 Eqs.(28.2) and (28.3) reveals that $\widehat{d}_{ij}$ is inversely proportional to ρ_{ij}. That is to say, the smaller the correlation coefficient is, the greater the UD is. For this season, we can use KA to derive an exclusive MaxST instead of MinST, and get an exclusive SUS of n assets then. The SUS is still topological, corresponding to an exclusive categorization.

S5 Finally, based on the MaxST, the exclusive IHST of n assets is ultimately derived.

Those steps infer that the UD's between assets are closely related to their correlations, and the corresponding categorization determined by IHST accurately reflects the quantity of correlation. So, if the categorizations in different periods are consistent, then we can use IHSA to construct different optimal portfolios containing various industries and stocks. Programming the above steps, the computation efficiency can be improved.

28.3 Application and Cause Analyses

28.3.1 Sample Selection and Data Collection

We choose the total 28 primary industries of SWS Research as samples. Removing 8 industries containing less than 60 stocks (including banking, home appliances, leisure service, steel, national defense and military, non-banking financial institutions, mining and comprehensive industry), 20 sample industries are left, including media(CM), electrical equipment(DQ), electronics(DZ), real estates(FC), textile (FZ), utilities(GY), chemicals(HG), building materials(JC), computer(JJ), machinery(JX), transport device(JY), architectural decoration(JZ), agriculture(NL), motor (QC), light-manufacturing(QG), commercial trade(SM), food beverages(SP), communications(TX), nonferrous metals(YS), pharmaceuticals(YY).

To avoid the non-tradable shares reform in 2005–2006 and the sharp fluctuation of stock index in 2007–2008 influencing test results, 2009–2015 is taken as sample period. Then the closing subscription prices of 20 industry indexes and 33,960 observation data of daily return rate are derived. All data are downloaded from the CSMAR, and computed by Eviews6.0.

28.3.2 Calculation Example and Categorization Discussion

Using the algorithm in Sect. 28.2, we draw out the SUS and IHST of 20 sample industries. The results are exhibited on Table 28.1 and Fig. 28.1.

Inspecting the IHST form left to right, we can gain the following conclusions. (1) The UD between FC and the rest 19 industries is the biggest, i.e. the correlation between them is the smallest. The result accurately reflects the reality that FC has been over-invested and developed independently in China. (2) The JJ, YY and SP are non-cyclical industries, showing relative weak independence and correlations with the rest industries. (3) In contrast, the SM, FZ, QG and HG show relative smaller UD's and stronger correlations with the rest industries, which are accord with their properties. In general, the FC is the most independent. The consumer goods industry is less affected by national economy and weak independent. Manufacturing and raw materials industries are affected by upstream and downstream industries, lacking of independence. The investment practice supported by the CSMAR also shows the fact that the proportion of FC listed companies held by institutional investors has been almost always maintained at more than 15% in the past decade. It confirms investors' affection for the FC indirectly.

Table 28.1 SUS of 20 sample industries (2009–2015)

Ind.	DQ	DZ	FC	FZ	GY	HG	JC	JJ	JX	JY	JZ	NL	QC	QG	SM	SP	TX	YS	YY
CM	0.854	0.854	0.839	0.854	0.851	0.854	0.851	0.854	0.854	0.851	0.784	0.854	0.851	0.854	0.854	0.851	0.854	0.851	0.854
DQ		0.854	0.743	0.854	0.854	0.854	0.854	0.854	0.854	0.854	0.854	0.854	0.854	0.854	0.854	0.854	0.854	0.854	0.854
DZ			0.772	0.854	0.854	0.854	0.854	0.854	0.854	0.854	0.854	0.854	0.854	0.854	0.854	0.854	0.854	0.854	0.854
FC				0.854	0.854	0.854	0.854	0.854	0.854	0.854	0.854	0.761	0.854	0.854	0.854	0.851	0.854	0.854	0.853
FZ					0.854	0.854	0.854	0.854	0.854	0.854	0.854	0.854	0.854	0.854	0.854	0.682	0.854	0.854	0.854
GY						0.854	0.854	0.854	0.854	0.854	0.854	0.854	0.854	0.854	0.854	0.733	0.854	0.854	0.854
HG							0.854	0.854	0.854	0.854	0.854	0.854	0.854	0.854	0.854	0.718	0.854	0.854	0.854
JC								0.854	0.854	0.854	0.854	0.854	0.854	0.854	0.854	0.759	0.854	0.854	0.854
JJ									0.854	0.854	0.854	0.854	0.854	0.854	0.854	0.854	0.854	0.854	0.854
JX										0.854	0.854	0.854	0.854	0.854	0.854	0.726	0.854	0.854	0.854
JY											0.854	0.854	0.854	0.854	0.854	0.759	0.854	0.854	0.854
JZ												0.854	0.854	0.854	0.854	0.854	0.854	0.854	0.854
NL													0.854	0.854	0.854	0.854	0.854	0.854	0.854
QC														0.854	0.854	0.744	0.854	0.854	0.854
QG															0.854	0.716	0.854	0.854	0.854
SM																0.664	0.854	0.854	0.854
SP																	0.809	0.835	0.854
TX																		0.854	0.854
YS																			0.851

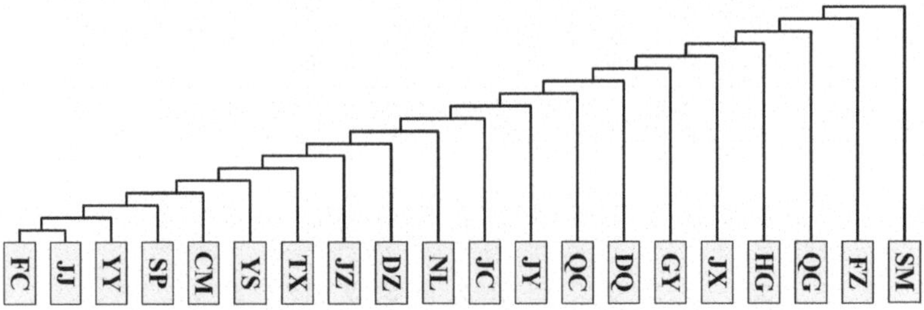

Fig. 28.1 IHST of 20 sample industries (2009–2015)

Table 28.2 Industry categorizations in four sample periods

Periods / Ranking	2009–2015	2009–2014	2009–2013	2009–2012
1	FC	FC	FC	FC
2	JJ	YY	YY	SP
3	YY	SP	SP	YY
4	SP	YS	YS	YS
5	CM	JJ	TX	TX
6	YS	CM	JJ	JJ
7	TX	TX	CM	JC
8	JZ	DZ	JC	QC
9	DZ	JC	QC	NL
10	NL	NL	NL	DQ
11	JC	JY	JY	JY
12	JY	QC	DZ	CM
13	QC	JZ	DQ	JZ
14	DQ	DQ	GY	GY
15	GY	GY	JZ	DZ
16	JX	HG	HG	HG
17	HG	JX	JX	JX
18	QG	QG	QG	QG
19	FZ	FZ	SM	SM
20	SM	SM	FZ	FZ

28.3.3 Stability Test and Analysis

To check whether the industry categorizations by IHSA are stable or not, we keep the beginning of the initial sample period fixed, and shorten forward 1 year from the end each time. Then we repeat the above steps in Sect. 28.2 in every new period and get another three results. The all results are all exhibited on Table 28.2 orderly.

It reveals that though the order of categorization in four periods is slightly different from each other, the difference is not significant. That’s to say, the correlations of 20 industries are slightly various in each sample period, but not distinct. Approximately, among all the independences, the FC's is always the highest, followed by the YY or SP's, and the JZ or SM's is the lowest. The tests prove that the industry categorizations are indeed stable, i.e. the IHSA is available and applicable.

There may be several reasons for the relative stable categorizations, but the main one is the industrial relevancy. Industrial relevancy refers to the technical and economic relations and dependences between industries, and is mainly reflected by input and output structure. Yu et al. [14] have classified the industrial relevancies, finding that the relevancies of FC and YY are relatively lower, while the relevancies of FZ and HG are relatively higher. Our results confirm their conclusions. In addition, the industry characteristics, such as cyclicity or non-cyclicity, stage of life cycle and so on, will also affect its correlation with other industries. Overall, the results calculated by IHSA comprehensively reflect the industries' internal and external factors.

28.4 Availability Test and Analyses

28.4.1 Study Design

Since the IHSA can accurately categorize financial assets according to their UD's and correlations, it can be applied to dynamically manage and maintain optimal portfolio theoretically. In order to test the availability of IHSA, we select a group of open-end funds for performance comparison. The study designs are as follows.

1. Constructing industry portfolios. According to the industry categorizations by IHSA in Fig. 28.1, we construct a series of equal-weighted industry portfolios. We assume that the stock position in each portfolio is 80%, cash and bond position is 20%.
2. Evaluating the monthly return rates of those given industry portfolios. Assuming that the monthly return rate of stock is calculated by each industry index and the monthly return rate of cash and bond is constantly equal to 1%. The monthly return rate of each industry portfolio will be obtained by adding the two rates proportionally.
3. Selecting sample funds for performance comparison. Selecting all the equity open-end funds established before 2014 from the CSMAR, removing the type of ETF, QDII, and structured mutual fund, we get 369 sample funds, including 263 growth funds and 106 index funds. The performance of every sample fund is approximately calculated by its monthly net growth rate.
4. Selecting comparison period. As industry categorization period ends at 2015, we selecting the first 6 months of 2016 as comparison period.

28.4.2 Comparison and Analyses

The performance rankings of different size industry portfolio are exhibited orderly on Table 28.3. We can achieve the following discoveries by observing the rankings.

Vertically, the return rate of different size industry portfolio ranks always among the top at the early period, indicating their performance are better than most sample funds. However, those rankings gradually decline over time, suggesting that the performance advantages of the industry portfolios are slowly weakened. It proves that the availability of IHSA is significant during the first half of comparison period, and decreases during the rest period. The main reason for the performance variability is that the industry portfolios are optimal when initial being constructed and constant throughout all the 6 months. On the contrary, the sample funds have been optimized continuously as the market changes. So the performance of sample funds is relatively better after 3 or 4 months. In addition, industry rotation will hinder those fixed industry portfolios to maintain stable outstanding performance. Horizontally, when the portfolio contains 4 industries, its performance shows the highest ranking in every month. Then with more and more industries being added, their performance rank in a certain range. When the portfolio contains more than 10 industries, its monthly performance ranking begin declines. This means that the optimal industry size is about 4, after that the industry diversification is no longer significant.

Table 28.3 Monthly performance rankings of industry portfolios

	Ranking in all 369 sample funds						Ranking in 106 index funds					
Industry size	Jan	Feb	Mar	Apr	May	Jun	Jan	Feb	Mar	Apr	May	Jun
2	21	172	249	315	223	214	10	48	63	98	77	17
3	15	179	262	255	222	206	6	53	69	72	76	15
4	9	144	235	172	193	171	3	37	52	40	64	8
5	11	196	216	219	193	230	5	64	43	62	64	25
6	14	90	214	176	173	216	7	13	42	42	50	19
7	16	123	219	169	182	203	8	26	45	39	57	14
8	13	130	218	181	213	224	6	30	45	46	75	24
9	19	135	208	181	176	190	11	34	38	46	52	13
10	15	106	211	199	181	202	8	19	41	53	56	14
11	18	92	208	182	210	214	10	14	38	47	75	20
12	19	100	214	206	216	240	11	17	43	57	78	32
13	20	95	209	184	218	237	12	15	39	49	80	32
14	24	111	201	178	227	220	16	24	36	44	83	23
15	25	112	209	180	238	237	17	25	39	45	91	32
16	26	122	203	183	240	227	18	30	37	48	92	28
17	28	120	200	172	251	220	19	29	36	41	100	23
18	29	121	197	167	262	211	20	30	35	37	104	20
19	30	114	194	166	266	229	21	27	34	37	106	30
20	30	122	195	169	274	243	21	31	35	39	111	38

Furthermore, we divide the sample funds into two groups of growth funds and index funds, compare each industry portfolio with them, and find that the performance rankings of industry portfolios in index funds are generally higher than in growth funds. This is because the industry portfolio is a negative one compared with growth funds, showing asset diversification at one hand, but a positive one compared with index funds, showing industry concentration at another hand. The balance between diversification and concentration in industry portfolio fully demonstrates its characteristics and advantages.

Overall, the return rates of industry portfolios have evident advantages at the early stage, especially higher than the rates of index funds. If we optimize the composition and the weight of industry assets according to M-V rule over time, the advantage of industry portfolio is expected to be significant consistently.

28.5 Conclusions

The main conclusions obtained by theoretical and empirical analysis in this paper are as follows. Firstly, the IHSA based on US is available and applicable, and can be applied to get stable industry categorizations. Secondly, the categorizations are consistent with the industrial relevancies mapped by input and output structure, suggesting that the ISHA can be used for industry selection. Lastly, the performance of industry portfolios is better than that of most sample funds. Our research is a useful supplement to the theory of dynamic assets allocation. We broaden the application field of IHSA, and provide a reliable basis for quantitative investment. Programming the process of IHSA, we will provide a scientific tool for quantitative investment here.

The IHSA can be applied to other aspects besides industry selection. For example, conducting regional or style assets selection, this is important for SWF, QFII and FOF. Identifying sector rotation and fund style drift when market fluctuating, this is useful for performance evaluation. Implementing industry support or supervision according to industry importance principle, this is significant for adjustment of economic structure and transformation and upgrading of traditional industries.

Acknowledgments This work is financially supported by the Guangdong Province-sponsored Philosophy and Social Science Funding Program (GD14XYJ16) and the Guangdong Province Education Department—sponsored Platform Funding Program for Humanities and Social Sciences (2014WQNCX074).

References

1. Bernanke, B. S., & Kuttner, K. N. (2005). What explains the stock market's reaction to federal reserve policy? *Journal of Finance, 60*, 1221–1257.
2. Kotter, J., & Lel, U. (2011). Friends or foes? Target selection decision and performance effects of sovereign wealth funds. *Journal of Financial Economics, 101*, 360–381.

3. Chen, X. X., Chen, W. Z., & Jin, Y. P. (2007). Country and industry factors on the performance of international asset allocation. *Finance & Trade Economics, 2*, 93–97.
4. Brands, S., Brown, S. J., & Gallagher, D. R. (2005). Portfolio concentration and investment manager performance. *International Review of Finance, 5*, 149–174.
5. Sapp, T., & Yan, X. (2008). Security concentration and active fund management: Do focused funds offer superior performance? *Financial Review, 43*, 27–49.
6. Liao, C. Y. (2013). Industry selection for fund asset allocation and fund performance: New evidence about the investment ability of fund managers. *Journal of Chong Qing University (Social Science Edition), 19*, 53–60.
7. Benzécri, J. P. (1984). L'analyse des données 1′, La Taxonomie Dumand. Paris.
8. Song, G. H., & Liu, G. (2013). An empirical study of dynamic asset allocation based on index hierarchical structure algorithm. *Soft Science, 27*, 32–37.
9. Mantegna, R. N. (1999). Information and hierarchical structure in financial markets. *Computer Physics Communications, 121*, 153–156.
10. Miccich, S., Bonanno, G., Lillo, F., et al. (2002). Degree stability of a minimum spanning tree of price return and volatility. *Physica A, 324*, 66–73.
11. Brida, J. G., & Risso, W. A. (2008). Multidimensional minimal spanning tree: The Dow Jones case. *Physica A, 387*, 5205–5210.
12. Gilmore, C. G., Lucey, B., & Boscia, M. (2008). An ever-loser union? Examining the evolution of linkages of European equity markets via minimum spanning trees. *Physica A, 387*, 6319–6329.
13. Tabak, B. M., Serra, T. R., & Cajueiro, D. O. (2010). Topological properties of stock market networks: The case of brazil. *Physica A, 389*, 3240–3249.
14. Yu, D. F., Gan, C. H., & Zheng, R. G. (2011). Industrial linkages in China——an empirical research based on structural decomposition technique of input-output model. *China Industrial Economics, 11*, 5–15.

Chapter 29
Forecasting Stock Price Index Volatility with LSTM Deep Neural Network

ShuiLing Yu and Zhe Li

Abstract In strong noisy financial market, accurate volatility forecasting is the core task in risk management. In this paper, we apply GARCH model and a LSTM model to predict the stock index volatility. Instead of historical volatility, we select extreme value volatility of Shanghai Compos stock price index to conduct empirical study. By comparing the values of four types of loss functions, we illustrate that LSTM model has a better predicting effect.

Keywords LSTM · Volatility forecasting · Extreme value volatility · GARCH

29.1 Introduction

Stock price volatility has been playing a significant rule in financial risk and investment. Stock price index is highly volatile financial time series, because financial market is a strong noise and complicated dynamic system. There exits various factors in financial market and any variable can be influenced by others in some way. The interaction of factors makes the stock price volatility difficult to be forecasted.

There are numbers of models to forecast time series volatilities, the most traditional models are GARCH-type model which is developed from ARCH (Autoregressive Conditional Heteroscedasticity) model [1]. ARCH model was first proposed by Engle [2], ARCH model can't portray leverage effect of financial asset return rate series, so it is generalized to GARCH model which is proved to have strong prediction ability. As to now, GARCH model contains many different types, such as TGARCH, EGARCH, FIGARCH and so on [3], but the most classical and widely applied one is still GARCH model. Recently, financial market environment becomes more complicated, researchers begin to focus on hybrid predicting model.

S. Yu · Z. Li (✉)
School of Science, Changchun University of Science and Technology, Changchun, China
e-mail: lizhe@amss.ac.cn

M. Tavana, S. Patnaik (eds.), *Recent Developments in Data Science and Business Analytics*, Springer Proceedings in Business and Economics,
https://doi.org/10.1007/978-3-319-72745-5_29

Hybrid model has been proved to be more effective than single GARCH-type model. In [4], the authors used ARIMA-SVM model to predict stock price. In [5], Li proposed ARIMA-GARCH to predict the volatility of WTI index.

Facing with the age of big data, traditional financial time series models are not accurate enough when forecast the volatility. Many studies now have primarily focused on the estimation of stock price index using machine learning methods. Deep learning neural networks such as recurrent neural networks are well-suited to this task. However, there are many issues plagued not only the training of net but the forecast effect, the most common difficulty is overfitting problem [6] and increased degree of the freedom. These difficulties have paved the way for more effective training novel architectures, one of the architectures is Long Short-Term Memory (LSTM) which is proposed in [7]. Lately, Gers [8] added forget gate and peephole connections to the architecture, so LSTM was completed. LSTM is a particular and advanced type of recurrent neural network, it is applied in numbers of area, for example artificial handwriting generation, speech synthesis, language forecasting and speech recognition [9–11].

In this paper, we attempt to model China stock price index volatility using an LSTM deep learning network. At the same time, we specify GARCH model as a benchmark model. While training LSTM model, we use extreme value volatility proposed by Parkinson [12]. When evaluating the models, except the predicting figure we also calculate the NMSE, NMAE, LL, LINEX of each model based on Hansen's suggestion [13]. By comparing the four values of each model, we find that LSTM model had the best predicting performance.

29.2 Models

Historical volatility assumes that the future is an extension of the past, and there will be no dramatic changes. So we can make reasonable predictions that future volatility is based on past and present data. However, due to many noises that are caused by policy, economy and psychology, the stock volatility shows a nonlinear change in high frequency which is hard to be described by accurate models. We can use the abstract mathematical expression below to describe the stock volatility:

$$h_t = f\left(h_{t-1}, h_{t-2}, \cdots, h_{t-p}\right),$$

Where h_t is the volatility of period t, $t = 1, 2, \ldots, Z$, $h_{t-1}, h_{t-2}, \cdots, h_{t-p}$ are volatilities of the pre-p periods we have already known, f denotes the nonlinear and linear relationship between input and output.

29.2.1 GARCH Model

To evaluate the performance of the LSTM model, we have developed one autoregressive model (GARCH) as benchmark models. The major idea of GARCH model is that the residual of the regression model relies on the square error of the previous information. The residua is not independent, although it is normally distributed. The conditional variance is the main object of interest, and it is considered as the prediction indicator for future variance, which is the square of volatility. The GARCH model is presented as followed:

$$\text{GARCH}: h_i^2 = \omega + h_{i-1}^2\left[\alpha + \beta\varepsilon_i^2\right], \varepsilon \sim \text{N}(0, 1) \qquad (29.1)$$

h_i^2 is the conditional variance, that is the square of volatility. ε is the error term of the function, and it is supposed to be normally distributed.

In this paper, the GARCH model is trained by a maximum likelihood estimator. And the coefficients of GARCH model were obtained by R software.

29.2.2 LSTM Model

Neural network usually exits error back-flow problem, vanishing gradients problem problems and only can remember short-time information. Long Short-Term Memory (LSTM) is one of the effective solutions. LSTM is a novel recurrent network architecture in conjunction with an appropriate gradient-based learning algorithm. A single LSTM hidden layer consisting of one LSTM block, assuming that there is no other hidden layers. The structure of this kind of neural network is shown in Fig. 29.1. Hochreiter, Schmidhuber [7] and Gers, Cummmins [8] indicate that the key of LSTM

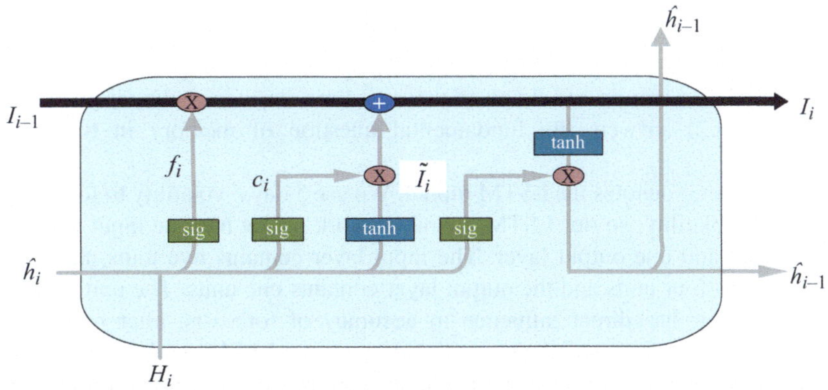

Fig. 29.1 Structure of the long short-term memory layer

is cell state I_i. Due to a "gating" mechanism, I_i can remember the information of past time and more importantly it has the ability to forget through Eq. (29.2).

$$I_i = f_i \cdot I_{i-1} + c_i \cdot \widehat{I_i} \tag{29.2}$$

Here f_i is the fraction of past-time information passed over to the present, $\widehat{I_i}$ measures the current information flow passed down from the last step and c_i weight the importance of this current information . At every time step i, the input of this layer consists of three parts: the volatility of the past five time steps which is denoted as vector H_i, volatility estimation $\widehat{h}_t$ passed down from the last step, the information flow I_{i-1} passed down from the last step. The functions that control the gates are either the sigmoid (sig) function or the hyperbolic tangent (tanh) function, the scalar multiplication are denoted by × and the addition is denoted by + operators. Equation (29.2) is the update of linear memory, and it is prominent together with the cell state flow. Both I_i and volatility prediction $\widehat{h}_{i+1}$ for the next time stamp are passed down to the next epoch. f_i, c_i and I_i, all these three quantities are functions of the input H_i and last-time's estimation of volatility $\widehat{h}_i$.

$$f_i = sigmoid\left[\left(\widehat{h}_l, H_i\right) \cdot W_f + b_f\right] \tag{29.3}$$

$$c_i = sigmoid\left[\left(\widehat{h}_l, H_i\right) \cdot W_c + b_c\right] \tag{29.4}$$

$$\widetilde{I}_l = tanh\left[\left(\widehat{h}_l, H_i\right) \cdot W_{\tilde{I}} + b_{\tilde{I}}\right] \tag{29.5}$$

To make a prediction of the next volatility value $\widehat{h}_{i+1}$, a linear activation function is used.

$$\widehat{h}_{i+1} = \alpha + \beta \cdot o_i tanh[I_i] \tag{29.6}$$

where o_i is also a function of H_i and $\widehat{h}_i$ tunes the output.

$$o_i = sigmoid\left[\left(\widehat{h}_l, H_i\right) \cdot W_O + b_O\right] \tag{29.7}$$

where I_i and $\widehat{h}_{i+1}$ are passed down to the next time step for continual predictions. Equation (29.2) answers the fundamental question of memory in time series forecasting.

In this paper, f denotes the LSTM model. We use 5 days' volatility to forecast the sixth day's volatility, so our LSTM neural network model has one input layer, one hidden layer and one output layer. The input layer contains five units, the hidden layer contains four units and the output layer contains one units. The unit's number of hidden layer has direct influence to accuracy of forecasts, after many times training and adjusting parameter, we decide that our LSTM model contains four LSTM blocks. LSTM model's coefficients here were learned through training with the python deep learning library Keras [14].

29.3 Empirical Research of China's Stock Index Volatility

29.3.1 Data Sources and Data Pre-processing

In this work, the empirical data are Shanghai Composite index (SSEC), Shenzhen Component Index (SZSE) and Industrial and Commercial Bank of China stock price index (ICBC), obtained by R software. The dataset which consists of opening price, closing price, ceiling price, the lowest price and the volume ranges from 2011–01-01 to 2015–12-31. For experiments, we use extreme value volatility proposed by Parkinson to denote real volatility. Traditionally, real volatility is unknown and is often substituted by square of yield rete, which often produce obvious error. Otherwise extreme value volatility contains the information that produced during the generating process of stock index's price, so it is a more effective substitution. The extreme value volatility is given by:

$$h_t = \left(\frac{1}{4\ln 2}\right)(\ln P_{t,h} - \ln P_{t,l})^2 \tag{29.8}$$

where $P_{t,\ h}$ is the intra-day ceiling price, $P_{t,\ l}$ is the intra-day lowest price.

In order to improve the training effects of our LSTM model, volatility series was normalized to [0,1]:

$$z_t = \frac{h_t - min\{h_t\}}{max\{h_t\} - min\{h_t\}} \tag{29.9}$$

Additionally, we split the whole data set into the training datasets with 70% of the observations to train and build our model, leaving the remaining 30% for testing the model.

29.3.2 Result

In Figs. 29.2, 29.3 and 29.4, we plot the predicted volatility together with the observed values of the test set. It can be observed visually that the trend and fluctuation of SSEC, SZSE and ICBC were predicted well by both GARCH model and LSTM model. But when there is a sharp change, the forecasting volatility of LSTM model is more close to real volatility.

When evaluating the model accuracy, we selected four types of loss functions as benchmark: normalized mean squared error (NMSE), normalized mean absolute error (NMAE), logarithmic loss function (LL) and exponential loss function (LINEX). They were defined as:

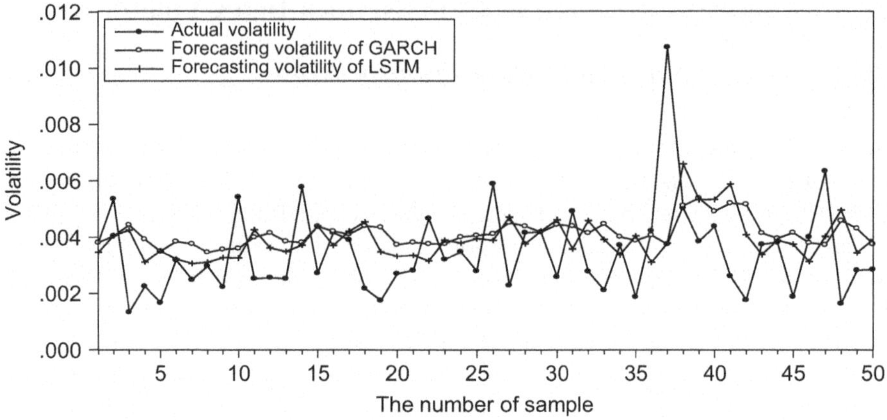

Fig. 29.2 SSEC volatility comparative results

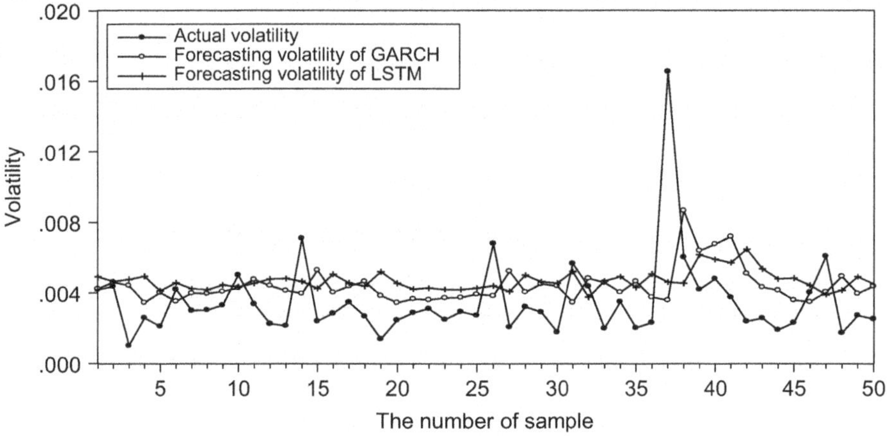

Fig. 29.3 SZSE volatility comparative results

$$NMSE = \left[\frac{\sum_{t=1}^{N} \left(\widehat{h_t} - h_t \right)^2}{\sum_{t=1}^{N} \left(h_{t-1} - h_t \right)^2} \right]^{1/2} \tag{29.10}$$

$$NMAE = \frac{\sum_{t=1}^{N} \left| \widehat{h_t} - h_t \right|}{\sum_{t=1}^{N} \left| h_{t-1} - h_t \right|} \tag{29.11}$$

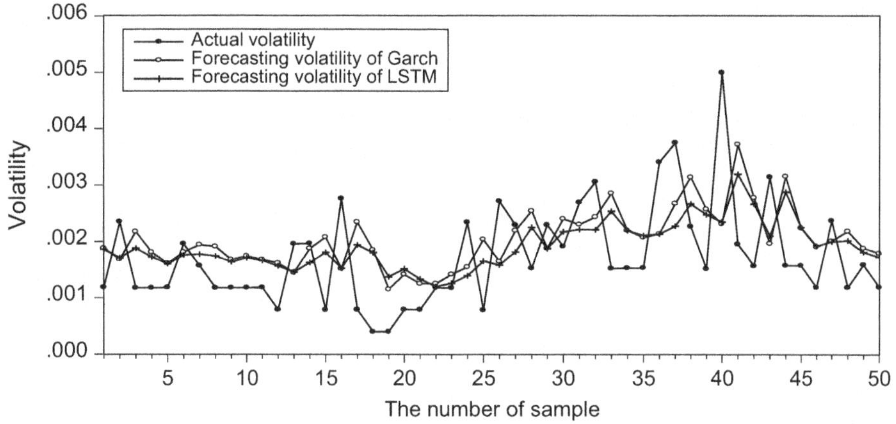

Fig. 29.4 ICBC volatility comparative results

Table 29.1 The estimating results of GARCH and LSTM model

Stock index	Model	NMSE	NMAE	LL	LINEX
SSEC	LSTM	0.6257	0.7525	0.8631	5.1346
	GARCH	0.9377	1.0647	1.1879	7.9324
SZSE	LSTM	0.7251	0.7968	0.8094	7.1968
	GARCH	1.0635	1.2053	1.3492	8.1872
ICBC	LSTM	0.8357	0.9345	0.9763	7.8592
	GARCH	1.1523	1.0549	1.2621	8.5837

$$LL = N^{-1} \sum_{t=1}^{N} \left[\ln\left(\widehat{h_t}\right) - \ln\left(h_t\right) \right]^2 \tag{29.12}$$

$$LINEX = N^{-1} \sum_{t=1}^{N} \left\{ exp\left[a\left(\widehat{h_t} - h_t\right) \right] - a\left(\widehat{h_t} - h_t\right) - 1 \right\} \tag{29.13}$$

Where N is the number of forecast sample, $\widehat{h_t}$ is the forecasting value of volatility, h_t is extreme value volatility which is a proxy variable of the real volatility. The smaller the value is, the higher the model accuracy is. Table 29.1 shows the values of our models when forecasting SSEC, SASE and ICBC volatility.

We can see from Table 29.1 that there are little difference between benchmark values when we use GARCH model to forecast volatility. But when LSTM model was used, the benchmark values of ICBC are larger than values of SSEC and SZSE. We conclude that LSTM has a better performance on synthetic stock index and a relatively poor performance on single stock index.

29.4 Conclusion

In this work, we consider extreme value volatility to be real volatility and we use deep learning LSTM model to forecast the stock price index volatility. We choose SSEC index, SZSE index and ICBC index as the object of empirical study, forecast them use LSTM model and GARCH model respectively. By comparing the loss functions' value we have verified the forecasting effect of each model. The result shows that LSTM with its long-term memory ability, has better predictive power than GARCH model. LSTM is an effective intelligent volatility prediction method, and it has certain practical value.

References

1. Bollerslev, T. (1986). Generalized autoregressive conditional heteroscedasticity. *Journal of Econometrics, 31*, 307–327.
2. Engle, R. F. (1982). Autoregressive conditional heteroscedasticity with estimator of the variance of United Kingdom inflation. *Econometrica, 50*(4), 987–1008.
3. Tray, R. S. (2012). *Analysis of financial time series*. Bei Jing: Posts & Telecom Press.
4. Cheng Chang-Pin, Chen Qiang, Jinag Yong-Sheng. (2012). Stock price forecasting based on ARIMA-SVM combination model. *Computer simulation, 06*, 343–346.
5. Li, L. (2017). An empirical study on WTI index based on ARIMA-GARCH model. *Value Engineering, 02*, 38–39.
6. Huang, Z., Chen, H., Hsu, C., Chen, W., & Wu, S. (2004). Credit rating analysis with support vector machines and neural networks, a market comparative study. *Decision Support Systems, 37*, 543.
7. Hochreiter, S., & Schmidhuber, J. (1997). Long short-term memory. *Neural Computation, 9*, 1735–1780.
8. Gers, F., Shmidhuber, J., & Cummmins, F. (2000). Learning to forget: Continual prediction with LSTM. *Neural Computation, 12*, 2451–2471.
9. Kang, S., Qian, X., & Meng, H. (2013). *Multi-distribution deep belief network for speech synthesis[C]//Acoustics, Speech and Signal Processing (ICASSP),* 2013 I.E. international conference on IEEE, 8012–8016.
10. Sundermeyer, M., Schluter, R., & Ney, H. (2010). *LSTM neural networks for language modeling*, International conferenceon spoken language processing, inter speech.
11. Graves, A., Mohamed, A., & Hinton, G. (2013). Speech recognition with deep recurrent neural networks. *Acoustics, Speech and Signal Processing (ICASSP), 10*, 6645–6649.
12. Parkinson, M. (1980). The extreme value method for estimating the variance of the rate of return. *Journal of Business, 53*, 61.
13. Hansen, P. R. (2005). A test for superior predictive ability. *Journal of Business and Economic Statistics, 23*(4), 365–380.
14. Chollet, F. (2015). *Keras*, Https://github.com/keras-team/keras.

Chapter 30
Improvement of Hedging Effect Based on the Average Hedging Ratio

Yang Liu and Chuan-he Shen

Abstract This paper is aimed at exploring the improvement of hedging effect based on the theory of portfolio hedging, with multiple groups of CSI300 stock index futures and spot sample data as the analysis object. The minimum variance method is employed to estimate the optimal hedging ratio under the OLS and GARCH hedging models and calculate the average of the hedge ratios. By comparing the hedging effects of the constructed portfolio outside of samples based on different hedging ratios, the empirical analysis displays that the hedging effect of the average hedge ratio was superior to the hedging effect of the estimated hedge ratio of most individual historical samples. Therefore, the methodology supposed is deeply improved by considering the average value of the hedging ratio in order to optimize the optimal hedging ratio.

Keywords The average hedging ratios · Hedging effect · Minimum variance method · OLS · GARCH model

30.1 Introduction

The application of financial derivative stock index futures in capital market is more and more extensive, mainly because of its hedging function in the solution of systemic risk has a superiority [1]. By combing the development of hedging theory, we found that the researchers have conducted a thorough and comprehensive study of the optimal hedging model and the effectiveness of hedging. The research theory has had a profound impact. Ederington [2] (1979) used the General Least Squares

Y. Liu
College of Mathematics and Systems Science of Shandong University of Science and Technology, Qingdao, China

C. -h. Shen (✉)
Institute of Financial Engineering of Shandong Women's University, Jinan, China
e-mail: schuanhe@sina.com; 15376401230@163.com

M. Tavana, S. Patnaik (eds.), *Recent Developments in Data Science and Business Analytics*, Springer Proceedings in Business and Economics,
https://doi.org/10.1007/978-3-319-72745-5_30

Estimator (OLS) to do the hedging of the US Treasury futures market and found that the hedging ratio was less than 1. He also raised the degree of reduction in the variance of the yield before and after hedging as a criterion for hedging. Engle [3] (1982) proposed an autoregressive Heteroscedastic Model (ARCH) and a generalized autoregressive Heteroscedastic Model (GARCH). And compared with the traditional valuation model, he found that GARCH model is more suitable for stock index futures hedge rate estimates. Holmes [4] (1996) conducted a study on the effectiveness of stock index futures contracts after hedging operations, and the results show that the risk of hedging portfolio without hedging is far greater than the hedging effect of the optimal hedging rate derived from the regression of OLS model. Wu Xianzhi [5] (2008) employed the hedging model to estimate the hedging ratio within a sample data, and on the basis of which the effectiveness of hedging is checked. However, this static analysis method is defective, mainly due to the existence of time-varying and dynamic characteristics of the impact index and the fluctuation of the stock index. In view of the multi-sample and dynamic hedging ratio, Zhang Hua et al. [6] (2014) proposed the method of estimating the dynamic hedging rate of grain futures based on GARCH model. The results show that both the data in the sample and the data outside the sample are better than the static estimation method in the dynamic estimation.

Follow the development of this theory, we assume that the joint distribution of the spot market and the futures market is fixed, thus taking into account the typical OLS model of static analysis. Simultaneously, the impact of new information on the market and expected changes will change the joint distribution of futures and spot markets. Then the time-dependent GARCH model can capture the dynamic characteristics of futures and spot market more accurately [7]. Therefore, this paper introduces OLS and GARCH hedging model. First of all, we lies in the selection of the longest futures and spot sample data from the last 4 years CSI 300 stock index futures contract as the analysis object, and a total of four sets of sample data were obtained. Next, we estimate the hedge ratio in the OLS and GARCH models and calculate the average of the hedge ratios. Then we compare the hedging ratio between the average of the hedge ratio and the hedging ratio of the single sample on the basis of the hedging portfolio outside the construction sample. The purpose is to avoid the problem that the hedging rate caused by the single sample is not significant in the actual hedging operation. Finally, the paper summarizes the full text and analyzes the future research direction.

30.2 Hedging Ratio Determination and Model Construction

30.2.1 Optimal Hedging Ratio Determination

The hedge ratio based on the minimum variance is the hedging ratio that the volatility minimizes of the constructed hedging portfolio by hedging tools [8].The investor adopts a long or short hedging strategy, where the price of the hedged asset portfolio changes to $\Delta S - h\Delta F$ or $h\Delta F - \Delta S$. No matter whether the investor adopts long hedging strategy or short hedging strategy, the variance of the hedging portfolio price is:

$$\sigma^2 = \sigma_s^2 - 2h\rho\sigma_s\sigma_f + h^2\sigma_f^2 \tag{30.1}$$

In (30.1) formula, $\sigma_s{}^2$ is the variance of spot asset price variation $\triangle S$, $\sigma_F{}^2$ is the variance of futures asset price variation $\triangle F$. Simultaneously, σ_{sf} is the covariance of ΔS and ΔF, ρ is the correlation coefficients of ΔS and ΔF. According to the minimum variance of the hedging ratio connotation, we make $\frac{d\sigma^2}{dh} = 0$ to get the hedging ratio of the minimum variance of portfolio return volatility. Therefore, optimal hedge ratio is:

$$h = \rho\frac{\sigma_S}{\sigma_F} \tag{30.2}$$

Therefore, the key of estimate the hedging ratio is using the hedging model to estimate *h*.

30.2.2 Hedging Model Construction and Effect Evaluation

30.2.2.1 Hedging Model Construction

1. OLS Hedging model Construction

When estimating the optimal hedging ratio based on the spot and Futures Portfolio, the general practice is to minimize the variance of the portfolio volatility (risk) of the spot and futures assets [9]. We assume that there is a linear relationship between the price of the goods and the price of the futures asset, and a linear function is introduced. Then, this paper construct the bivariate linear regression model:

$$\ln\left(\frac{S_{t+1}}{S_t}\right) = \alpha + \beta\ln\left(\frac{F_{t+1}}{F_t}\right) + \mu_t \tag{30.3}$$

In (30.3) formula, $\ln\left(\frac{S_{t+1}}{S_t}\right)$ is the daily logarithmic yield of spot index stock at t time, $\ln\left(\frac{F_{t+1}}{F_t}\right)$ is the daily logarithmic yield of t time stock index futures. Simultaneously, α is the intercept of the equation, μ_t is the disturbance term, and β is slope of the equation (that is, the optimal hedge ratio under the OLS model).

2. GARCH Hedging model Construction

The implicit condition of the OLS model is that the variance of the random error is constant, but the actual financial time series mostly has heteroscedasticity. Therefore, the autoregressive conditional heteroscedasticity model is used to solve heteroscedasticity problems and to capture more time series financial information. Then, the generalized autoregressive conditional heteroscedasticity model is derived on the basis of the ARCH model. In addition, the GARCH model is not only related to the squared u^2_{t-q} of the random error in the previous Q period, but also related to the variance σ^2_{t-p} of the random error in the previous P period [10]. Therefore, the GARCH (p, q) model of under the daily logarithmic yield sample data of the CSI300 stock index futures and spot is:

$$\begin{cases} \ln\left(\frac{S_{t+1}}{S_t}\right) = C + \beta \ln\left(\frac{F_{t+1}}{F_t}\right) + \varepsilon_t \\ \sigma_t^2 = \alpha_0 + \sum_{i=1}^{q} \alpha_i u_{t-i}^2 + \sum_{j=1}^{p} \gamma_j \sigma_{t-j}^2 \end{cases} \tag{30.4}$$

In the (30.4) formula, ε_t is a random error and obeys the GARCH (p, q) process, $\ln\left(\frac{S_{t+1}}{S_t}\right)$ is the daily logarithmic yield of spot index stock at t time, and $\ln\left(\frac{F_{t+1}}{F_t}\right)$ is the daily logarithmic yield of at time stock index futures. In addition, σ_t^2 is the variance of the random error, ε_{t-j}^2 is the square lag term of the lag J period residual, and σ_{t-i}^2 is the variance lag term of the lag I period residual.

Here consider the leverage effect is to solve the future multi-step forecast when the volatility is negative situation, and the asymmetric effect of market information is reflected in the EGARCH model [11]. Therefore, a common EGARCH (1,1) model is established here:

$$\begin{cases} \ln\left(\frac{S_{t+1}}{S_t}\right) = C + \beta * \ln\left(\frac{F_{t+1}}{F_t}\right) + \varepsilon_t \\ \ln(\sigma_t^2) = \omega + \alpha\left|\frac{u_{t-1}}{\sigma_{t-1}}\right| + \gamma\frac{u_{t-1}}{\sigma_{t-1}} + \lambda \ln \sigma_{t-1}^2 \end{cases} \tag{30.5}$$

EGARCH model of leverage effect can be judged by γ is not equal to 0. When $u_{t-1} > 0$ represents the presence of good information can bring $\alpha + \gamma$ times the impact; and $u_{t-1} < 0$ on behalf of the negative information can bring $\alpha - \gamma$ times the impact [12]. So, β is the optimal hedging ratio for GARCH and EGARCH models.

30.2.2.2 Evaluation Criteria for Hedging Effect

The most common way to measure the hedging effect under different models is to examine the increase or decrease of the variance of the portfolio return before and after the hedging operation [13]. The formula for calculating the yield of our stock portfolio without risk avoidance is as follows:

$$Var(X_t) = Var\left(\ln\left(\frac{S_t}{S_{t-1}}\right)\right) = Var(\ln S_t \text{-} \ln S_{t-1}) = \sigma_x^2 \tag{30.6}$$

In the (30.6) formula, X_t and $\sigma_x{}^2$ represent the variance of the daily logarithmic yield of stock index spot and the daily logarithmic yield. The variance of the yield of the portfolio after hedging operations is:

$$\begin{aligned} Var(\Pi) &= Var\left[\ln\left(\frac{S_t}{S_{t-1}}\right)\right] + h^2 Var\left[\ln\left(\frac{F_t}{F_{t-1}}\right)\right] \\ &\quad - 2hCov\left[\ln\left(\frac{S_t}{S_{t-1}}\right), \ln\left(\frac{F_t}{F_{t-1}}\right)\right] \\ &= \sigma^2{}_\Pi \end{aligned} \tag{30.7}$$

In the (30.7) formula, Π stands for the constructed hedging portfolio, the σ_Π^2 hedging portfolio's yield variance, and the h as the hedging ratio. The measure of the minimum variance hedging effect is the percentage reduction in the risk of the return on the portfolio after the hedging rate relative to the non-hedged spot rate of return [14]. In combination with (30.6) and (30.7), the measure of hedging effectiveness is:

$$e^* = \frac{\sigma_x^2\text{-}\sigma_\Pi^2}{\sigma_x^2} = \frac{2hCov\left[\ln\left(\frac{S_t}{S_{t-1}}\right), \ln\left(\frac{F_t}{F_{t-1}}\right)\right] - h^2 Var\left[\ln\left(\frac{F_t}{F_{t-1}}\right)\right]}{Var\left[\ln\left(\frac{S_t}{S_{t-1}}\right)\right]} \tag{30.8}$$

In the formula, e^* represents the proportion of the risk of the hedging portfolio to the original risk. The greater the e^*, the better the effect of the hedging [15].

30.3 Empirical Analysis

30.3.1 Sample Data Selection and Processing

30.3.1.1 CSI 300 Stock Index Sample Data Selection

This paper lies in the selection of the longest futures and spot sample data from the last 4 years CSI 300 stock index futures contract as the analysis object, and a total of four sets of sample data were obtained. The contract varieties are IF1412, IF1512, IF1612 and IF1703. Moreover, a set of spot price sequences corresponding to the futures price series is generated. Through the CSI 300 stock index futures trading chart and the correlation coefficient of 0.95566, we can get a high consistency between the daily logarithmic price of CSI300 stock index futures and the daily logarithmic price of the spot, so it will avoid the basis risk to a certain extent.

30.3.1.2 Descriptive Statistics Analysis of Sample Data

This paper uses Eviews software to make a descriptive statistical analysis of the daily logarithmic yield of the CSI 300 stock index futures(See Figs. 30.1 and 30.2). As can be seen from the figure, first of all, the average daily yield of CSI300 stock index futures and spot contracts is −0.000182 and −0.000173, the results are less than 0 and fluctuate around 0. Secondly, the standard deviations are 0.015195 and 0.018430, indicating that both fluctuations are not large, that is, the degree of dispersion is relatively stable. Then, the *P* value of the *J-B* statistic is close to 0, indicating that the daily return of the spot is rejected to the normal distribution. At last, by the kurtosis 8.468, 10.904 and skewness −0.706, −0.313, we can see that the daily logarithmic yield of CSI300 stock index futures contracts and share index spot follows the left skewed spike tail distribution.

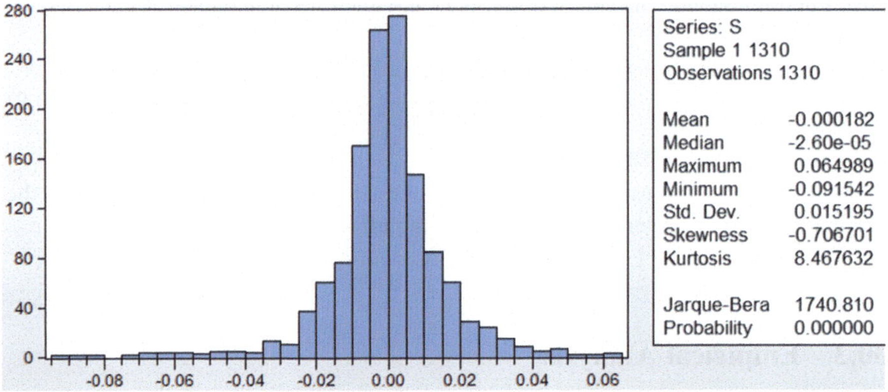

Fig. 30.1 Descriptive statistics of stock index spot

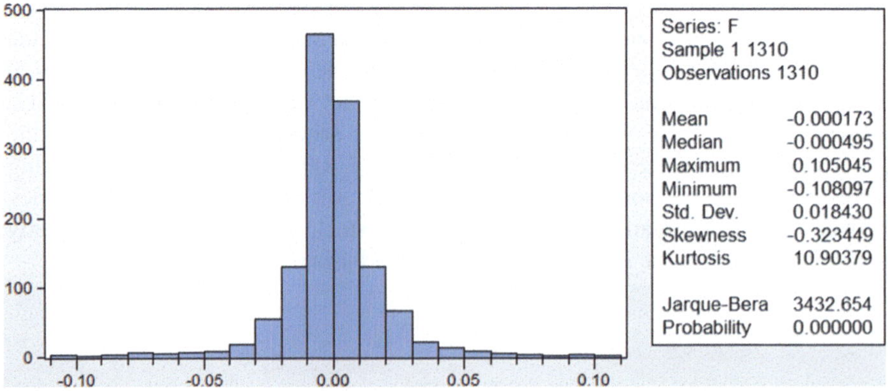

Fig. 30.2 Descriptive statistics of stock index futures

30.3.1.3 Processing and Checking of Sample Data

In order to avoid the phenomenon of pseudo regression, this paper uses Eviews software to do the unit root test for the sample data of CSI 300 stock index futures and spot stock. CSI 300 stock index futures and spot daily closing price logarithm sequence and first-order difference sequence ADF test results shown in Table 30.1.

Table 30.1 shows the original data by ADF test *P* value in the range of 0.0488–0.9894 and more than 0.01 significant level under, so accept the null hypothesis, that is, indicating that these contracts data is not stationary. In addition, the *P* values of the ADF tests for first difference sequences are all 0 and are less than the significant level of 0.05 and 0.01, that is, rejecting the original hypothesis. Therefore, the two column data are a stationary sequence and are the first order single integral sequence, that is, $\ln(S_t){\sim}I(1)$ and $\ln(F){\sim}I(1)$. This paper selected the sample data for cointegration test and Granger causality test (See Figs. 30.3 and 30.4).

As can be seen from Fig. 30.3, the *P* value of the *T* statistic is 0 in the cointegration test and less than the significant level of 0.05. Therefore, the original hypothesis is rejected and there is a long-term equilibrium relation between the two variables. According to Fig. 30.4, "spot is not the cause of futures," the hypothesis that the *P* value of the *F* statistic is 0.2843 and it is greater than 0.05 of the significance level, that is, inspection passed. However, "futures do not cause spot", the hypothesis that the *P* value of the assumed *F* statistic is 0.0261 and less than the significant level of 0.05, that is, the test does not pass. To sum up, the CSI 300 stock index futures is "because", the spot is "fruit".

30.3.2 Hedging Ratios Analysis

30.3.2.1 OLS Model Empirical Analysis

This paper use the OLS model to do empirical analysis and find that the *P* value of the *T* statistic of the constant term in the partial sample fitting is greater than 0.05, that is, accepting the original hypothesis shows that the constant term is not significant. Therefore, the regression of the OLS model does not include the constant term. Then the regression results of OLS model were tested by residual difference test and autocorrelation test, and found that there was no heteroskedasticity in the regression results of all samples. However, we found that the regression results of some samples had residual first-order autocorrelation. Therefore, we use the Cochrane Orcutt method to eliminate the original model of the first-order autocorrelation [16]. Finally, we do a residual autocorrelation test on the adjusted fitting results and found that there is no autocorrelation, as shown in Table 30.2 below.

From Table 30.2, The *P* values of the *T* statistics of the 4 sets of samples under the OLS model are 0, and R^2 are close to 1, which shows that the OLS model fits well. The *P* values of the White test and the LM test were greater than the significant level of 0.05, and indicating that there was no residual heteroscedasticity and

Table. 30.1 Unit root test results

		IF1412		IF1512		IF1612		IF1703	
Contract variety		spot	Futures	spot	Futures	spot	Futures	spot	Futures
Pricelogaritmic	ADF	−1.01	−0.32	−1.44	−1.46	−3.44	−3.32	−2.54	−2.36
	P	0.93	0.98	0.55	0.83	0.04	0.06	0.40	0.39
First order difference	ADF	−6.91	−6.66	−10.2	−11.2	−12.2	−12.2	−12.5	−13.5
	P	0.00	0.00	0.00	0.00	0.00	0.00	0.00	0.00

Null Hypothesis: RESID01 has a unit root
Exogenous: Constant
Lag Length: 0 (Automatic - based on SIC, maxlag=7)

		t-Statistic	Prob.*
Augmented Dickey-Fuller test statistic		-15.49409	0.0000
Test critical values:	1% level	-3.470934	
	5% level	-2.879267	
	10% level	-2.576301	

Fig. 30.3 Cointegration test results

Pairwise Granger Causality Tests
Date: 06/04/17 Time: 11:55
Sample: 1 166
Lags: 2

Null Hypothesis:	Obs	F-Statistic	Prob.
S does not Granger Cause F	163	1.26772	0.2843
F does not Granger Cause S		3.73247	0.0261

Fig. 30.4 Granger causality test results

Table. 30.2 Empirical analysis results and their tests based on OLS model

	OLS model fitting regression results						
Contract variety	β	*p value*	R^2	*SC*	*AIC*	White test	LM test
IF1412	0.88	0.00	0.90	−8.33	−8.36	0.65	0.15
IF1512	0.63	0.00	0.75	−5.67	−5.69	0.33	0.14
IF1612	0.76	0.00	0.92	−9.09	−9.07	0.10	0.53
IF1703	0.75	0.00	0.88	−9.15	−9.17	0.16	0.53

autocorrelation in the fitting results. Therefore, the β value is the optimal hedging ratio and calculate the average hedge ratio under the OSL model is 0.7532.

30.3.2.2 Empirical Analysis Based on GARCH Model

According to the requirements of the GARCH model, the ARCH effect of the sample data is checked and the independent Q test is used to determine the lag order. Moreover, a common GARCH (1,1) GARCH (1,2) GARCH (2,1) GARCH (2,2) model is established here. We selected the appropriate model by comparing the fitting degree, coefficient significance, AIC and SC criteria, and found that the IF1612 and IF1703 sample data did not meet the GARCH modeling requirements. Therefore, the EGARCH model is established and then the ARCH LM test of the fitting results ensures that there is no autocorrelation of the fitting residuals.

Table. 30.3 Fitting regression results based on GARCH model

Contract variety	ARCH test (*P* value)	Independent *Q* test(lag order)	β	*T* statistic (P value)	R^2	Sum of coefficient of variance equations	Residual ARCH LM test
IF1412	0.00	1 order lag	0.89	0.00	0.89	0.98	0.67
IF1512	0.00	2 order lag	0.81	0.00	0.69	0.94	0.58

Table. 30.4 Fitting regression results based on EGARCH model

Contract variety	ARCH test (*P* value)	Independent*Q* test(lag order)	β	*T* statistic (P value)	AIC	SC	R^2	Residual ARCH LM test
IF1612	0.01	2 order lag	0.80	0.00	−9.20	−9.20	0.91	0.61
IF1703	0.00	2 order lag	0.85	0.00	−9.23	−9.10	0.85	0.32

According to the sample data, the GARCH and EGARCH models are established, and the fitting results are shown in Tables 30.3 and 30.4.

As can be seen from Table 30.3, the *P* values of the ARCH effect tests for the contract IF1412 and the IF1512 samples are less than the significant level of 0.05 and indicates the existence of ARCH effect. Here, according to the nature of the coefficient significance test and the sum of the variance equation coefficients of the GARCH model is less than 1 and the coefficient is greater than 0, the GARCH (1,1) model (the mean equation does not include the constant term) is established. In the fitting result, the *P* value of coefficient significance is 0 and shows that the coefficient is remarkable. Moreover, the sum of coefficients of the variance equation is less than 1 and the R^2 is close to 1, which shows that the model fitting is better. In this paper, the residual value of GARCH (1,1) model to do ARCH LM test and we found that the *P* values of the *F* statistics are greater than the significant level of 0.05, that is, there is no autocorrelation of the residuals. Therefore, the β value can be the optimal hedging ratio.

As can be seen from Table 30.4, the *P* values of the ARCH effect tests for the contract IF1612 and the IF1703 samples are less than the significant level of 0.05 and indicates the existence of ARCH effect. Since the variance equation coefficient is negative and the GARCH model can not be established. Therefore, the EGARCH (1,1) model is established and the *P* value of the coefficient *T* statistic is 0, and indicating that the coefficient is significant. The values of AIC and SC are relatively small and R^2 is close to 1, which shows that the model fits well. Finally, the residual of the EGARCH model is tested by ARCH LM and the *P* value of the *F* statistic is greater than 0.05, which indicates that the original assumption is accepted, that is, there is no autocorrelation. Therefore, the β value is the optimal hedging ratio and calculate the average hedge ratio under the GARCH model is 0.838.

30.3.3 Hedging Effects Comparative Analysis

In this paper, we use the OLS and GARCH model to estimate the hedge ratio and construct the futures and spot investment portfolio outside the sample. Here with the hedging performance formula for hedging the effect of comparative analysis(See Fig. 30.5). As shown in Fig. 30.5 and Table 30.5, the average of the hedging ratios under the GARCH model constructs the portfolio outside the sample and the hedging effect is better than the hedge ratio under a single sample.Similarly, this paper using the average of the hedging ratios under the OLS model to build a portfolio outside the sample and the hedging effect is better than that the hedge ratio obtained under IF1512 and IF1703 samples, but slightly lower than the hedge ratio of IF1412 and IF1612 samples. Thus, the average of the hedging ratios in the two models is superior to the estimated hedging ratio estimated in most individual historical samples in terms of hedging effectiveness.

This paper is based on the theory of portfolio hedging, the minimum variance method is used to estimate the average of the optimal hedging ratio under the OLS and GARCH hedging models. The results show that the average of the hedging ratios is better than the hedging ratios estimated by most individual historical samples in terms of hedging effectiveness. Therefore, the analysis processing method by calculating the average hedge ratio optimization can adapt to different market conditions.

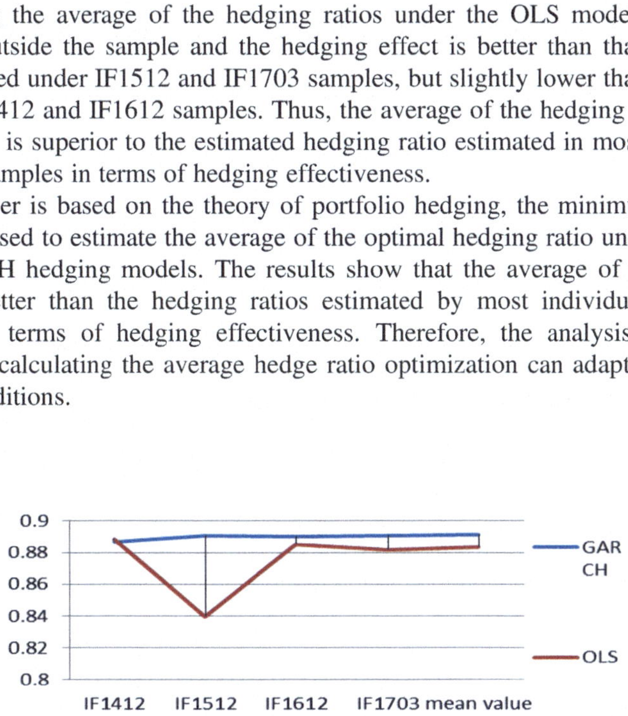

Fig. 30.5 Hedging effect of different hedging ratios outside the sample

Table. 30.5 Hedging effects of different hedging ratios outside of sample

Contract variety	Hedging ratio under OLS model	Hedging effect under OLS model	Hedging ratio under GARCH model	Hedging effect under GARCH model
IF1412	0.88	88.87%	0.89	88.69%
IF1512	0.63	83.98%	0.81	89.06%
IF1612	0.76	88.53%	0.80	89.02%
IF1703	0.75	88.20%	0.85	89.06%
Average value	0.75	88.36%	0.84	89.12%

This paper suggests that investors can select several continuous historical contracts at the moment of hedging as the object of study. Then, we estimate the hedge ratio in the optimal hedging model and calculate the average of the hedge ratios. At the same time, this paper only uses the average value of the hedging ratio to construct a hedge portfolio outside the sample and the future needs a lot of empirical data validation. In addition, the average of the hedging ratio should be dynamically adjusted on the basis of multiple tests as the future research direction.

Acknowledgments This work was supported by Social Science Research Foundation of Ministry of Education of China (15YJA790051), National Social Science Fund Project of China (17BGL058) and Shandong Province Natural Science Foundation (ZR2016GM20).

References

1. Crosby, J. (2014). Optimal hedging of variance derivatives. *The European Journal of Finance, 20*(02), 150–180.
2. Ederington, L. H. (1979). The hedging performance of new futures inarkets. *Journal of Finance, 34*, 157–170.
3. Eegel, R. F. (1982). Autoregressive conditional heteroscedastucity with estimates of variance of U.K inflation. *Econometrica, 50*, 987–1008.
4. Holmes, P. (1996). Stock index futures hedging: Hedge ratio estimation, duration effects, expiration effects hedge ratio stability. *Journal of Business Finance and Accounting, 23*, 63–78.
5. Wu, X.-z. (2008). The optimal hedging ratio of index futures: An empirical study based on CSI300 futures. *Journal of Shanghai Lixin University of Commerce*, *22*(04), 78–84.
6. Ping, Z., Han, D.-p., & Mei-jia, G. U. O. (2014). Estimation of dynamic hedging rate of grain futures based on GARCH model. *Statistics and Decision*, *30*(02), 149–152.
7. Jonathan Dark, A. (2005). Critique of minimum variance hedging. *Accounting Research Journal, 18*(01), 40–49.
8. Olgun, O., & Hakan Yetkiner, I. (2011). Determination of optimal hedging strategy for index futures: Evidence from Turkey. *Emerging Markets Finance and Trade, 47*(06), 68–79.
9. Xie, J.-q., & Zhao-jun, Y. A. N. G. (2005). An empirical analysis on EMT of stock market based on GARCH model. *Statistics & Information Forum*, *20*(03), 57–60.
10. Lien, D., Tse, Y. K., & Tsui, A. K. C. (2002). Evaluating the hedging performance of the constant-correlation GARCH model. *Applied Financial Economics, 12*(11), 791–798.
11. Koutmos, G. (2012). Modeling interest rate volatility: An extended EGARCH approach. *Managerial Finance, 38*(06), 628–635.
12. Kao, W.-S., Lin, C.-H., Changchien, C.-C., & Chien-Hui, W. (2016). Return distribution, leverage effect and spot-futures spread on the hedging effectiveness. *Finance Research Letters, 22*(08), 158–162.
13. Lee, M., Chiou, J.-S., Wu, P.-S., & Chen, C.-D. (2005). Hedging with S& P500 and E-mini S& P500 stock index futures. *Journal of Statistics and Management Systems, 8*(02), 275–294.
14. Hong-feng, P., & Yong-gang, Y. (2007). The evaluation and comparison research of dynamic optimal hedging ratios based on modified ECM-GARCH. *Chinese Journal of Management Science*, *24*(05), 29–35.
15. Zhen-nan, Z. (2016). Be based on GARCHS tudy on the performance of model futures hedging of agricultural products. *West Forum on Economy and Management*, *27*(04), 43–58.
16. Shen, C.-h., & Wang, X.-r. (2015). Nonlinear analysis on the pattern structures of connection between financial markets. *Journal of Management Sciences, 18*(2), 66–75.

Chapter 31
Finding the Lenders of Bad Credit Score Based on the Classification Method

Haifeng Li and Yuejin Zhang

Abstract The online P2P lending is a creative and useful finance way for tiny enterprises who can conduct by the internet. To exclude the risk of this method, we make a study on predicting the potential lenders that may have a bad credit score. We use a classification method to perform this detection. Our experimental results show that this method can achieve a high precision.

Keywords Trust model · Credit score · Classification · P2P

31.1 Introduction

The online peer-to-peer (P2P) lending is a creative and useful finance way for the small enterprises. To finance certain tiny enterprises with a valid method has been focused on by many attentions. This problem is very important especially in China. With using the advantages of the information techniques, a new-coming finance method, the online P2P lending is becoming very important to replace the traditional ones. It allows us to distribute the funds online with no financial mediations.

When a lender wants to acquire capitals from the online P2P companies, a risk will be raised. Traditional bank can audit the background of a lender with his application document, which, for the P2P companies or the borrowers, is an impossible task. Since a lender is never known has a good credit score or a bad one. Thus, how to find the lenders with bad credit score is a very challengeable question. Many researches have focused on this problem and proposed some useful method.

In [1], the authors represented the credit risk based on the corporate bond and securities risk pricing literature, and analyzed the concept drift of credit scores, also, the authors used the residential mortgages data to do the experiments and observe the credit score migration post loan origination, which, indicated the credit scores

H. Li (✉) · Y. Zhang
School of Information, Central University of Finance and Economics, Beijing, China
e-mail: mydlhf@cufe.edu.cn

M. Tavana, S. Patnaik (eds.), *Recent Developments in Data Science and Business Analytics*, Springer Proceedings in Business and Economics,
https://doi.org/10.1007/978-3-319-72745-5_31

providing data to the users in a similar manner to the commercial credit ratings. Soner [2] proposed a three stage hybrid system ANFIS, whose performance was compared to the conventional utilized models and was tested by empoying a cross-validation method on the credit card data, which is got from the Turkey bank. The analyzed results showed that the ANFIS method has a better performance than the Linear Discriminant Analysis methods (LDA), as well the Logistic Regression methods (LR) and the Artificial Neural Network (ANN) approaches. Arya et al. [3] focused on the problem of the factors that may determine a bad score, which is compared by measuring some important factors over the data generated from a decision-making laboratory, as well certain financial reviews. The score was evaluated with the online estimator distributed by the FICO. In addition, the preferences were assessed by the same factors. Chen et al. [4] developed an useful trusting module of the P2P lending context based on trust theories The module was evaluated by employing the data of PaiPaiDai lenders, which showed that not only the trust of borrowers, but also the intermediaries were useful factors that will affect the lending intention. In [5], Emerkter et al. used the Lending Club data to discover the borrowing features, and estimate the risk, as well measure the borrowing efficiencies. The authors showed that debt ratio, credit grade, FICO score and some other factors are very important factors in loan, which was consistent to the Cox Proportional Hazard test. Plus, they detected that the larger interest rates for the bad borrowers cannot compensate enough to the loan default. Finally, Harris [6], who has explored the use of scores and employed a method of the clustered SVM, which has been extensively used for large credit scoring datasets. The author compared the clustered SVM with other nonlinear ones and presented that the it can get better performance.

In this paper, we also addressed this problem and proposed a classification method by the online documents of the lenders. The organization for the rest paper is as follows: Section 31.2 shows the data related the lenders. Section 31.3 proposes our predicting method. Section 31.4 concludes this paper.

31.2 Data Preparation

We use the data crawled from the website, which is a BBS that provide the lenders to discuss the issues related to P2P lending. We preprocess the data and get the dataset with 18 properties. We describe it with Table 31.1. In this dataset, the title and the descriptions are string information, which are not useful in our method. In addition, we transform the continously changed property values, such as age, to the discrete values with an aequilate method. Also, we convert the credit rate and other string type properties to integer properties.

Since not all the properties are valid in our problem, we employ the randomized logistic regression to filter certain the properties that have little impacts, and get the final properties. As shown in Fig. 31.1, the age, membership score, group, amount has a very little percentage on our prediction; thus, we remove these properties. Also, we can see that the failed loan number, the payback time and the borrowed credit score may have a relatively much larger impact on the final predicting results.

Table 31.1 The characteristics of the dataset

Properties	Record Count
Title, Amount, Annual interest rate, Repayment Time, Descriptions, Credit rate, Successful loan number, Failed loan number, Gender, Age, Borrowed credit score, Lending credit score, Overdue, Membership score, Prestige, Forum currency, Contribution, Group	20,000

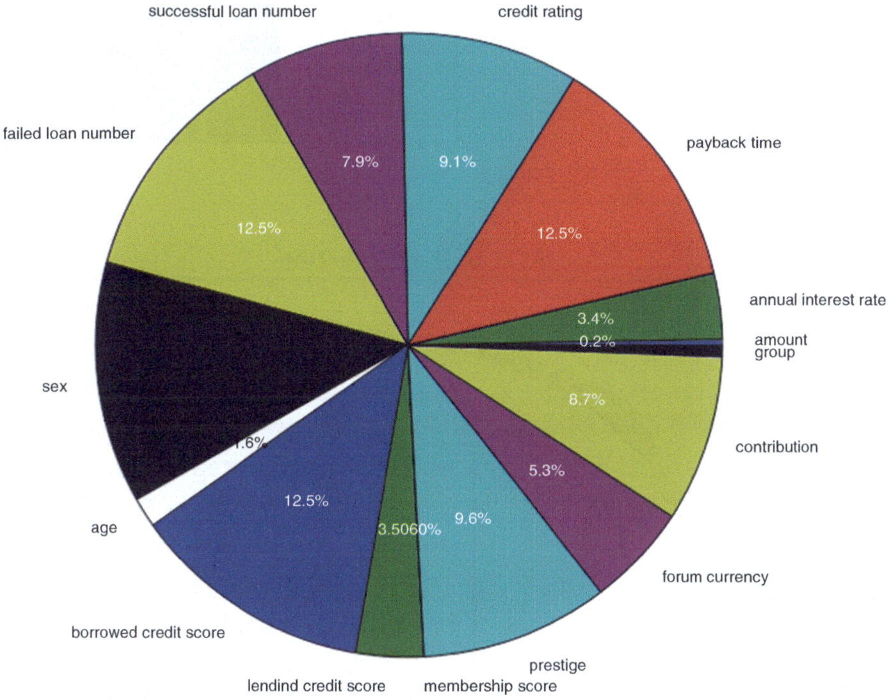

Fig. 31.1 The impacts of the properties

31.3 Predicting Model

In this section, we use a logistic regression method to build the predicting model. In this model, we try to find the probability of a lender having a bad credit score w.r.t. all our mentioned properties shown in Fig. 31.1. As can be seen in Fig. 31.2, we use the confused matrix to represent the accuracy. The light green section and the dark green section show the right predictions, and the other two sections show the wrong predictions; thus, the average predicting accuracy is about 76%. We also use the sixfold cross validation to multi-test the accuracy. Figure 31.3 is the ROC curve of the result of the validation. As can be seen, all this sixfolds have the consistent accuracy.

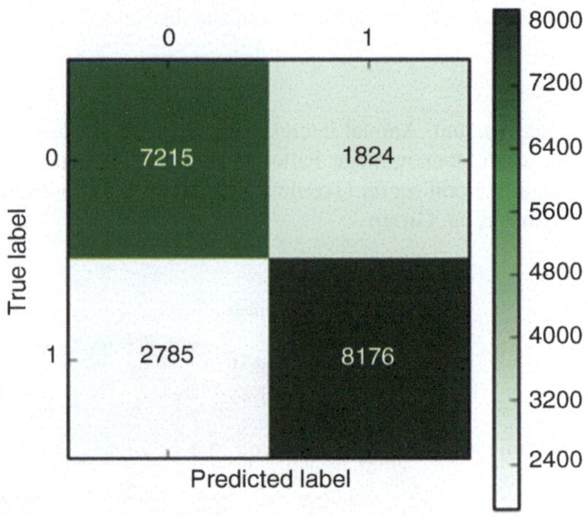

Fig. 31.2 The confused matrix

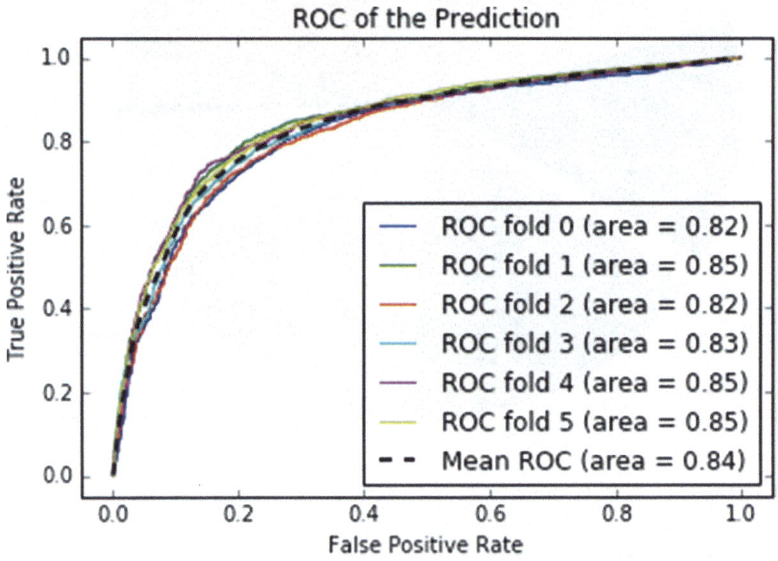

Fig. 31.3 ROC of the cross validation

31.4 Conclusions

In this paper, we used the social network data to decide whether a lender has a bad credit score when he is applying the P2P loan, and we employed a classification method, which achieved a 76% accuracy. This presented that the social network data has a very intimate relation to the credit score, and provided an effective method for P2P companies.

Acknowledgements This research is supported by National Natural Science Foundation of China (61100112, 61309030), Beijing Higher Education Young Elite Teacher Project (YETP0987), Key project of National Social Science Foundation of China(13AXW010).

References

1. Smith, B. C. (2011). Stability in consumer credit scores: Level and direction of FICO score drift as a precursor to mortgage default and prepayment. *Journal of Housing Economics, 20*(4), 285–298.
2. Soner, A. (2012). An empirical comparison of conventional techniques, neural networks and the three stage hybrid Adaptive Neuro Fuzzy Inference System (ANFIS) model for credit scoring analysis: The case of Turkish credit card data. *European Journal of Operational Research, 222* (1), 168–178.
3. Arya, S., Eckel, C., & Wichman, C. (2013). Anatomy of the credit score. *Journal of Economic Behavior & Ornanization, 95*, 175–185.
4. Chen, D., Lai, F., & Lin, Z. (2014). A trust model for online peer-to-peer lending: A lender's perspective. *Information Technology and Management, 15*(4), 239–254.
5. Emerkter, R., Tu, Y., Jirasakuldech, B., & Lu, M. (2014). Evaluating credit risk and loan performance in online peer-to-peer(P2P) lending. *Applied Economics, 47*, 57–40.
6. Harris, T. (2015). Credit scoring using the clustered support vector machine. *Expert Systems with Applications, 42*(2), 741–750.

Part IV
Predictive Modeling and Data Analytics

Chapter 32
Research Status and Prospect of Data Extraction and Cleaning Technology in Large Environment

Mingzhe Wang and Zhaochan Li

Abstract In the era of big data, how to get effective data from the massive data, the data obtained by the relevant analysis and processing is particularly important. This paper first introduces the importance of data cleaning and data extraction technology, secondly from different angles, introduced the two technology, and then summarizes the current domestic and international data cleaning and data extraction technology research, and finally describes the data extraction and data cleaning technology development prospects. It has a certain guiding role in the research of data extraction and cleaning technology in the future.

Keywords Data cleaning · Data fetch · Research Status · Developing Prospect

32.1 Introduction

With increased data growth, increasing data volumes, the redundancy of the data will also be a greater, the traditional model of data analysis software can no longer meet the needs of mass information processing and analysis of needs, how effective information is pulled from the data, to data for effective analysis and decision making, creating new data models, data analysis and processing is particularly important. Data throughout its lifecycle, through human interaction, computational and communication procedures, each link can introduce errors,resulting in abnormal data, data quality issues, consists of the following four types of data-quality issues source: (1) data input / update; (2) measurement; (3) simplify; (4) data integration.

M. Wang (✉)
Department of Economic Management, China Institute of Industrial Relations, Beijing, Chao yang District, China
e-mail: wangmingzhe@ciir.edu.cn

Z. Li
Beijing Wuzi University, Beijing, Tong zhou District, China

M. Tavana, S. Patnaik (eds.), *Recent Developments in Data Science and Business Analytics*, Springer Proceedings in Business and Economics,
https://doi.org/10.1007/978-3-319-72745-5_32

The instance-level view of the data from the data, detect and eliminate errors and inconsistencies in the data quality problems, to improve the quality of data. It is one of the 19 research topics on data quality. Data cleaning (Data Cleansing or Data Scrubbing) to detect data errors and inconsistencies, remove or correct them, to improve the quality of data [1].

32.2 Data Cleaning Technique

Major headings should be typeset in boldface with the first letter of important words capitalized.

32.2.1 Introduction

Data cleaning, also known as Data cleansing or Data scrubbing. Due to the different areas of application to data cleaning, its definition is slightly different. Data cleansing is mainly used for data warehousing (DW), data mining and overall data quality management 3 areas.

32.2.2 Data Cleaning Objects

Data cleaning objects can be classified according to the source area and the cause of data cleaning. The former belongs to the macro level and the latter belongs to the micro level.

32.2.2.1 Source Fields

Many areas are related to data cleaning, such as digital documents [1], search engines, financial services, government agencies and others, data information system for the purpose of cleaning is to provide accurate and valid data.

32.2.2.2 The Causes

In microscopic aspect, the object of data cleaning can be divided into pattern layer data cleaning and instance level data cleaning. Data cleaning tasks are filtered or modified data that do not meet the requirements. Do not meet the requirements of the data is mainly incomplete data, error data and duplicate data of these 3 categories.

32.2.3 Data Cleaning Model the Principles and Framework

32.2.3.1 The Basic Principle

The principle of data cleaning is to make use of the related technologies, such as statistical method, data mining method, pattern rule method, etc. Data cleaning in accordance with the realization of the scope and can be divided into the following four: manual implementation; write a specialized application; solve the problem of a particular application domain; independent of the specific application data cleaning. Either way, it is made up of 3 phases: data analysis, definition, search, identification, error correction, error correction.

32.2.3.2 Data Cleaning Framework

At present, there have been a lot of data cleaning framework models, the following 3 representative framework model: Trillium Model; Bohn Model; AJAX Model. The data cleaning process is divided into 5 steps: mapping; matching; clustering; merging; viewing.

32.2.4 Data Cleaning Algorithms

For the property cleaning and repeat record cleaning from the detection and cleaning of the two angles to discuss the related algorithms.

1. Property cleaning

   ①Method for detecting attribute error: Statistical based approach; Clustering method; Method of association rules.
   ②Method for cleaning property: Method for cleaning blank value; Method for cleaning noise data; Method for cleaning inconsistent data.

2. Duplicate record cleaning

   ①Algorithm for detecting duplicate records: Basic field matching algorithm; Recursive field matching algorithm; Smith-Waterman algorithm; Matching algorithm based on edit distance; Cosine similarity function.
   ②Algorithm for duplicate record cleaning; Priority queue algorithm; Nearest neighbor sorting algorithm SMM; Multiple nearest neighbor sorting MPN.

32.3 Data Pump

32.3.1 Data Analysis Confirms

Before the data pump must be the nature of the data, but also to understand how the data collected is used. The complexity of the source data, rules model, completeness of data warehousing of larger effect than other factors. Which data source to pay special attention to the data type, size, and content are compatible. The extraction of data depends on how the data is stored in the source system.

32.3.2 Data Extraction Technology

After you know how the data is stored in the source system, to discuss general technical data pump now. Extracts the data from the source operating system, there are two main types, static data and revised data, respectively.

32.4 Data Cleaning Technology Based on Data Extraction and Research at Home and Abroad

Data cleaning is mainly in 3 areas: data warehouse, database knowledge discovery (also known as data mining) and overall data quality management. In the field of research and application of data warehouse, data cleaning process is the first step in the construction of data warehouse, due to the huge amount of data, not artificial processing, so the automatic data cleaning is widely concerned by industry.

32.4.1 Domestic Research

At present, the domestic research on data cleaning technology is still in the preliminary stage. Directly for data cleaning, especially for Chinese data cleaning research results are not many. Most of the data warehouse, decision support, data mining research, to make some simple exposition. Banking, insurance and securities such as accuracy of customer data demanding industry, are doing the cleaning work of the respective customer data, according to their specific application and development of software, and there is little theoretical results reported.

32.4.2 Foreign Research Status

Foreign study of data cleaning first appeared in the United States, from the start for America's social security number error correcting. The development of information technology and business in the United States, has greatly stimulated the data cleaning technology research, mainly concentrated in the following four aspects.

1. To detect and eliminate the abnormal data

   Using statistical methods to test the numeric attributes, calculate the mean and standard deviation of the field values, considering the confidence interval of each field to identify abnormal fields and records. Data cleansing, introduced the method of data mining, such as the clustering method is used to detect abnormal record, model method found that do not conform to the existing pattern of abnormal records, association rules method is found that the data set is not in conformity with the abnormal data with high confidence and support rules [2].

2. To detect and eliminate the approximately duplicate records

   Duplicate records for cleaning. Eliminate the approximately duplicate records of the data set is one of the most current research in the field of data cleaning. In order to eliminate duplicate records from a data set, the first question is how to determine whether two records similar to repeat.

3. Data integration

   In the applications of data warehouse, data cleaning must first consider the data integration, mainly structure in the data source and data mapping and domain to the target structure. In this respect has carried out a lot of research work.

4. Specific areas of data cleaning

   A lot of data cleaning solution and algorithm are for a specific application question, applies only to a smaller range. Generic, has nothing to do with the application fields of the algorithm and scheme is less [3].

32.5 Cleaning Technology of Data Extraction and Data Development Prospects

32.5.1 Research and Development of Chinese Data Cleansing Tools

At present, data cleaning is mainly concentrated in the western, Western Chinese data cleaning and data cleaning are different (such as many matching algorithm does not apply to Chinese). Chinese data cleaning has not attracted attention.

32.5.2 Application of Data Mining Method in Data Cleaning

Numbers and strings is the main focus of the main fields. The relationship between the identification number field is not yet mature and practical. The application of data mining algorithm in data cleaning [3].

32.5.3 Duplicate Recognition Efficiency Need to Be further Improved

Duplicate records identification has received the widespread attention, and take a lot of measures, but the recognition efficiency and the accuracy is not satisfactory. Especially in the recorded data is very long, and time consuming too much, that wants to study better.

32.5.4 Cleaning of Non-Structured Data

Previous data cleaning are mainly concentrated in structured data, now paid unstructured or semi-structured data (such as XML) are constantly by more and more people pay close attention to. Especially due to the universality and the descriptive characteristics of the XML itself, should be more attention in the data cleaning.

32.5.5 The Interoperability Between Data Cleaning Tools

Although according to user friendliness, many tools or systems provide descriptive language, but basically is through some existing languages (such as SQL, XML) according to their own needs through expansion, can't meet the need of data cleaning, generally do not have interoperability.

32.5.6 The Generality of Data Cleaning Scheme

The problem of data cleaning in a specific domain is still the focus of research and application, but more general cleaning schemes will receive more and more attention.

32.5.7 Future Research Areas

Future research should focus on integrating a variety of different methods for error detection. Methods based on the analysis of a set of related fields (e.g. statistical correlations) have also been shown to be powerful methods. It is necessary to design a set of common operators and theories (similar to relational algebra) to solve the problem of data cleaning. This formal foundation is essential to the design and construction of high quality and wide range of software tools to support data cleaning.

32.6 Summary

In the development of technology, the data extraction of a single technology has been relatively mature, some of which can't be separated from the programming, but the overall integration is not enough. Currently available on the market are mostly data extraction tools. These tools automatically generate the code for data extraction by selecting the corresponding relationship between the source data and the target data. But the type of data extraction tool support is limited; and the data extraction process involves data conversion, it is closely related with the practical application of the part, its complexity can't be extraction tools embedded in user programming often cannot meet the requirements [4].

In the field of computer science and information systems, little basic research is directly related to error detection and data cleaning, no in-depth comparison of data cleaning techniques and methods, resulting in many methods of data cleaning has a strong theoretical basis. Future research should focus on the integration of a variety of different methods for error detection. Methods based on the analysis of a set of related fields (e.g. statistical correlations) have also been shown to be powerful methods. It is necessary to design a set of common operators and theories (similar to relational algebra) to solve the problem of data cleaning. This formal foundation is essential to the design and construction of high quality and wide range of software tools to support data cleaning [5].

References

1. Rahm, E., & Do, H. H. (2000). Data cleaning problems and current approaches. *IEEE Data Engineering Bulletin, 23*(4), 3–13.
2. Rifen, W., & Chengzhi, Z. (2007). Review of data cleaning research. *J Modern book information technology, 158*(12), 50–57.
3. Harte-Hanks Trillium Software[EB/OL]. (2007). http://www.trillium software.com. [2007-01-09].
4. Li, Y. (2013). *Study on data extraction technology based on network*. Harbin.
5. Congjian, B. (2007). *Research on key techniques of data extraction*. Jiangsu university.

Chapter 33
Research on Intelligent Sales Platform of Automobile Industry Based on Large Data Mining

Jinzi Lee

Abstract In the context of Internet + and large data, the traditional automotive industry sales platform has been unable to meet the requirements of users of high quality services. How to use the online and offline large data, tap out the potential of potential customers, accurately capture the current needs of each customer, the accurate information through accurate channels to each other, as the current smart sales platform goals. Based on the current development of large data mining technology and the existing problems of automobile industry sales system, this paper gives the suggestions of intelligent sales platform based on large data mining.

Keywords Big data · Data mining · Sales platform · Automobile industry

Big data age has arrived. Big data (big data) This concept was first proposed by the four well-known consulting firm McKinsey. Big to space, science and technology, military, economic, small to eat and let us, all with large data are closely related. Big data has brought great changes to our lives, with the people for the massive data mining and use, a new wave of productivity and consumption level of the wave of growth is coming.

I during the graduate school, Changchun FAW office system for the development of MIS, the system functions for the repetitive labor automation. All the data in the database need to be added manually. Late maintenance workload. Dr. stage has also been involved in the study of the status quo of cross-border automobile mergers and acquisitions, before 2010, the automotive industry sales model in the basic customer relationship management (CRM) model. Then the "big data" era will give car marketing what kind of opportunity? For the traditional car companies, how to use large data trends, the establishment of intelligent, accurate sales platform, you can accurately target the target crowd, monitor brand influence, improve product quality,

J. Lee (✉)
East Campus Administrative Building, ShenZhen Polytechnic, ShenZhen, China

M. Tavana, S. Patnaik (eds.), *Recent Developments in Data Science and Business Analytics*, Springer Proceedings in Business and Economics,
https://doi.org/10.1007/978-3-319-72745-5_33

improve production technology, and gradually enhance market sales and improve after-sales Quality of service, so as to contribute to business decision-making.

33.1 The Status Quo of Large Data Analysis and Sales Platform in the Automobile Industry

33.1.1 User Behavior Characteristics and Data Storage Type

33.1.1.1 User Characteristics of the Automotive Industry

2015, China's auto industry continues to flourish, car sales and ownership remains double up trend. The cumulative annual production of 23.7729 million vehicles, an increase of 7.3%, the annual total sales of 23.4919 million vehicles, an increase of 6.9%. Car ownership of 154 million, the new registered vehicles 21.88 million, a net increase of 17.07 million, a record high level. Over the past few years, with the development of Internet technology, led the auto industry sales model changes. Whether the new car or used car, sales model to electricity and O2O mode, driven by sales under the main form, has accounted for 10% of the overall sales.

In such a big trend, the user behavior is divided into pre-sale behavior, purchase behavior and after-sales behavior. There are two general forms: First, online behavior, such as to 4 s shop direct inquiry, advice and so on. Second, online behavior, common car vertical website, portal car channel, 4S shop/vendor official website and the major car forum. The proportion of common user behavior is shown in Fig. 33.1 [1].

Visible, with the vertical car website, portal car channel and 4S point official website/vendor official website and other vertical automotive media efforts, the Internet has become the primary channel for users to obtain car information. Word-of-mouth communication and experience-related communication is the decisive factor that affects the user's purchase. The network has developed into a set test drive experience, word of mouth and price comparison of the most effective way, so the Internet is essential to car decision-making. Who can quickly and accurately extract valuable information, has become the car sales industry, the most important factor.

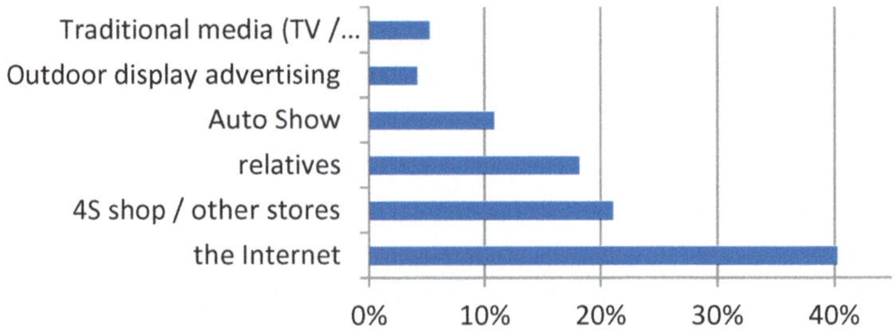

Fig. 33.1 Comparison of major vehicle access channels

33.1.1.2 Data Structure Types Stored in the Automotive Industry

China Internet Information Center (CNNIC) the latest statistical report shows that as of December 2016, China's Internet presence on the Internet outlets 711 million, a total of 42.99 million new users throughout the year. Internet penetration rate of 53.2%, compared with the end of 2015 increased by 2.9% points. It is important to link the online and offline data of all the people involved in the vehicle, that is, to summarize the online, offline attributes and network behavior data of such people. Including the user in the car forum, the site published comments, upload pictures, video and users use the type of network equipment, the number of users use the location, movement trajectories and other semi-structured and unstructured data, the total data to EB level Even ZB level development. Therefore, the data volume is large, the data type is rich, the value density is low, the data acquisition is quick, the data processing is quick and becomes the main characteristic of the big data of the automobile industry.

33.1.2 Automotive Industry Intelligent Sales Platform Facing the Problem

Prior to the automotive industry has been used in customer relationship management system CRM due to their own technical capacity constraints, did not achieve the full integration and utilization of all data sources. Most of the large data on the network exists in decentralized form. These large numbers of scattered, different sizes, different structures of the data source contains a wealth of valuable user behavior data. The combination of automobile sales system and large data is the trend, the car sales system into large data environment research and construction, relying on large data technology, build intelligent sales platform, mining and analysis of user behavior data, for users to carry out personalized information, Service is imminent.

33.2 Research on the Platform of Automobile Industry Sales Platform Under Large Data Environment

33.2.1 The Goal and Framework of Building Intelligent Sales Platform

Car industry consumers are characterized by a long process of decision-making, from the school car, car, car to the back of the sale and second-hand car trading, upgrading, the whole cycle is very long. How can we accurately grasp the information of each stage of the customer, will be suitable for the customer's preferred information, with the fastest and most effective way to accurately send to the customer, is a smart sales platform, a major goal [2]. The key to this technology is

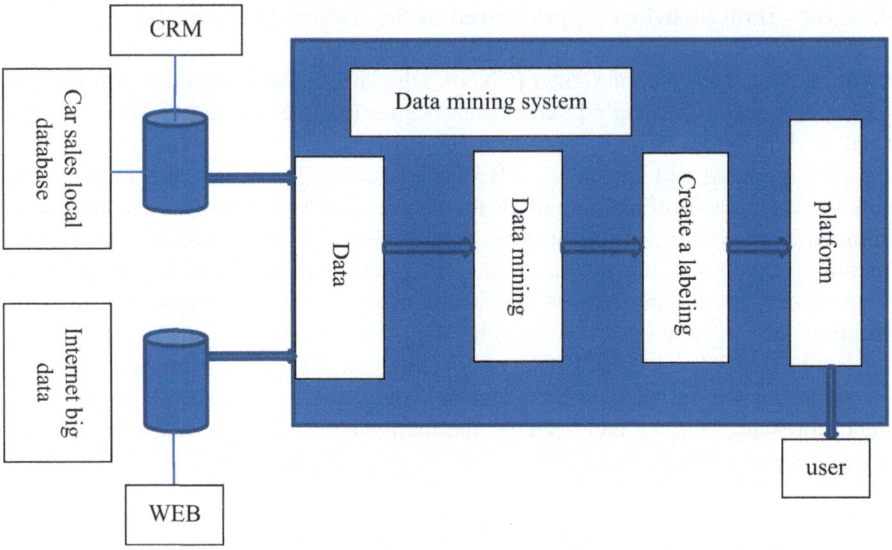

Fig. 33.2 Automotive industry intelligent sales platform system framework

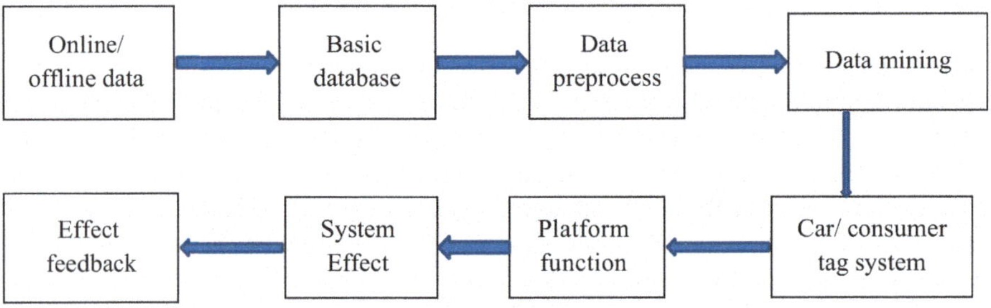

Fig. 33.3 Automotive industry intelligent sales platform flow chart

how to dig out the valuable data in the complex data of the structure. Thus forming a decision-making information, and even become an effective and feasible platform for the industry chain. Intelligent sales platform framework shown in Fig. 33.2.

33.2.2 Intelligent Sales Platform for the Automotive Industry

The Internet data and the automotive industry sales company data together, allows us to accurately capture the consumer from the interest to the end of the car, and even the latter part of the aftermarket, upgrading and other characteristics of the process, combined with the actual product positioning, market positioning, For accurate marketing decision-making to help. Platform flow chart shown in Fig. 33.3.

33.2.2.1 Data Collection

Large data mining and traditional mining Although the purpose is to extract valuable information, the difference lies in the excavation of the scope of the depth. With the cloud computing, Internet of Things, mobile intelligent terminal a large number of popular, traditional data mining technology and cloud computing technology how to form a complete system, is imperative. According to the characteristics of the automotive industry, data collection range comes down to two categories, consumer data and automotive data. The consumer data source may be the registration information of the website, the click rate of the website, the time period of access, the frequency, the app data of the mobile terminal, and so on. Automotive data sources can be GPS, RFID, camera image processing, sensors and other devices.

The problems are: large amount of data, massive, data structure is complex, there are comments posted on the website, there are upload pictures, audio, video, and the user's network equipment type, quantity, location, movement trajectory and other semi-structured And unstructured data. These data are multiplied by the order of magnitude. So in order to make better use of the mining model, we have to extract valuable information, we must do a good job before the data preprocessing.

33.2.2.2 Data Preprocessing and Mining Technology Analysis

Because of the differences in the structure, mobility, real-time and other aspects of traditional data mining and large data mining, traditional data storage and mining algorithms are not practical for large data mining. Large data mining has a wide range of active and automatic, complex and diverse, semi-structured and unstructured data-based, high data redundancy, the overall data volume is large, the accuracy is not high, the processing efficiency real-time, fast and so on [3].

The core of massive data mining, including distributed storage and distributed parallel computing. Distributed storage includes distributed file storage and distributed database storage. Distributed file storage to Google Labs proposed GFS and Hadoop proposed HDFS as the representative. Google and Hadoop main technology compared to Table 33.1.

Large data mining platform, HDFS, MapReduce, HBase as the core of Hadoop with high reliability, high scalability, high fault tolerance and high efficiency, open source and so on. It is recommended that the intelligent sales platform for the automotive industry be used in conjunction with the Hadoop system. The MapReduce model is suitable for dealing with large volumes of data, updating low-speed static

Table 33.1 Google, Hadoop main technical comparison

	Google	Hadoop
Distributed Parallel Computing Technology	MapReduce	MapReduce
Distributed file system	GFS	HDFS
Large-scale distributed database	BigTable	HBase

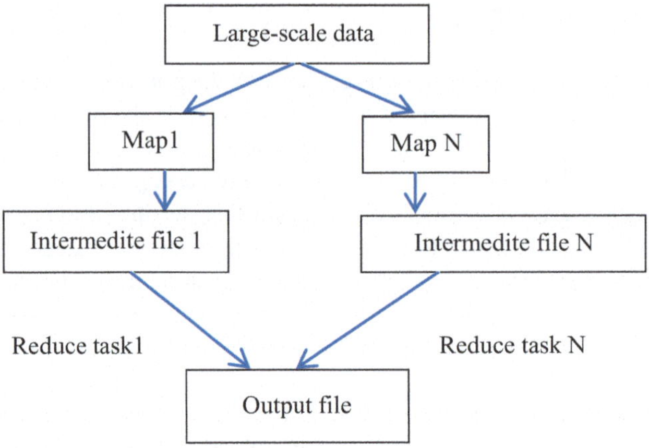

Fig. 33.4 MapReduce operating mechanism

data, such as the automotive industry's local sales database [4–9]. MapReduce main mechanism shown in Fig. 33.4.

Flume, Pig, etc. are more effective for dynamic processing of data streams, such as dynamic data on the network. So the combination of traditional data mining and large data mining industry intelligent vehicle sales platform mining system architecture shown in Fig. 33.5.

Between the massive data source and the data mining algorithm, the data is preprocessed. Function is to remove the data independent of the calculation, reduce all kinds of noise information (such as filling out incomplete data), complete the data type conversion, eliminate overlapping information. In order to dig out valuable information, the prerequisite is that there must be a good data source. The data source structure of the MapReduce model is consistent. So the face of a variety of structured, semi-structured, unstructured massive data must be pretreated. Commonly used preprocessing methods are: data extraction, data conversion, data cleaning and integration, data protocol, attribute concept of hierarchical automatic generation [10].

Regardless of the model, the commonly used algorithm is: C4.5 algorithm, Apriori algorithm, maximum expectation (EM) algorithm, clustering analysis algorithm, information system reduction clustering algorithm. C4.5 algorithm from the ID3 decision tree algorithm, the advantage is easy to understand the classification rules, high accuracy. Disadvantages are: low efficiency. The Apriori algorithm is also called mining algorithms that boolean Boolean association rules with frequent itemsets. The advantages are easy to understand and simple. The disadvantage is the consumption of memory, space and time. The maximum expectation method (EM) can handle relational data, the disadvantage is that the processing speed from fast to slow, easy to fall into local priority. The clustering analysis algorithm is divided into K-Means algorithm and two-stage clustering analysis method, which divides the research goal into KDD technology of relatively homogeneous groups.

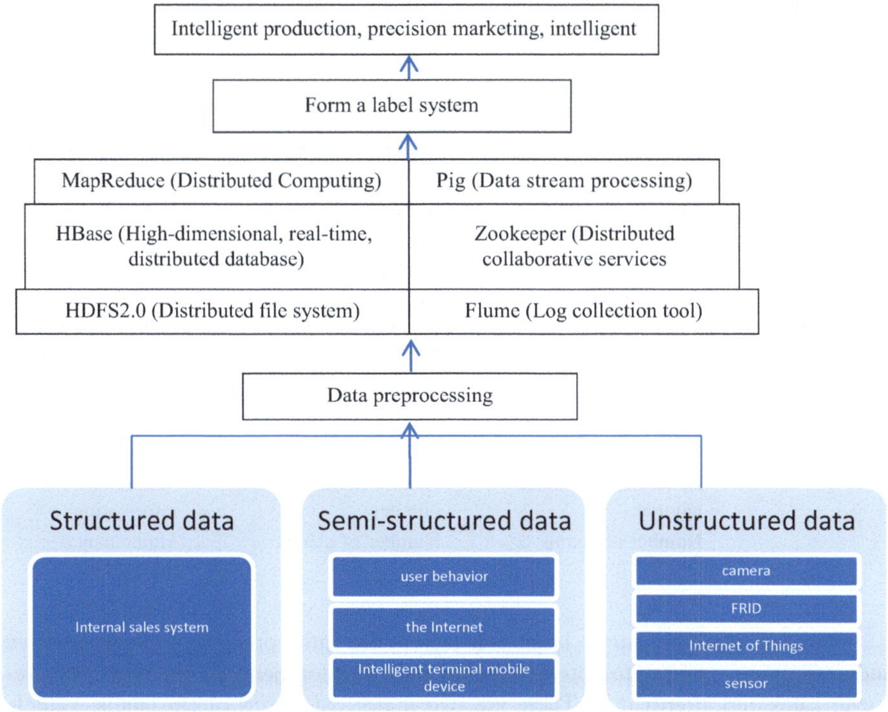

Fig. 33.5 Intelligent sales platform Large data mining system framework

The advantage of K algorithm is that the algorithm is simple, efficient and effective. The disadvantage is that the outliers are sensitive to the initial center point. The advantages of the two-stage algorithm are mass data analysis, complex class structure recognition. The disadvantage is that the main component of the factor selection is more difficult. The advantages of information reduction clustering algorithm are strong anti-noise and high accuracy. The disadvantage is that the attribute range is more complex, increasing the complexity of the calculation time [11].

Commonly used data mining software are: spss modeler, SAS, RapidMiner, MineSet, OracleDarwin, Weka, Orange. Commonly used large data processing platform: Apache Hadoop, Tencent large data platform, Ali cloud ODPS and so on.

33.2.2.3 Label System to Establish --------- Platform Basis

Data collection, the establishment of a good data source, data cleaning, finishing, access to data mining technology for the data structure, the ultimate goal is to establish for our decision-making reference customer label, car label set. We can form each customer from these labels and the portrait of each model.

Table 33.2 Car tag content

Label category	Label set	Information item	Type of data
Vehicle preference	Model preference	Vehicle model	Basic information
	Geographical distribution	On the card city	Ownership
	Engine capacity	Engine capacity	Basic information
	Interior color	Interior color	Basic information
	Appearance color	Appearance color	Basic information
	Car use	Vehicle use	Basic information
Car characteristics	Motivation of car	Family structure	Basic information
	Daily mileage	Monthly mileage	Travel information
	Information sources	Customer source	Basic information
	Yield concerns	Service category	Basic information
	Then into the factory stage	Then into the factory stage	Maintenance information
	History maintenance number	History maintenance number	Maintenance information
	Number of claims	Number of claims	Maintenance information

The automotive industry's large data covers the entire process of the automotive industry chain, from automotive research and development, production, procurement, sales and aftermarket. There are two major categories of car labels, vehicle preferences and car characteristics. Vehicle preferences include the model, the area, displacement, interior color and so on. Car features include daily mileage, historical repairs and so on. See Table 33.2. These data can be obtained by means of on-board recorder, sensor, GPS, camera and other devices.

Early automotive industry CRM has all the customer's information. Now we are based on large data not only get the information of old customers, but also to obtain information on potential customers. Customer labels can be divided into population attributes, owner attributes, customer value, after-sales service, brand value, driving characteristics and marketing preferences. See Table 33.3.

33.2.2.4 Platform Function Module Design and Implementation Environment

The greatest value of large data is that it can provide strong support for decision making. The intelligent sales platform of the automotive industry should run through the intelligent production of the automotive industry, precision marketing, after-sales service and some intelligent applications throughout the process. See Fig. 33.6.

Intelligent production module can help enterprises to improve product quality, improve the production process, product optimization. For example: vehicle status analysis, vehicle condition analysis, driving behavior analysis. The labels used are: basic data, maintenance, maintenance data, driving data, etc. of the vehicle. For

Table 33.3 Consumer tag sets

Label category	Label set	Information item	Type of data
Population attributes	User basic information	Name, age, gender	Basic information
	General life radius	Occupation, education	Basic information
	User preferences	Interest	Basic information
Owner attribute	Car length	Car information	Basic information
	Customer active state	Customer information	Basic information
	Customer type	Car information	Basic information
	Pit stop frequency	Driving habits information	Basic information
Value of customer	The user's own value		Value information
	Customer lifetime value		Value information
After-sales service	Satisfaction	After sale, feedback	Aftermarket information
	Loyalty		Aftermarket information
	Blacklist		Aftermarket information
Brand value	User opinion information	Website, intelligent terminal	Basic information
	User behavior information	Website, intelligent terminal	Basic information
Driving characteristics	Driving habits	Driving habits	Basic information
	Driving age type	Driving habits	Basic information
	Travel mileage	Driving mileage	Basic information
Marketing preferences	Activity preference	Website, intelligent terminal	Basic information
	Consumption characteristics	Website, intelligent terminal	Basic information
	Financial characteristics	Website, intelligent terminal	Basic information

example, we can use the car repair data, and each car portrait, for auto repair shop, manufacturer inventory adjustment, quality improvement and so on.

Accurate marketing module based on customer labels, from the customer form of life and characteristics of the two aspects of the customer market positioning, from the customer life cycle at all stages, according to the customer at different stages of the product and service demands and attention to the formation of preliminary Of the automotive customer market positioning, in order to establish a car customer base tag library, to enhance the quality of car brand service basis.

Predictive decision-making modules are designed to help companies achieve intelligent planning and support as a reference for strategic forecasting and decision making in strategic decision-making and analysis. For example, we can use the basic data and ownership of the vehicle, brand communication, brand crisis monitoring and management support, discovery of new markets and new trends, market forecasting and decision analysis support.

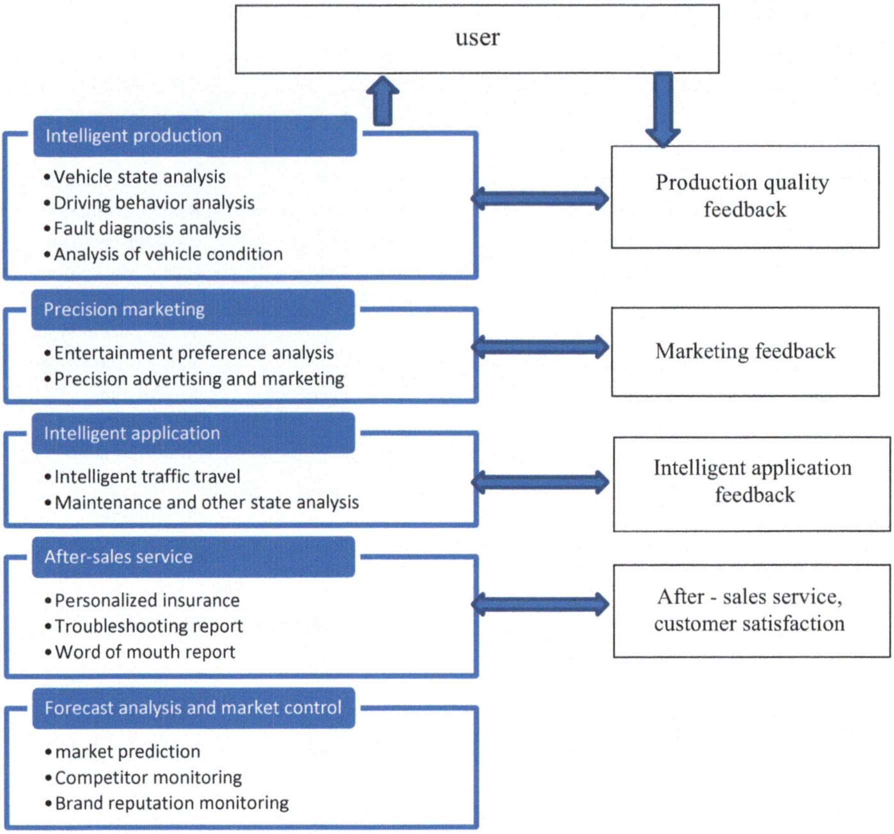

Fig. 33.6 Intelligent sales platform architecture

The results of data mining show the visualization of human-computer interaction visual display to the user. At present, there are more popular cosmic planets, tag clouds, cluster analysis and visualization techniques that show cluster members' distribution [3].

33.3 The Prospect and Suggestion of Intelligent Sales Platform in the Automobile Industry under Large Data Environment

Based on large data mining intelligent sales platform is an efficient marketing model that can be achieved in the accurate time, through accurate ways to provide customers with accurate service to meet the personalized service needs of consumers to help enterprises more accurate of the market positioning. The platform is to customers as the goal, the marketing concept and the perfect blend of modern science and technology together, is the inevitable trend of future business development. Suggestions on the platform:

33.3.1 Data Source Analysis to Be Comprehensive

First of all business needs to do research, assess the scale of operation of the system, the development of data warehouse structure and data sources. Review first-party data, such as product library, customer information, sales records, etc. to fully meet the demand. If the data is incomplete, develop third-party data supplements, such as the introduction of user social platform data.

33.3.2 Technical Advice, Flume as a Data Acquisition Channel, a Data Source Real-Time Transmission to the Hadoop Cluster Storage

Hadoop clusters include Zookeeper cluster, Hbase cluster, Storm cluster. Hive, Pig as ETL tools, providing real-time computing, offline computing two modes. For the real-time requirements of high business Xu Xiu, Redis and other Nosql program can be added as a supplement [12–15].

33.4 End

With the large data technology and machine learning technology mature, large data system based on existing data to predict the future behavior of customers. By forecasting models, salespeople can better understand customer needs. Combined with CRM forecasting model can be more in-depth interpretation of customer needs, predict when customers will be interested in what products. Combined with reptile tools, can reveal the customer needs the product, and other customers of the product evaluation. From price to quality to customer service, the monitoring and response of these information is essential. Large data support under the intelligent sales platform should be able to the company's Web site and the evaluation of the community together to ensure that the customer dynamic to make a timely response.

References

1. https://wenku.baidu.com/view/71ac244af111f18583d05a97.html
2. Hu, J. (2016). *Research on accurate marketing strategy of Chang'an automobile in large data*. Beijing: Central University for Nationalities.
3. Deng, Z., Liu, W., & Lu, Y. (2015). Research on content and solution of large data mining based on cloud computing. *Information Theory and Practice, 07*(020), 103–107.
4. Li, C., Zhang, X., & Jin, H. (2011). MapReduce: A new distributed parallel computing programming model. *Computer Engineering and Science, 33*(3), 129–135.
5. Li, J. (2010). *Research on map redize of cloud computing and some data mining algorithms*. Chengdu: University of Electronic Science and Technology.
6. Lu, J., Jin, J., & Song, A. (2011). Chengdu: University of electronic science and technology. *The Journal of Communication, 32*(7), 3–21.
7. Tuo, S. (2010). Research on cloud computing and cloud data storage technology. *Computer development and application, 23*(9), 1–3.
8. Todd, D. P., Yung, R. C., & Yoshimura, A. (2012). Using performance measurements to improve map reduce algorithms. *Procedia Computer Science, 9*, 1920–1929.
9. Liu, P., Wu, Z., & Hu, G. (2013). Big data – is undergoing profound changes. *ZTE Technologies, 19*(4), 1–7.
10. He, Y., Wang, W., & Xue, F. (2013). Research on massive data mining based on cloud computing. *Computer Technology and Development, 23*(2), 69–72.
11. Liu, S. (2016). *Analysis and application of automobile sales state based on data mining technology*. Nanjing: Southeast University.
12. Han, J. W., & Micheline, K. (2012). *Concept and technology of data mining*. Beijing: Mechanical Industry Press.
13. Wang, Y., Jin, X., & Cheng, X. (2013). Network big data: Current situation and prospect. *Journal of Computer Science, 36*(6), 1125–1138.
14. Li, J., & Liu, X. (2013). An important aspect of large data: Data availability. *Computer Research and Development, 50*(6), 1147–1162.
15. Lazer, R. D., Kenndy, R., King, G., & Vespignani, A. (2014). The parable of google flu: Traps in big data analysis. *Science, 343*(6176), 1203–1205.

Chapter 34
A Local Neighborhood Constraint Method for SIFT Features Matching

Qingliang Li, Lili Xu, Pengliang Zheng, and Fei He

Abstract For improving the accuracy of the SIFT matching algorithm with low time cost, this paper proposes a novel matching algorithm which is based on local neighborhood constraints, that is, SIFT matching feature is optimized by the local neighborhood constraint method in the SIFT algorithm. We optimize the matching results by using the information of SIFT feature descriptor and the relative position information of SIFT feature, then the final matching result obtained by RANSANC algorithm to filter the false matched pairs. The experimental results show that our method can improve the accuracy of the matching feature pairs without affecting the time cost.

Keywords Image matching · SIFT algorithm · Local neighborhood constraints

34.1 Introduction

Image matching is an analysis and processing technology that matches the image of the overlapping area and obtains the position relation between the images. However, due to differences between two matching images including translation, rotation, scale and perspective in practical application, therefore, there have many problems that need to be solved through the matching process. The predominant research tendency of image matching is focused on feature-based. According to [1–3], several representative feature-based matching algorithms have been introduced. In conclusion, SIFT algorithm is the most robust feature matching algorithm compares to other.

SIFT algorithm has been successfully applied in many fields [1–3], however, it has possible of prodigious improvement in matching accuracy in application. In

Q. Li · L. Xu · P. Zheng · F. He (✉)
Changchun University of Science and Technology, School of Computer Science and Technology, Changchun, China
e-mail: liqingliang@cust.edu.cn

M. Tavana, S. Patnaik (eds.), *Recent Developments in Data Science and Business Analytics*, Springer Proceedings in Business and Economics,
https://doi.org/10.1007/978-3-319-72745-5_34

Reference [4], in order to remove the unreliable matching points, the SIFT algorithm combined with RANSAC algorithm. It is unfortunately that the RANSAC algorithm can filter out most of the mismatched points, but it is not guarantee that all the error matching points can be eliminated. References [5, 6] proposed a bidirectional matching filtering strategy, but the bidirectional matching requires an artificial custom threshold. In Reference [7], the more color information added to the image, the more matched pairs will be filtered out, however, the image may not match better with less feature points.

This paper proposes a filtering algorithm which is based on local neighborhood constraints, filtering the false matched pairs by computing the number of other matched feature points in the local neighborhood of matched features. The proposed filtering algorithm not only uses the information of SIFT feature descriptor, but also uses the relative position information of the SIFT features, it could optimize the matching accuracy, and the final result obtained by adding RANSAC algorithm for filter the false matched features. Compared to other filtering algorithms, our method eliminates fewer features and remain more positive matched pairs, so that the accuracy is greatly improved.

34.2 SIFT Algorithm

The SIFT matching algorithm consists of four steps: the construction of Gaussian scale space, feature points detection, the generation of feature descriptor and feature matching [8, 9].

We choose the SIFT algorithm [10] for image feature extraction. Firstly, constructing image pyramid by Gaussian kernel function, then, using Difference-of-Gaussian (DoG) detector, the key points of image are detected. Secondly, for accurate location, scale, and ratio of principal curvatures, model fitting and Hessian are performed to the nearby data of key points. As a result, the low contrast key points (sensitive to noise or are poorly localized along an edge) are eliminated. Calculating the magnitude and orientation for the gradient histogram for every pixel, it is a remaining key point as heart point, in a neighboring region. The peak in gradient histogram corresponds to dominant orientations. Finally, using the gradient direction distribution characteristic of neighborhood pixels of key points, the direction parameters of each key point can be computed. a set of orientation histograms is created on 4×4 pixel neighborhoods with 8 bins. What's more, these histograms are computed from magnitude and orientation values of samples in a 16×16 region around the key points. Then, concatenating all the 16 histograms for 128-dimension feature vector. In conclusion, the extracted SIFT features are invariant to modification of translation, rotation and scale.

34.3 Filtering Algorithm Based on Local Neighborhood Constraints

In Sect. 34.2, we obtained the SIFT matched pairs, then we adopts the local neighborhood constraints method to optimize the SIFT feature matching result. Figure 34.1 shows the process of the local neighborhood constraint method. Figure 34.1a denotes the five matching feature pairs of two images. The red circular area in Fig. 34.1b shows that the matching feature (a1,b1) is used as a reference pair, and the local neighborhood area is obtained based on its scale parameter (black arrow). Figure 34.1c shows that the local neighborhood matched pair is defined as the correlation matching feature, and acts on the reference pair (a1,b1). In Fig. 34.1, (a3,b3) referrers to the point in local neighborhood of reference pair, regarded as the relevant matched pairs, other matched pairs (a1,b1), (a4,b4), (a5,b5)) are regarded as irrelevant matched pairs.

In Fig. 34.1, (a3,b3) referrers to the point in local neighborhood of reference pair, regarded as the relevant matched pairs, other matched pairs (a1,b1), (a4,b4), (a5,b5)) are regarded as irrelevant matched pairs.

$$D_{q_i} = \sigma_{constant} * scl_i \quad (34.1)$$

$$D_{d_i} = \sigma_{constant} * scl'_i \quad (34.2)$$

First, select a matching feature pair (q_i, d_i) as a reference pair. In order to verify that the matching is correct, the relevant matching feature is obtained by the local neighborhood of reference pair. The local neighborhood is defined as below:

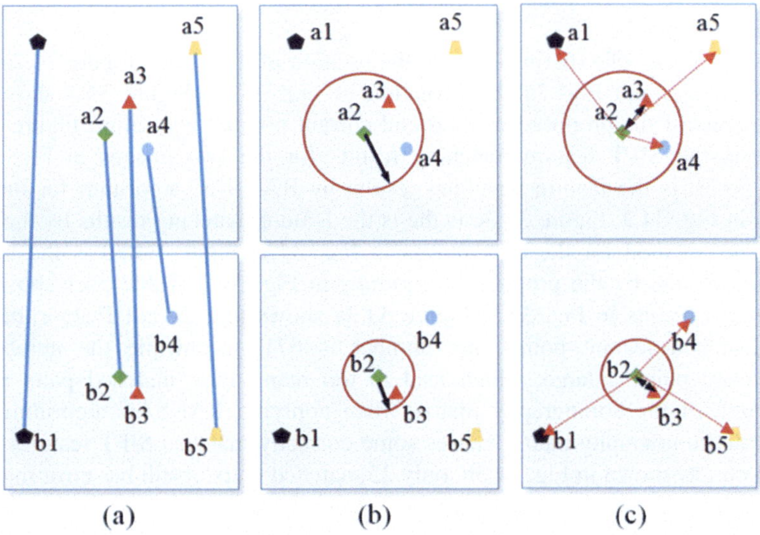

Fig. 34.1 A diagram of local neighborhood constraints method

In the formula, scl_i, scl'_i separately represent the scale parameters of the matching pair (q_i, d_i). The size of $\sigma_{constant}$ decides the scope of the local neighborhood of the reference pair, it is set to 10 in this experiment. Then, the relevant matched pairs in the local neighborhood are remained. The relative matched pairs are defined as below:

$$CorMatch_{(q_i,d_i)} = \left\{ (q_j, d_j) \middle| \begin{array}{l} (q_j, d_j) \in M(Q, D) \\ dist(q_i, q_j) < D_{q_i} \\ dist(d_i, d_j) < D_{d_i} \end{array} \right\} \quad (34.3)$$

In the above formula, *dist*(,) represent the Euclidean distance of two feature points in the spatial position of the image. Finally, the local neighborhood constraint value of the reference pair is calculated, that is, the ability to spatial constraint.

If the *CorMatch* of matched pairs was 0, it is regarded as an error matched feature, and filtered its matched pair, or vice versa. In the optimized matched pair, based on the RANSAC algorithm, the position information of the small matching feature pair is randomly selected as the observation data. And the transformation matrix is calculated to obtain the model. In final, the matched feature pairs which dissatisfy with the transformation matrix are eliminated, while the final matched pairs are obtained.

34.4 Experimental Results and Analysis

In order to verify the adaptability of the algorithm to image rotation and image scaling, the two experiments are separately verified. The experimental image size is 1334×1001 pixels.

The first group of experiments tests the rotation of the image. Figure 34.2a is an original image, and Fig. 34.2b is rotation of Fig. 34.2a. Figure 34.3 shows the comparisons of the proposed method and current popular algorithm. Figure 34.3a represents the SIFT feature matching results for the two images in Fig. 34.2. Figure 34.3b is the feature matching results by RANSAC algorithm for the two images in Fig. 34.2. Figure 34.3c is the is the feature matching results by the local constraints method for the two images in Fig. 34.2. Figure 34.3d is the feature matching results by the proposed algorithms in Fig. 34.2. Table 34.1 shows the quantitative results in Fig. 34.3. Figure 34.3a shows that the SIFT algorithm has more matched feature points, the number is 671, meanwhile the number of mismatched pairs is large, which lead to too many false matched pairs in the calculation of the homography matrix when applying RANSAC algorithm. The inaccurate homography matrix causes some correctly matched SIFT features to be filtered out, as shown in Fig. 34.3b, only 25 matched pairs, it still has error matched pairs are left. Figure 34.3c shows that the SIFT algorithm based on local neighborhood constraints can optimize the matching results in the phase matching the SIFT feature (352 points), which lead to more correct matched pairs when calculating the

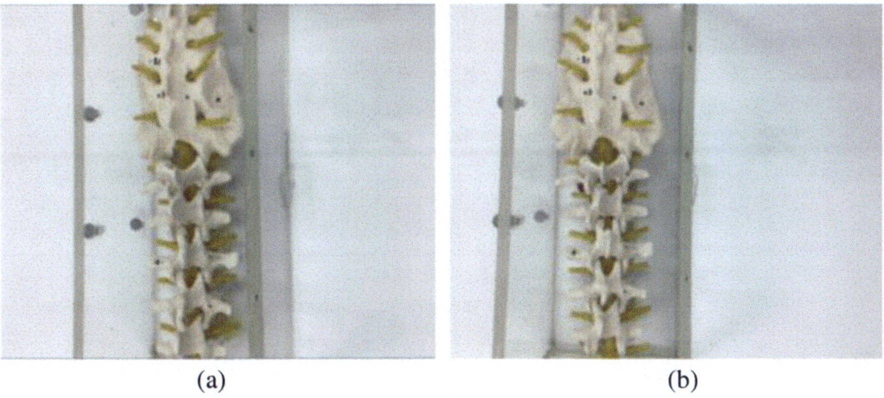

Fig. 34.2 Image Rotation of Vertebra Modal (a) Original Image (b) Rotation Image

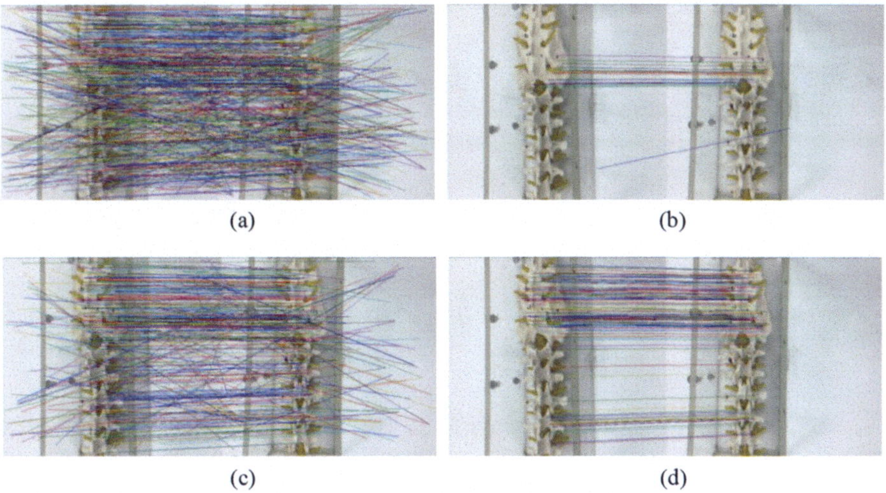

Fig. 34.3 Matched features of Vertebra Model. (a) Based on SIFT algorithm; (b) Based on RANSAC algorithm; (c) SIFT algorithm based on Local Neighborhood Constraints method; (d) Based on our method in this paper

Table 34.1 Quantitative results of the above algorithms in Fig. 34.3

Matched algorithms	The number of total matched points	The number of filtered points	Matched time/s
SIFT	671	–	2.3402
RANSAC	25	646	2.3487
Local + SIFT	352	–	2.7041
Proposed algorithm	148	204	2.7793

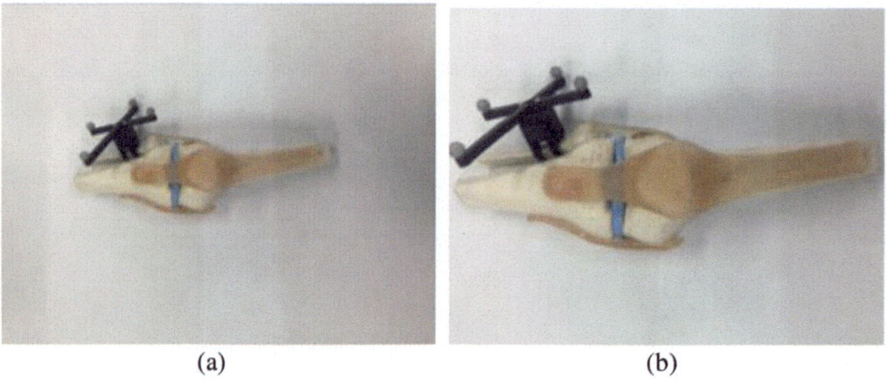

Fig. 34.4 Image size of Knee-joint Modal (**a**) Original image (**b**) Enlarged image

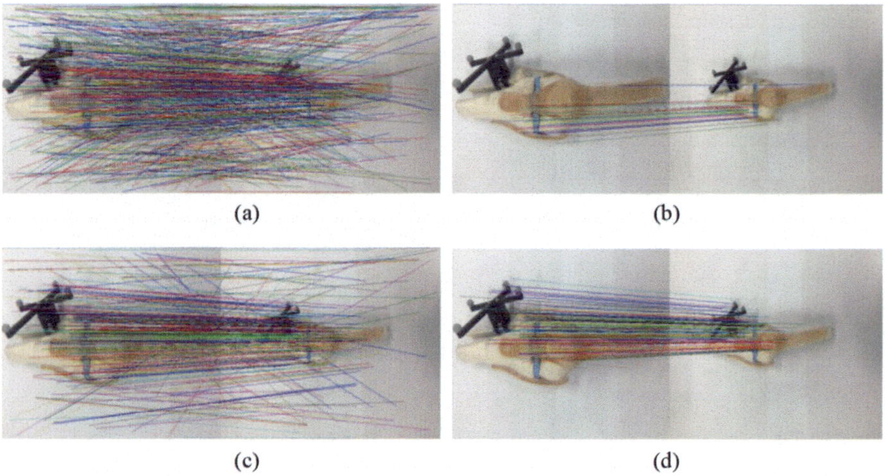

Fig. 34.5 Matched features of Knee-joint Model. (**a**) Based on SIFT algorithm; (**b**) Based on RANSAC algorithm; (**c**) SIFT algorithm based on Local Neighborhood Constraints method; (**d**) Based on our method in this paper

homography matrix when using RANSAC algorithm. And It can also obtain more accurate homography, so that the remained matched pairs after filtering is more accurate, as shown in Fig. 34.3d, the matched features which are all positive is 148. In computing time, our method takes much time, it is only 0.07–0.3 s more than other algorithms, however the accuracy is greatly improved.

The second group of experiments tests the scaling of the image. Figure 34.4a is an original image, and Fig. 34.4b is enlarged by Fig. 34.4a. Figure 34.5a is represents the SIFT feature matching results for the two images in Fig. 34.4. Figure 34.5b is the

Table 34.2 Quantitative results of the above algorithms in Fig. 34.5

Matched algorithms	The number of total matched points	The number of filtered points	Matched time/s
SIFT	641	–	2.4392
RANSAC	32	609	2.5041
Local + SIFT	293	–	2.6512
Proposed algorithm	129	164	2.7023

feature matching results by RANSAC algorithm for the two images in Fig. 34.4. Figure 34.5c is the is the feature matching results by the local constraints method for the two images in Fig. 34.4. Figure 34.5d is the feature matching results by the proposed algorithms in Fig. 34.4. Table 34.2 shows the quantitative results in Fig. 34.5. From the experimental result, it can get the same conclusion with the first group of experiments.

From the experimental results, the matching results of our method are more accurately under the almost same time consuming in computation compared with other algorithms. But in this paper our method only depends on the matched pairs of the SIFT feature descriptor. If the SIFT feature matched points are less, the accuracy of our algorithm will not be significantly improved.

34.5 Conclusion

The proposed algorithm in this paper is an improved filtering algorithm, which is based on the SIFT algorithm. In SIFT algorithm, the algorithm adopts the local neighborhood constraint method to optimize the SIFT matching feature. The experimental results show that the proposed algorithm can improve the accuracy of the matching feature pairs without affecting the time cost. However, this paper only optimizes from the matched SIFT feature pairs. If the SIFT feature matched points are less, the accuracy of this algorithm will not be significantly improved.

Acknowledgements This work was supported by the Science & Technology Development Program of Jilin Province, China (Nos. 20140101182JC, 20150101060JC, 20150307030GX, 2015Y059 and 20160204048GX), and by the International Science and Technology Cooperation Program of China (Nos. 20140101182JC, 20150101060JC, 20150307030GX and 20160204048GX).

References

1. Vourvoulakis, J., Kalomiros, J., & Lygouras, J. (2016). FPGA accelerator for real-time SIFT matching with RANSAC support. *Microprocessors and Microsystems, 49*, 105–116.

2. Chen, Y., & Shang, L. (2016). Improved SIFT image registration algorithm on characteristic statistical distributions and consistency constraint. *Optik - International Journal for Light and Electron Optics, 127*(2), 900–911.
3. Jin, R., & Kim, J. (2015). Tracking feature extraction techniques with improved SIFT for video identification. *Multimedia Tools & Applications, 76*, 1–10.
4. Wei, W., Hong, J., & Tang, Y. (2008). *Image matching for geomorphic measurement based on SIFT and RANSAC methods*. International Conference on Computer Science and Software Engineering. IEEE Computer Society (pp.317–320).
5. Qian, S., & Zhu, J. (2007). Improved SIFT-based bidirectional image matching algorithm. *Mechanical Science & Technology for Aerospace Engineering, 26*(9), 1179–1182.
6. Huo, C. L., Zhou, Z. X., Liu, Q. S., et al. (2007). Remote sensing image registration based on SIFT and the distance between generalized tight pair-wise prototypes. *Remote Sensing Technology & Application, 22*(4), 524–530.
7. Zhang, R. (2008). Study on Color Image Registration Technique based on CSIFT. *Acta Optica Sinica, 28*(11), 2097–2103.
8. Jiang, Y., & Wang, J. (2010). *Research on multi-source remote sensing image registration base on SIFT algorithm of window segmentation*. International Conference on Wireless Communications Networking and Mobile Computing. IEEE (pp. 1–4).
9. Yi, Z., Zhiguo, C., & Yang, X. (2008). Multi-spectral remote image registration based on SIFT. *Electronics Letters, 44*(2), 107–108.
10. Lowe, D. G. (2004). Distinctive image features form scale-invariant key-points. *International Journal of Computer Vision, 60*(2), 91–110.

Chapter 35
A Wine Consumption Prediction Model Based on L-DAGLSSVM

Xiao Wang, Sijie Lu, and Zhijian Zhou

Abstract With the increasing demand of wine consumption, the marketing of wine consumption is expanding. In this paper, we do a research about the decision behavior of Chinese wine consumers in order to grasp the consumption demand of wine at different prices better. We acquire 774 questionnaires finally, and the 528 of which are valid. According to the consumption prices, we divide wine consumers into three types. Then we propose a multi-class classification method named L-DAGLSSVM for constructing prediction model of consumption types, which is based on LDA and the directed acyclic graph least squares support vector machine (DAGLSSVM). The numerical experiments demonstrate that our algorithm gains better performance compared with other algorithms. And the prediction model plays an important role in commercial fields that it can provide an effective reference for the wine production, purchase and marketing strategies etc.

Keywords LSSVM · The decision directed acyclic graph (DDAG) · LDA · Prediction model of consumption types

35.1 Introduction

Wine has a history of more than 2000 years in China, and it is a kind of alcoholic beverage which is healthy, nutritious and elegant. In recent years, with the improvement of the Chinese income and the spread of the wine culture, the wine has become more and more popular. The marketing of wine consumption is expanding, while the competition in wine marketing is becoming more and more intense and the wine marketing is facing huge opportunities and challenges. In order to seize the opportunity better and grasp the marketing demand more accurately, marketers and enterprises pay more attention to the prediction of consumption types. Therefore, taking a quantitative analysis with the data about the wine consumption is necessary,

X. Wang · S. Lu · Z. Zhou (✉)
College of Science, China Agricultural University, Beijing, China

M. Tavana, S. Patnaik (eds.), *Recent Developments in Data Science and Business Analytics*, Springer Proceedings in Business and Economics,
https://doi.org/10.1007/978-3-319-72745-5_35

which can not only realize the value of the data, but also improve the prediction accuracy of consumption types.

The support vector machine (SVM) [1, 2] is proposed by Vapnik. It is a famous method for classification problems. A large number of improvements to SVM have been proposed in the past few years, such as LSSVM [3] whose solution is simple by solving a set of linear equations. Now, they have been successfully applied in various areas, for example, disease detection, speech recognition and etc. SVMs are originally designed for binary classification, however, in our real life, multi-class classification problems are more commonly seen. As for multi-class problems, we usually build a multi-class function through combining several two-class classifiers. Generally, the split strategies include one-verse-one (OVO) [4] and one-verse-all (OVA) [2, 5, 6] algorithms. But there may exist inseparable areas in the above strategies. DDAG introduced by Platt et al. [7] avoids above problem. The DAGLSSVM [8] which combines DDAG and LSSVM, is an effective solution for multi-class problem. Meanwhile LDA [9] is an excellent method for dimension reducing and has a good performance for many classification problems. So in our paper, we propose a multi-class classification approach named L-DAGLSSVM which is based on LDA and DAGLSSVM. We collect relevant data about wine consumption through an investigation of the Chinese wine consumers' decision-making behavior. According to the collected data, we do quantitative analysis and then construct prediction model of the consumption types based on L-DAGLSSVM.

35.2 The Basic Theories

35.2.1 DAGLSSVM

The DDAG is a learning architecture based on OVO algorithm. Assuming a K multi-class classification problem, the training set is marked as $\mathbf{X}_1$, $\mathbf{X}_2$, $\cdots$, $\mathbf{X}_K$, respectively and testing set is marked as $\mathbf{X}_t$. It is similar to the OVO algorithm by solving $K(K\text{-}1)/2$ binary functions at training step. Then it predicts $\mathbf{X}_t$ by a rooted binary DAG which includes K leaves and $K(K\text{-}1)/2$ internal nodes at testing step. Every node presents a binary function of every two classes between ith and jth. For $\mathbf{X}_t$, evaluate the node's binary function starting from root node. Next it turns to either right or left through the output. Finally, we can predict the $\mathbf{X}_t$ by a leaf node which we acquire from the path. DAGLSSVM is a multi-class algorithm combining learning architecture DDAG with binary classification algorithm LSSVM. First, we introduce LSSVM in the following:

Consider a classification problem with the dataset: $S = \{(\mathbf{x}_1, y_1), \cdots, (\mathbf{x}_l, y_l)\}$ where $\mathbf{x}_i \in R^n$, $y_i \in \{-1, 1\}$, $i = 1, 2, \cdots, l$. We finds a separating hyper-plane: $\mathbf{w}^T\mathbf{x} + b = 0$, whose original optimal problem is described as:

$$\begin{array}{lll} \min_{\xi, \mathbf{w}, b} & \| \mathbf{w} \|^2/2 + C\sum_{i=1}^{l} \xi_i^2 & \\ s.t. & y_i(\mathbf{w}^T\mathbf{x}_i + b) = 1 - \xi_i, & i = 1, \cdots, l, \end{array} \tag{35.1}$$

where ξ_i, $(i = 1, \cdots, l)$ are slack variables. $C > 0$ is a parameter chosen a priori. And then DAGLSSVM can be presented in the following Algorithm 35.1.

Algorithm 35.1: DAGLSSVM

Input: $\mathbf{X}_1, \mathbf{X}_2, \cdots, \mathbf{X}_K, \mathbf{X}_t$

Training phase:

Step 1: Train a binary classifier by LSSVM between every two classes $\mathbf{X}_i$ and $\mathbf{X}_j$, where $i < j$, then we can obtain $K(K\text{-}1)/2$ binary classifiers totally.

Step 2: Construct a learning architecture (prediction model) according to the DDAG based on the classifiers obtained by above training step.

Testing phase:

Step 3: Take $\mathbf{X}_t$ into prediction model above, and the outputs are the samples' labels.

35.2.2 *Linear Discriminant Analysis (LDA)*

LDA is a useful technique for dimension reduction and applied for various classification problems. Assume that there are K classes, marked as $\mathbf{X}_1, \mathbf{X}_2, \cdots, \mathbf{X}_K$, respectively, and each class has M samples. The objective function is maximizing the ratio of S_b against S_w and is described as follows:

$$\arg\max_{\mathbf{W}} \frac{|\mathbf{W}S_b\mathbf{W}^T|}{|\mathbf{W}S_w\mathbf{W}^T|} \tag{35.2}$$

where $S_b = \sum_{k=1}^{K}\left(\bar{\mathbf{X}}^k - \mathbf{X}^k\right)\left(\bar{\mathbf{X}}^k - \mathbf{X}^k\right)^T, S_w = \sum_{k=1}^{K}\sum_{m=1}^{M}\left(\mathbf{X}_m^k - \bar{\mathbf{X}}_m^k\right)\left(\mathbf{X}_m^k - \bar{\mathbf{X}}_m^k\right)^T$, $\bar{\mathbf{X}}^k = (1/M)\sum_{m=1}^{M}\mathbf{X}_m^k$, and $\bar{\mathbf{X}} = (1/KM)\sum_{k=1}^{K}\sum_{m=1}^{M}\mathbf{X}_m^k$. We resolve the problem by the method of singular value decomposition (SVD) and get the $\mathbf{W} = S_w^{-1}S_b$, which can be eigenvectors corresponding greatest eigenvalues of $S_w^{-1}S_b$.

35.3 L-DAGLSSVM

In this article, we introduced a new algorithm L-DAGLSSVM for multi-class classification which is based on LDA and DAGLSSVM. The L-DAGLSSVM is shown in the following Algorithm 35.2.

Algorithm 35.2: L-DAGLSSVM

Input: $\mathbf{X}_1, \mathbf{X}_2, \cdots, \mathbf{X}_K, \mathbf{X}_t$.

Data preprocessing:

Step 1: Normalize the data and reduce the dimensions of the samples by LDA for reducing the computing complexity and improve the classification accuracy. Meanwhile save the linear transformation matrix W. The samples' dimensions are reduced from 20 to 2 in this paper.

Training phase:

Step 2: Train a binary classifier by LSSVM between every two classes $\mathbf{X}_i$ and$\mathbf{X}_j$, where $i < j$, then we can obtain $K(K\text{-}1)/2$ binary classifiers totally.

Step 3: Construct a learning architecture (prediction model) according to the DDAG based on the classifiers obtained by above training step.

Testing phase:

Step 4: $\mathbf{X}_t$ is dealt with LDA the same as step1 by the linear transformation matrix **W**. Then take the data into prediction model above and the outputs are the samples' labels.

35.4 Prediction Model

35.4.1 Questionnaire and Data

The questionnaire is conducted for investigating the basic situation of wine consumption in China. The main contents of the questionnaire include the following items: (1) the basic situation of consumers, including gender, age, marital status, income, education and occupation. (2) the annual consumption, the purchase purpose, the brand (domestic or imported), the purchase place, etc. (3) screening problem: the frequency of consumer' purchase, whose main purpose is to eliminate the consumers who do not meet the requirements.

We acquire 774 questionnaires finally, and 528 of which are valid in this survey. We regard each questionnaire as a sample which contains 20 characteristics. According to the purchased price of wine consumers in real life, the consumers are separated into basic, middle, and high levels. (1) The basic level: these consumers are beginning to experience wine which is a kind of low and middle-end products and consumption price is in the range of 200 RMB. (2) The middle level: these consumers drink wine frequently and consumption price is between 200 and 500 RMB. (3) The high level: these consumers are professional drinkers with a high ability to payment and consumption price is 500 RMB and above. After statistics, there are 344 people at the basic level, 116 people at the middle level and 68 people at high level.

Table 35.1 the testing accuracy (%) on algorithms with the change of *C*

	DAGSVM		L-DAGSVM		DAGLSSVM		L-DAGLSSVM	
	Acc (%)	Std	Acc (%)	Std	Acc (%)	Std	Acc (%)	Std
$C = 2^{-4}$	65.5	5.58	65.19	1.58	65.96	1.75	66.54	2.39
$C = 2^{-3}$	61.92	11.27	65.19	1.85	63.08	5.71	66.35	2.72
$C = 2^{-2}$	62.12	9.04	64.04	8.03	65.96	3.51	67.12	5.33
$C = 2^{-1}$	55.77	5.18	63.65	6.85	63.27	4.48	65.00	4.64
$C = 2^{0}$	55.00	8.11	57.88	5.83	64.23	6.21	64.62	3.15
$C = 2^{1}$	53.08	16.10	53.65	14.90	59.62	10.53	61.92	6.72
$C = 2^{2}$	57.50	13.23	60.00	8.15	63.46	4.71	65.77	4.06
$C = 2^{3}$	54.04	10.23	59.04	7.59	63.27	2.30	64.81	2.60
$C = 2^{4}$	60.96	6.07	63.46	7.16	**69.23**	2.39	67.88	2.11
$C = 2^{5}$	**67.50**	3.15	**67.12**	3.81	67.50	3.79	69.23	3.72
$C = 2^{6}$	51.54	9.34	58.08	7.40	62.69	5.01	**70.19**	4.66
$C = 2^{7}$	51.92	12.48	57.88	10.84	59.81	7.21	63.08	5.71
$C = 2^{8}$	62.69	10.95	66.15	5.15	65.38	3.85	65.00	3.01
$C = 2^{9}$	55.77	17.05	64.23	5.37	63.46	10.06	67.69	1.75
$C = 2^{10}$	48.65	16.01	51.92	16.97	59.23	11.02	61.92	8.56

35.4.2 Experiments and Discussion

We construct the prediction model by L-DAGLSSVM on the wine consumption data that we acquire from questionnaires. Then we conduct experiments on the wine consumption data to verify the validity of prediction model. All the experiments are operated in MATLAB. We take 80% of the dataset as the training samples, and 20% of the dataset as the testing samples. Parameters are obtained from $\{2^i \mid i = -4, -3, \ldots, 10\}$ with 5-fold cross-validation method. We also take DAGSVM, DAGLSSVM and L-DAGSVM into experiments. The performance comparison with the change of *C* is in Table 35.1. Where "Acc" presents the average value of the five test results and "Std" presents standard deviation of five testing results. For the sake of comparing the performance of four models more intuitively, we plot results as a line chart shown in Fig. 35.1.

From the perspective of prediction accuracies, the algorithm L-DAGLSSVM performs the best among four models on our wine consumption data. Obviously, we can see that DAGLSSVM has a better performance compared with DAGSVM and L-DAGSVM. What's more, the algorithm L-DAGLSSVM not only inherits DAGLSSVM's excellent performance, but also improve the performance because of the feature extraction by LDA. What's more, our method is more stable compared with other algorithms. Therefore, the algorithm we proposed is more suitable to the prediction model and performs better on wine consumption data.

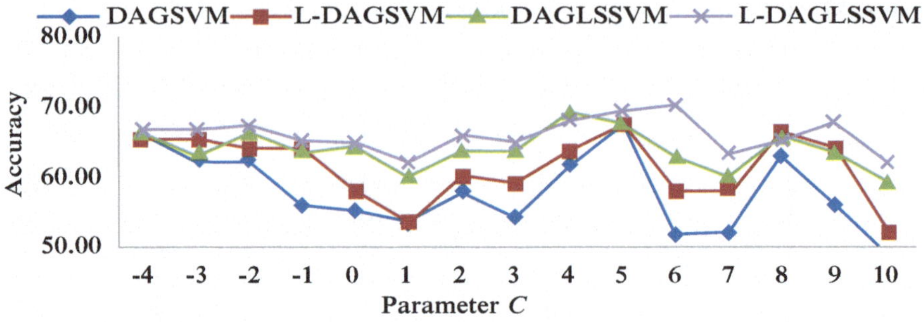

Fig. 35.1 Testing accuracies (%) on algorithms with the change of C

35.5 Conclusions

In this article, we propose a multi-class algorithm L-DAGLSSVM on wine consumption data. We first use a feature extraction technique which is linear discriminant analysis for dataset's dimensionality reduction and then use a prediction model by the acyclic graph least squares support vector machine algorithm. Finally we analyze and compare our prediction model with other algorithms on the wine consumption data, and we find that our algorithm performances better. Therefore marketers and enterprises can predict the types of wine consumers by the prediction model, which can provide an effective reference for the wine production, purchase and marketing strategies, etc. So, it has great practical significance and is a potential application for constructing the prediction model in commercial fields.

References

1. Naiyang, D., & Yingjie, T. (2009). *Support vector machine: theory, algorithm and development*. Beijing: Science Press (in Chinese).
2. Cristianini, N., & Shawe-Taylor, J. (2004). *An introduction to support vector machines and other kernel-based learning methods*. Beijing: Publishing House of Electronics Industry.
3. Suykens, J. A. K., & Vandewalle, J. (1999). Least squares support vector machine classifiers. *Neural Processing Letters, 9*(3), 293–300.
4. Kressel, B. U. (1999). *Pairwise classification and support vector machines, advances in kernel methods: Support vector learning*. Cambridge: MIT Press.
5. Hsu, C., & Lin, C. A. (2002). Comparison of methods for multiclass support vector machines. *IEEE Transactions on Neural Networks, 13*(2), 415–425.
6. Yang, X., Yu, Q., He, L., et al. (2013). The one-against-all partition based binary tree support vector machine algorithms for multi-class classification. *Neural Computing, 113*, 1–7.
7. Platt, J. C., Cristianini, N., & Shawe-Taylor, J. (2000). Large margin DAGs for multiclass classification. *Advances in Neural Information Processing Systems, 12*(3), 547–553.
8. Wu, H. (2015). Least squares DAGSVM for multiclass classification. *Journal of Information & Computational Science, 12*(18), 6863–6871.
9. Chen, L. F., Liao, H. Y. M., Ko, M. T., et al. (2000). A new LDA-based face recognition system which can solve the small sample size problem. *Pattern Recognition, 33*(10), 1713–1726.

Chapter 36
Fuzzy Control and Network System Design for Time Series Prediction Model

X. L. Lu, H. X. Wang, and Z. X. Zhao

Abstract This paper proposed and developed a set of fuzzy time series prediction model FTSFM (Fuzzy Time Series Forecasting Model) based on the historical data, the concepts of fuzzy number function and inverse fuzzy number function and predictive function, which the basic theory of FTSFM was initially established. The general elements of FTSFM and the prediction function are FTSFM (μ). FTSFM (0.0004) is one of the commonly used prediction models of FTSFM. Based on the forecast of tourism revenue of Sanya city in 2006~2014, this paper introduces the whole process of the application of FTSFM (0.0004). FTSFM (0.0004) provides a new way of thinking for the research of time series prediction.

Keywords Fuzzy time series forecasting model · FTSFM (0.0004) · Inverse fuzzy number · Predictive value · Tourism income

36.1 Introduction

In 1965, Zadeh [1] proposed the landmark Fuzzy set theory, which became an indispensable and powerful tool in the study of uncertainty problem. Song [2, 3] used the Fuzzy set theory to create the first fuzzy time series forecasting model in 1993 and 1994, and studied the classic case (the number of registration forecasting problems from the University of Alabama in 1971~1992). Saxena [4] invent the high precision forecasting model in 2012, it is based on the fuzzy time series forecasting model of inverse fuzzy number. In the study of the classic case, it get the average forecasting error rate AFER = 0.3406% and mean square error MSE = 9169, which is an unprecedented forecasting accuracy. Proposed the definition 3 of fuzzy time series forecasting model with high forecasting accuracy by applying these data [5–8]. Therefore, the forecasting model created by Saxenais the first fuzzy time series

X. L. Lu (✉) · H. X. Wang · Z. X. Zhao
Hainan Tropical Ocean University, Sanya, Hainan, China
e-mail: vingo0314@163.com

M. Tavana, S. Patnaik (eds.), *Recent Developments in Data Science and Business Analytics*, Springer Proceedings in Business and Economics,
https://doi.org/10.1007/978-3-319-72745-5_36

forecasting model with high forecasting accuracy [2]. Wang and other people [9] improve the forecasting model, and establish several new fuzzy time series forecasting model in 2014. Based on these concepts, the concepts of fuzzy number function, inverse fuzzy number function and predictive function are presented, and the convergence theorem 2 is proved. Therefore, develop a set of fuzzy time series prediction model (FTSFM). The proof of theorem 2 marks the basic theory of FTSFM. The general element of FTSFM is FTSFM (μ), and it is also the forecasting formula of the model [10, 11].

Although the name of FTSFM is a set of fuzzy time series prediction model, it is actually a set of time series prediction model [12, 13]. FTSFM (0.0004) is one of the commonly used prediction models of FTSFM. Because AFER = 0.000006% and MSE = 0.095238 are obtained in the study of classical cases, so it is a fuzzy time series forecasting model with high prediction Accuracy [14, 15]. By applying the prediction model to study the prediction of tourism revenue in Sanya city 2006~2014 years, it can achieve a very high prediction accuracy of AFER = 0.0086% and MSE = 0.000238. When take the 2014 as unknown years using FTSFM (0.0004) and FTSFM (3-4-5) prediction rules to predict unknown of Sanya City 2014 years of tourism income, the prediction value and the real value of the prediction error rate of the absolute value of the first ordination is H_2 predictive value, prediction error rate of the vast of value is only 0.0408%, prediction accuracy is very high. Therefore, when predict the forecast value of tourism revenue in Sanya city in 2015, the forecast error rate should be expected to be Good [16–18].

The structure of this paper: the second section introduces the basic concept; the third section establishes a fuzzy time series prediction model of the set FTSFM; In the fourth section, study the classic case "registration number prediction problem", which proves that the commonly used model FTSFM (0.0004) is a prediction model with high prediction accuracy; The sixth section is the application of FTSFM (0.0004) to forecast the whole process of the application steps of the tourism revenue in Sanya, including the simulation of historical data and the prediction of the data for unknown years; The seventh section is the concluding remarks.

36.2 Application Steps of Prediction Model Ftsfm (μ)

Assuming there is a time series prediction problem. For the determination of membership degree $\mu \in (0,1)$, using the prediction formula FTSFM (μ) the steps of time series analysis are as follows:

Step 1: establish the table of historical data as in Table 36.1;

Step 2: establish the historical data field H = {H1971 = 13055..., H1992 = 18876};

Step 3: establish prediction formula FTSFM (0.0004), Fi is the predictive value of t years historical data, Hi−1 is i−1 years historical data, Hi is t years historical data, 0.0004 is (historical data Hi−1) degree of membership.

$$F_i = (0.0004 + 1)/(0.0004/H_{i-1} + 1/H_i) \quad (36.1)$$

Table 36.1 application of FTSFM (0.0004) to Predict the history registration number of Alabama university

Year	H_i	F_i	F_i-H_i	$(F_i-H_i)^2$	$\lvert F_i-H_i\rvert/H_i$
1971	13,055	–	–	–	–
1972	13,563	13,563	0	0	0.000000
1973	13,867	13,867	0	0	0.000000
1974	14,696	14,696	0	0	0.000000
1975	15,460	15,460	0	0	0.000000
1976	15,311	15,311	0	0	0.000000
1977	15,603	15,603	0	0	0.000000
1978	15,861	15,860	0	0	0.000000
1979	16,807	16,807	0	0	0.000000
1980	16,919	16,919	0	0	0.000000
1981	16,388	16,388	0	0	0.000000
1982	15,433	15,434	1	1	0.000065
1983	15,497	15,497	0	0	0.000000
1984	15,145	15,145	0	0	0.000000
1985	15,163	15,163	0	0	0.000000
1986	15,984	15,984	0	0	0.000000
1987	16,859	16,859	0	0	0.000000
1988	18,150	18,149	−1	1	0.000055
1989	18,970	18,970	0	0	0.000000
1990	19,328	19,328	0	0	0.000000
1991	19,337	19,337	0	0	0.000000
1992	18,876	18,876	0	0	0.000000
AFER		0.3406%	3.11%	0.5701%	1.02%
MSE		9169	407,507	21,575	41,426

Step 4: Application of FTSFM (0.0004) to study the predictive value of historical data; use formula (36.1) to calculate the value in 1972~1992, and fill in Table 36.1.

Step 5; establish FTSFM (μ) prediction rules for predicting the unknown year data in sanya City;

Step 6: apply FTSFM (μ) to investigate the predictive value of unknown year data.

Because the formula FTSFM (μ) can be used for time series analysis, the prediction formula FTSFM (μ) is actually a time series forecasting model. That is to say FTSFM (μ) not only represents a time series forecasting model, but also indicates that it's prediction formula. The application of the formula FTSFM (μ) is the application of the time series prediction model FTSFM (μ) steps.

36.2.1 FTSFM Represents a Collection of Time Series Prediction Models

FTSFM is a set of time series forecasting model, but according to the theorem 2 and theorem 1, time series forecasting model FTSFM is based on the fuzzy inverse model to forecast the time series of the number of sets, so dubbed "fuzzy" word.

It can be seen from the Table 36.1, we can get AFER = 0.000006% < 0.3406%, and MSF = 0.095238 < 9169, so that according to the definition of 3 to know that FTSFM (0.0004) is a fuzzy time series forecasting model with high prediction accuracy.

36.3 Application of FTSFM (0.0004) to Study the Forecast of Tourism Revenue in Sanya City

This section use model FTSFM (0.0004) to study the forecast of tourism revenue in Sanya City in 2006–2014 years. Through this case, this paper introduces the application of FTSFM (0.0004) to predict the forecast value of unknown tourism revenue in the whole process.

Step 1. establish the Sanya city tourism income historical data field H = {H2006 = 65.40,H2007 = 80.11,. . ., H2013 = 233.33,H2014 = 269.73}.

Step 2. Set up FTSFM (0.0004) forecast formula (1), Ft is the forecast value of t years of tourism revenue, Ht−1 is t−1 years of tourism revenue, Ht is t years of tourism revenue, μ=0.0004 is (tourism income Ht−1) membership degree.

Step 3. Application of FTSFM (0.0004) to study the tourism revenue of Sanya city in 2006~2015 to calculate the Sanya city of tourism revenue forecast and inspection, fill in Table 36.2. From the Table 36.2 shows that the average prediction error rate AFER = 0.0086%, mean square error MSE = 0.000238, indicates that the prediction accuracy is very high.

Table 36.2 Using FTSFM (0.0004) to predict the tourism revenue of Sanya 2006~2014

Year	H_i	F_i	F_i-H_i	$(F_i-H_i)^2$	$\lvert F_i-H_i\rvert/H_i$
2006	65.4	–	–	–	–
2007	80.11	80.1	−0.01	0.0001	0.000125
2008	91.05	91.05	0	0	0
2009	103.77	103.76	−0.01	0.0001	0.000096
2010	139.64	139.62	−0.02	0.0004	0.000143
2011	160.71	160.7	−0.01	0.0001	0.000062
2012	192.22	192.2	−0.02	0.0004	0.000104
2013	233.33	233.31	−0.02	0.0004	0.000086
2014	269.73	269.71	−0.02	0.0004	0.000074
AFER					**0.0086%**
MSE				**0.000238**	

Step 4. Application of FTSFM (0.0004) to predict the tourism revenue of Sanya city in 2014~2015. Because the 2015 tourism revenue H2015 is still unknown, it cannot be directly applied to predict the formula (36.1) in 2015 to calculate the forecast value of tourism revenue. Assuming that the 2015 tourism revenue and previous years of data have the relationship (do not need to find a specific relationship), on this basis to set up a FTSFM (0.0004) forecast of unknown tourism revenue forecast rules.

36.3.1 FTSFM (3−d−1−d + 1) Prediction Rules

Assuming C years of the first 3 years C-3, C-2, C-1 annual tourism revenue is H_{c-3}, H_{c-2}, H_{c-1}. Calculate $\alpha=\{\max\{H_{c-3},H_{c-2},H_{c-1}\} - \min\{H_{c-3},H_{c-2},H_{c-1}\}\}/d$,

Use parameters separately H_{c-1} and $H_d = \max\{H_{c-3},H_{c-2},H_{c-1}\} + d\alpha$;

Application of FTSFM (0.0004) of the prediction formula (5) calculation, the resulting value from small to large array that is unknown years C years of tourism revenue . . . H_{-1} predictive value, H_0 predictive value, H_1predictive value, . . ., H_{d-1} predictive value, H_d predictive value, ... the parameter D value standard: with the predicted values, the minimum of the absolute value of the prediction error of the predicted value and the true data should be <1%., which is called the FTSFM (3−d−d + 1) prediction rule.

36.3.2 Decision Method of Predictive Value

Decision making method one: the predicted value of the unknown year in 2015..., H_{-1} type predictive value, H_0 type predictive value, H_1 type predictive value,..., H_{d-1} type predictive value, H_d type predictive value,... The ranking of the recommendations and the historical data of 2014..., type H_{-1} predictive value, H_0 type predictive value, H_1 type predictive value,..., H_{d-1} type predictive value, H_d type predictive value,... The absolute value of the prediction error rate of the real data is the same as that of the small to large. Decision method two: the decision makers making their decision by their own experience.

36.3.3 Application of FTSFM (3-4-5) Forecasting Rules to Predict the Tourism Revenue of Sanya City in 2014

Set 2014 is the unknown year, take d = 4, Apply FTSFM (3-4-5) forecast rules to predict Sanya's tourism revenue in 2014 unknown year. From the Table 36.2 in 2014 the first 3 years of 2011~ 2013 annual tourism revenue were $H_{2011} = 160.71$,

Table 36.3 Application of FTSFM (3-4-5) forecasting rules to predict the tourism revenue of Sanya in 2014 and 2015

	F_{2014}	H_{2014}	$F_{2014}-H_{2014}$	$(F_{2014}-H_{2014})/H_{2014}$	F_{2015}	$F_{2015}-H_{2014}$	$(F_{2015}-H_{2014})/H_{2014}$
H_{-1}	269.73	215.18	−54.55	−0.202239	250.36	−19.37	−0.071813
H_0	269.73	233.33	−36.4	−0.13495	269.73	0	0
H_1	269.73	251.48	−18.25	−0.06766	251.48	19.37	0.071813
H_2	269.73	269.62	−0.11	−0.000408	308.47	38.74	0.143625
H_3	269.73	287.77	18.04	0.066882	327.83	58.1	0.215401
H_4	269.73	305.91	36.18	0.134134	347.2	77.47	0.287213
H_5	269.73	324.05	54.32	0.201387	366.62	96.89	0.359211

$H_{2012} = 192.20$, $H_{2013} = 233.33$. Calculate $\alpha = \{\max\{H_{2011}, H_{2012}, H_{2013}\} - \min\{H_{2011}, H_{2012}, H_{2013}\}\}/4 = 72.62/4 = 18.155$. Use parameters separately $H_{2013} = 233.33$ and $H_{-1} = \max\{H_{2011}, H_{2012}, H_{2013}\} - \alpha = 215.18 \ldots H_{2013} = 233.33$ and $H_5 = \max\{H_{2011}, H_{2012}, H_{2013}\} + 3\alpha = 324.105\ldots$ as in Table 36.3. From Table 36.3, we can see that H2 predictive value 269.62, the absolute value of predictive error rate with actual value is 0.0408%, the prediction accuracy is high.

36.3.4 Application of FTSFM (3-4-5) Prediction Rules to Study the Forecast Value of Tourism Revenue of Sanya City in 2015

2015 is the truly unknown year, taking $d = 4$, using FTSFM (3-4-5) prediction rules to predict Sanya's tourism revenue in 2014 year. From the Table 36.2, $H_{2012} = 192.22$, $H_{2013} = 233.33$, $H_{2014} = 269.73$. Calculate $\alpha = \{\max\{H_{2012}, H_{2013}, H_{2014}\} - \min\{H_{2012}, H_{2013}, H_{2014}\}\}/4 = 77.51/4 = 19.3775$. Use parameters separately $H_{2014} = 269.73$ and $H_{-1} = \max\{H_{2012}, H_{2013}, H_{2014}\} - \alpha = 250.3525, \ldots$ $H_{2014} = 269.73$ and $H_5 = \max\{H_{2012}, H_{2013}, H_{2014}\} + 5\alpha = 366.6175\ldots$ as in Table 36.3. From Table 36.3, we can see that H_0 predictive value 269.73, the absolute value of predictive error rate with actual value is 0.0000%, the prediction accuracy is high. According to the decision-making method two, the decision maker can take the tourism economic situation as reference, and play their own initiative to select the reasonable decision-making sequence of the predicted value.

36.4 Conclusion

Based on the fuzzy number function S(μ), inverse fuzzy number function T(μ) and prediction function F(μ). FTSFM is a set of time series forecasting model, FTSFM (0.0004) is one of the commonly used prediction models of FTSFM. FTSFM (0.0004) has the function of predicting the historical data, and with the FTSFM (3−d−d + 1) prediction rule, it has the function of predicting the unknown data. For each forecast value of unknown year data, a sort of recommendation method is given, and a sort of self – decision – making function is also given. When applying FTSFM(0.0004) of FTSFM (3-4-5) prediction rules to predict the value of tourism revenue in 2014, The absolute value rate is only 0.0408%, So that people can have a good anticipation of the forecast error rate of H2 type forecasting value of the 2015 tourism income. FTSFM (0.0004) provides a new way of thinking for the research of time series prediction.

Acknowledgments This work is supported by Natural Science Foundation of Hainan Province (Grant No.714283), Scientific Research Foundation of Qiong zhou University (Grant No. QYXB201301).

References

1. Zadeh, L. A. (1965). Fuzzy set. *Fuzzy Sets and Systems, 8*, 338–353.
2. Song, Q., & Chissom, B. S. (1993). Forecasting enrollments with fuzzy time series – Part I. *Fuzzy Sets and Systems, 54*, 1–9.
3. Song, Q., & Chissom, B. S. (1994). Forecasting enrollments with fuzzy time series – Part II. *Fuzzy Sets and Systems, 62*, 1–8.
4. Saxena, P., Sharma, K., & Easo, S. (2012). Forecasting enrollments based on fuzzy time series with higher forecast accuracy rate. *Computer Technology & Applications, 3*, 957–961.
5. Hagawati, B., Joshi, P., & Kumar, S. (2013). A computational method for fuzzy time series forecasting based on difference parameters. *International Journal of Modeling, Simulation, and Scientific Computing, 4*, 1–12.
6. Uslu, V. R., Bas, E., Yolcu, U., & Egrioglu, E. (2014). A fuzzy time series approach based on weights determined by the number of recurrences of fuzzy relations. *Swarm and Evolutionary Computation, 15*, 19–26.
7. Kai, C., Ping, F. F., & Gang, C. W. (2010). A novel forecasting model of fuzzy time series based on K-means clustering. IWETCS, IEEE, pp. 223–225.
8. Wang, H. X., Guo, J. C., Feng, H., & Zhang, F. J. (2014). *A new model of forecast enrollment using fuzzy time series*. International Conference on Education Management and Management Science (ICEMMS 2014) (pp. 95–98) August 7–8, Tianjin, China.
9. Gao, Y., & Liu, Y. J. (2016). Adaptive fuzzy optimal control using direct heuristic dynamic programming for chaotic discrete-time system. *Journal of Vibration and Control, 22*, 595–603.
10. Cheng, C. H., Chen, T. L., Teoh, H. J., & Chiang, C. H. (2008). Fuzzy time series based on adaptive expectation model for TAIEX forecasting. *Expert Systems, 34*, 1126–1132.
11. Jilani, T. A., & Burney, S. M. A. (2007). M-factor high order fuzzy time series forecasting for road accident data. IEEE-IFSA 2007, June 18–21 Cancun, Mexico.

12. Liu, Y. J., Gao, Y., Tong, S. C., & Li, Y. M. (2016). Fuzzy approximation-based adaptive backstepping optimal control for a class of nonlinear discrete-time systems with dead-zone. *IEEE Transactions on Fuzzy Systems, 24*, 16–28.
13. Chen, S. M. (2002). Forecasting enrollments based on high-order fuzzy time series. *Journal of Cybernetics and Systems, 33*, 1–16.
14. Duan, M., Shi, J., Wang, J. H., & Lu, X. L. (2003). Control system based on fuzzy-neural network theory for an engine test bed. *Journal of Jilin University of Technology(Natural Science Edition), 33*, 107–109.
15. Shi, J., Duan, M., Chen, Y., Zhang, T., & Lu, X. L. (2006). A research on fuzzy neural network control for the measuring system of engine test bed. *Journal of Automotive Engineering, 28*, 1000–680x.
16. Liu, Y. J., Tong, S. C., Li, D. J., & Gao, Y. (2015). Fuzzy adaptive control with state observer for a class of nonlinear discrete-time systems with input constrain. *IEEE Transactions on Fuzzy Systems*, https://doi.org/10.1109/TFUZZ.2505088, (in press).
17. Jilani, T. A., Burney, S. M. A., & Ardil, C. (2007). Multivariate high order fuzzy time series forecasting for car road accidents. *International Journal of Computational Intelligence, 4*, 15–20.
18. Liu, Y. J., & Tong, S. C. (2015). Adaptive fuzzy control for a class of unknown nonlinear dynamical systems. *Fuzzy Sets and Systems, 263*, 49–70.

Chapter 37
Research on Data Storage Based on Cloud Platform

Xiaoman Zhang and Fangqin Xu

Abstract With the rapid development of mobile communication technology, "5G" era to speed up the pace, and large data age is surging from the times. Relational database is built on the relational model of the database, It uses mathematical concepts and methods to deal with the data in the database. In recent years. With the rise of web2.0, Data in the form of pictures, video and other forms of data storage was explosive growth, The existing data storage methods have been unable to meet the needs of the system. NoSQL's distributed large data storage technology is widely used in cloud computing because of its scalability. Cloud storage of unstructured storage technology can alleviate the current unstructured data storage problems and improve the quality of unstructured data storage services. The evolution of cloud computing makes the importance and value of cloud databases increasingly visible, let us see the future of data storage development.

Keywords Big Data · Relational Database · Non-relational Database Nosql · Cloud Storage

37.1 Introduction

37.1.1 Big Data Explosion Age

IT technology is developing rapidly, but also brought the pressure of data processing, and human intelligence has been unable to meet the needs of computational analysis. In 2012, the Obama administration launched a "Big Data Research and Development Initiative" on the White House website in order to enhance the use of massive data and access to knowledge, while also advocating government agencies access to digital data and mining technology. There are many types of data, a

X. Zhang (✉) · F. Xu
Shanghai JianQiao University, Shanghai, China
e-mail: 18221631802@163.com

M. Tavana, S. Patnaik (eds.), *Recent Developments in Data Science and Business Analytics*, Springer Proceedings in Business and Economics,
https://doi.org/10.1007/978-3-319-72745-5_37

large number of distribution is also wide, these data if the full analysis and mining can play its value [1].

Big data, which means a collection of data that can not be captured, managed, and processed with regular software tools within a certain timeframe. It needs a new processing model in order to have a stronger decision-making power, insight into the discovery of power and process optimization capabilities of information assets which is massive, high growth rates and diversified .

Big data stored in a huge amount of data, the initial unit of measurement is at least P, E or Z. It has many types, including the network log, audio, geographical location, its many types of features on the data processing capacity of a higher demand. With the extensive application of things, information gradually increased, but the value of Big data density is low, need to continue to tap and purify the value of the data. And the data growth rate is fast, so the processing speed of the data should also be required, otherwise it will lose the timeliness of the data. Big data age poses new challenges to human data, and people try to find their unprecedented space and potential.

37.2 Data Storage

37.2.1 Relational Database Overview

Relational database is based on the relational database model, which handles the data in the database by means of concepts and methods such as set algebra. It is also a form that is organized into a group and has a formal description. This form is essentially a special group of data items, and the data in the tables can be accessed or re-convened in many different ways without reorganizing the database tables.

There are more than one table in the relational database, and there is a need for a relationship between the tables to communicate their data. There are three types of relationships between tables and tables: one to one, one to many, many to many.

One to one example: Everyone has a unique ID number.

One to many example: A student belongs to only one class, but a class has more than one student.

Many to many example: A student can choose multiple classes, a class also has a number of students.

37.2.2 Non-Relational Database Nosql0

Non-relational database (NoSQL) is a new non-relational distributed storage technology, the database data not only includes the daily text data, including such as pictures, videos, animation and other data [2]. It is because of mass storage, flexible and easy to use, high concurrency and other characteristics, the system can provide a scalable data model. This model strictly abide by the CAP theorem, can be very good support for unstructured data storage to meet the high demand for read and write, it has good scalability [3].

Nosql data storage mainly includes:

1. key-value

By providing the key value for the data storage, each tuple can have a different field, each tuple can be added as needed to their own key pairs, so that it does not confine to a fixed structure and reduces some time and space Of the overhead, while a good system to meet the needs of reading and writing.

2. document store

It can efficiently meet the system's massive data storage and access needs and greatly improve the efficiency of mass data storage access. The current storage is mainly included MongoDB and couchDB, etc [4].

3. column-oriented

In a column-oriented storage system, the data is listed as a storage unit. The same column of data is stored together to support the dynamic expansion of the column, for a column or a few columns of the query has a clear I/O advantage.

NoSQL database is generated in order to solve the large-scale data collection of multiple data types of challenges, especially large data application problems.

37.3 Cloud Computing

37.3.1 Three Cloud Services

Software-as-a-Service(SaaS): Consumers can use the application, but do not control the operating system, hardware, or operational network infrastructure. This is the basis of a service concept, the software service provider to provide the concept of leasing customer service rather than purchase, the more common mode is to provide a set of account password.

Platform as a Service(PaaS): Consumers use the host to run the application. Consumers can control the environment in which the application is running, but does not control the operating system, hardware, or operational network infrastructure. A platform is usually an application infrastructure.

Infrastructure as a Service(IaaS): Consumers use basic computing resources such as processing power, storage space, network components, or middleware. Consumers can control the operating system, storage space, deployed applications and network components (such as firewalls, load balancers, etc.), but does not control the cloud infrastructure.

37.3.2 Cloud Computing

Cloud computing is one of the popular technologies for information technology development, which is an Internet-based computing. In this way, shared hardware and software resources and information can be provided to the computer on a variety of terminals and other equipment. It breaks through the limitations of time and space, as a new model of service computing, cloud computing on the IT industry and related industries have caused a huge impact. Cloud-based data storage services are also developing rapidly, which makes now more and more enterprises and individual users can enjoy the cloud computing and cloud storage technology to bring the convenience of services. To a certain extent, it has played a huge role for individual users and business users [5].

37.4 Cloud Platform Data Storage

Cloud storage is a network-based data storage model in which data is stored on multiple virtual servers and is typically managed by a third-party organization rather than a dedicated server.

Users use data storage center storage space to store their own data, the data center operators in the background according to the needs of users virtual resources. For users, it's like using a virtual server, so users can manage their own data. In fact, these resources may span many different physical servers.

The traditional relational database adopts the up-scaling way to improve the performance, and the Nosql database adopts the horizontal expansion mode to improve the performance. In other words, it is evenly distributed to each host. With the explosive growth of network data, Nosql database has been widely used, it has a large amount of data storage, cost-effective, flexible and scalable features. To a certain extent it is a substitute for traditional databases. Distributed databases are logical collections of databases on various sites or nodes in a computer network environment, logically that they belong to the same system, but physically they are scattered across multiple nodes of a computer network connection. And they are unified by a distributed database management system management. From the CAP theory, the distributed storage system is more suitable for nosql database. In some scenarios, nosql database and relational database can also be combined to compensate for each other's shortcomings to solve some of the web2.0 encountered some performance, scalability and other issues.

Distributed databases have been in existence for many years, it can be used to manage a large number of distributed storage of data, and usually use non-shared architecture, cloud databases and traditional distributed database has similarities [6]. For example, they all put data on different nodes. But the distributed database in the scalability can not be compared with the cloud database.

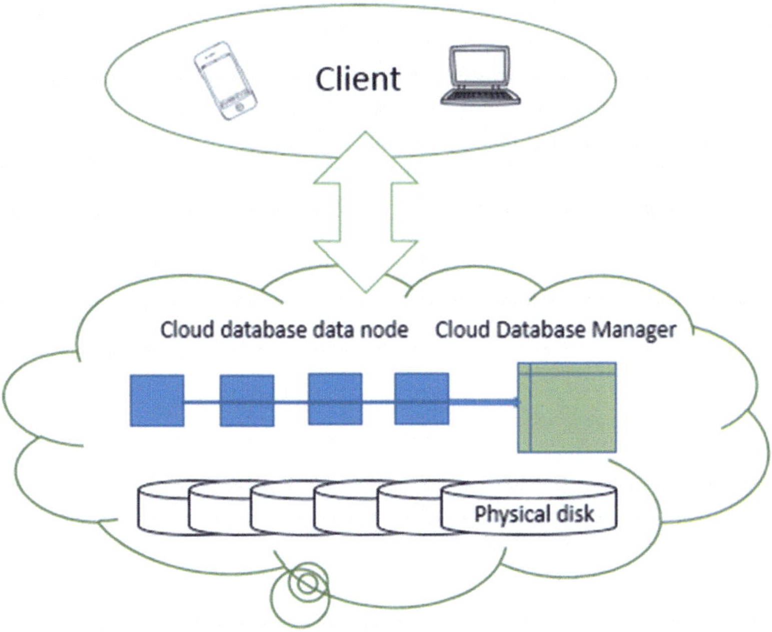

Fig. 37.1 Cloud database application diagram

Because of the need to consider data synchronization and partition failure and other overhead, the former with the increase in nodes will lead to performance degradation, while the latter is very good scalability, because the latter has been designed to avoid a lot will affect the scalable Sexual factors. Such as the use of a more simple data model, the metadata and application data separation, relaxation of the consistency requirements. In addition, the use of the cloud database is also different from the traditional distributed database. Cloud databases usually use multiple instances to share data storage problems, but also reduce the cost of the user to use the database, the tenant's data is both isolated and shared.

As shown in Fig. 37.1, in the cloud database application, the client does not need to understand the underlying details of the cloud data, all the underlying hardware has been virtualized, the client is transparent, it is like using a Running on a single server on the same database, it is very easy and easy, but also can get theoretically almost unlimited storage and processing power.

37.5 Conclusion

In the current era of this data explosion, cloud database has a very good application prospects. According to the research report, in the future business storage requirements for structured data will continue to increase. In the case of small-scale applications, the system load changes from the redundant resources of the system to deal with, but in large-scale applications, not only the existence of massive data storage needs, the application of the demand for resources is also dynamic changes, which means that a large number of virtual machines Increase or decrease. The traditional relational database has been unable to meet the needs of NOSQL database of massive data storage, cost-effective, flexible scalability and other characteristics, it replaced the traditional database. In the distributed database of scalability factors, the cloud database can better solve the problem of data storage, so the cloud database is an inevitable choice. Although cloud storage is not yet able to completely address our storage needs, but its emergence to ease the current explosive data growth pressure, and let us see the future of data storage development prospects.

References

1. Lohr S. (2012). The age of big data. New York Times, 11.
2. Stonebraker, M. (2010). SQL databases v. NoSQL databases. *Communications of the ACM, 53*(4), 10–11.
3. Moniruzzaman, A. B. M., & Hossain, S. A. (2013). NoSQL database: New era of databases for big data analytics – classification, characteristics and comparison. *International Journal of Database Theory & Application, 6*(4), 1–14.
4. Pokorny, J. (2011). *NoSQL databases:a step to database scalability in web environment.* Iiwas'2011 – the, International Conference on Information Integration and Web-Based Applications and Services (pp. 278–283), December 5–7 2011, Ho CHI Minh City, Vietnam. DBLP.
5. Hayes, B. (2008). Cloud computing. *Communications of the ACM, 51*(7), 9–11.
6. Zhang, J. H. (2014). Design and implementation of data mining based on distributed computing. *Applied Mechanics and Materials, 644-650*, 1702–1705.

Chapter 38
An Automatic Multi-Objective Clustering Based on Hierarchical Method

Chao Chen and Feng Qi

Abstract Just optimizing a single objective function or need to know the exact number of clusters in advance is the choice of most clustering methods. However, less knowledge of the data set to be clustered makes it difficult to select the appropriate number of clusters. Motivated by this, we propose an automatic multi-objective clustering based on hierarchical method (AMOH-Cluster), which can not only automatically calculate the optimal number of clusters but also divide all data sets properly based on intra-cluster data compactness and inter-cluster data connectivity. The proposed algorithm has advantages of providing higher clustering accuracy and requiring only a few parameters. As shown in the experiment, the comparison with the known multi-objective clustering algorithms proves that the proposed algorithm provides a solution with higher accuracy and optimal clustering number in various clusters of artificial data sets.

Keywords Multi-objective clustering · Automatic clustering · Hierarchical method

38.1 Introduction

Clustering is the process of separating a collection of data objects into clusters that consist of similar objects by a similarity evaluation criterion. In fact, the clustering process can be considered as one of the multi-objective optimization problems (MOOP) in which multiple complementary or conflicting objective functions need to be optimized simultaneously [1, 2]. The study of the field of automatic clustering has attracted public attention after the work of Handl and Knowles [3] entitled "Multi-objective clustering and K-determination", also known as MOCK. Prior to this, the same author proposed the first non-automatic multi-objective clustering algorithm based on PESA-II [4], called VIENNA [5]. MOCK fine-tunes the

C. Chen · F. Qi (✉)
Shandong Normal University, Jinan, China
e-mail: qfsdnu@126.com

M. Tavana, S. Patnaik (eds.), *Recent Developments in Data Science and Business Analytics*, Springer Proceedings in Business and Economics,
https://doi.org/10.1007/978-3-319-72745-5_38

objectives used in VIENNA to automatically determine the appropriate number of clusters. Previous papers have shown that MOCK is superior to traditional single-objective clustering in different ranges of reference data sets. However, it is suitable for super-spherical shapes or separated clusters and can provide better clustering results, but it is unsatisfactory on overlapping clusters [6]. VAMOSA, proposed in [6], utilizes the multi-objective optimization method based on simulated annealing and the center-based coding as the optimization strategy and use the newly proposed point-symmetric distance in [7] to replace the Euclidean distance that we often use for clustering. It provides overall performance better than MOCK. However, VAMOSA does not show good robustness to the clustering of overlapping data sets. In addition, Bandyopadhyay et al. [8] and Xia et al. [9] developed some automatic clustering methods based on NSGA-II [10], including MOKGA, MOEASSC.

The highlight of this paper is that we propose an automatic multi-objective clustering based on hierarchical method with automatic K-determination, called AMOH-Cluster. We use the hierarchical method to derive the optimal clustering and provide higher clustering accuracy. The algorithm can not only dynamically detect the number of the final clusters of the data points to be clustered, but also realize the rational distribution of the clusters.

The rest of this paper is generally organized as follows: some of the theoretical basis of AMOH-Cluster and the details of other algorithms involved is described in Sect. 38.2. Section 38.3 describes data sets, experiment platform and the clustering results of AMOH-Cluster compared with those of other algorithms. Finally, Sect. 38.4 concludes the paper and describes the ideas for future work.

38.2 AMOH-Cluster Algorithm

38.2.1 Clustering Algebraic Operations

In general, clustering can be considered as the process of partitioning a collection of data objects O = $\{O_1, \cdots, O_i, \cdots, O_n\}$, where each object $O_i = \{O_{i1}, \cdots, O_{ij}, \cdots, O_{id}\}$ into K disjoint clusters, C = $\{C_1, C_2, \cdots, C_K\}$, such that "similar" data objects are divided into the same cluster and "dissimilar" data objects are partitioned into different clusters.

In AMOH-Cluster algorithm, two clustering algebraic operators are introduced to define segmentation and aggregation of clusters respectively [11], namely '$\cup$' and '$\cap$', show as Fig. 38.1a and b.

Theorem 38.1 Let O_x, O_y be the d-dimensional data points to be clustered, and $O_x = (x_1, x_2, \cdots, x_d)$, $O_y = (y_1, y_2, \cdots, y_d)$. Then,

(1) $\cup(O_x, O_y) = \{O_x, O_y\}$;
(2) $\cap(O_x, O_y) = \{O_{xy}\}$, where O_{xy} is the mean of O_x, O_y.

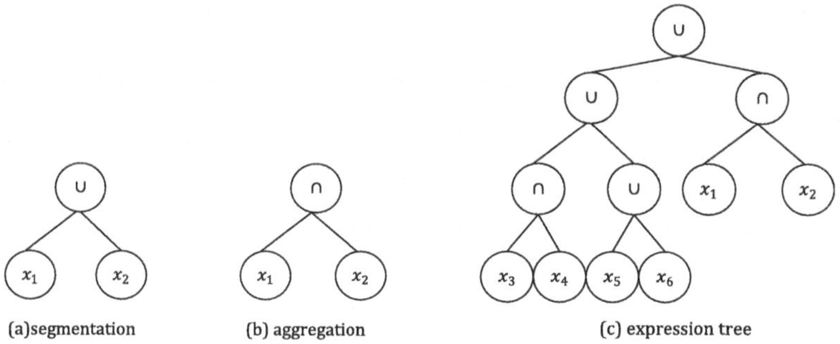

Fig. 38.1 (**a**, **b**) segmentation and aggregation of clusters defined by '∪' and '∩' and (**c**) expression tree

Table 38.1 An expression string

∪	∪	∩	∩	∪	x_1	x_2	x_3	x_4	x_5	x_6

38.2.2 *Encoding the Expression String and Decoding the Expression Tree(ET)*

In AMOH-Cluster, each expression string is composed of a head that contains the function and the tail that contains the terminal. The functions set is composed of $\{\cup, \cap\}$, and the terminals set is composed of $\{x_1, \cdots, x_i, \cdots, x_n\}$,where x_i is the *ith* data object in the data sets to be clustered. Table 38.1 shows an example of the representation of an expression string for AMOH-Cluster. Figure 38.1c shows its corresponding expression tree.

The expression string is converted to the corresponding expression tree by the sequence traversal method. The Algorithm 38.1 details the decoding process.

Algorithm 38.1 Cluster-ET Decoding

Input: The Cluster Expression Tree
Output: The initial cluster center-points
1: Cause '∪';
2: Pruning the sub-tree, placing the sub-tree into the set *i*;
3: For each set, recursively decode each sub-tree in it;
4: Cause '∩';
5: Recursively decode left and right sub-trees in turn;
6: Center-point ← calculates the mean of all the data in the left and right sub-trees;
7: Return all the cluster center-points;

38.2.3 *Objectives Functions*

For the automatic multi-objective clustering algorithm, the selection of the objective function determines the performance of the final clustering results. In this paper, we choose two complementary objectives to offset each other to achieve automatic clustering: compactness and connectivity.

In this paper, we use the global deviation of clustering results to represent compactness [3]. The overall deviation is measured by the overall Euclidean distance of the data objects in each cluster to its corresponding cluster center:

$$\mathrm{Dev}(\mathrm{C}) = \sum_{C_k \in C} \sum_{i \in C_k} \delta(i, \mu_k) \tag{38.1}$$

Where C is the all data objects to be clustered, μ_k is the center-point of C_k, and $\delta(i, \mu_k)$ is the Euclidean distance between the *ith* data object of C_k and μ_k .

As a clustering objective function, connectivity is used to assess the extent of adjacent data objects in the same cluster [3]:

$$\mathrm{Connc}(\mathrm{C}) = \sum_{i=1}^{k} \frac{1}{n} \sum_{h=1}^{|C_i|} \left(\sum_{j=1}^{L} x_{h,nn_{hj}} \right) \tag{38.2}$$

Where:

$$x_{h,nn_{hj}} = \begin{cases} \frac{1}{j}, & if \nexists C_i : x_h \in C_i \wedge x_{nn_{hj}} \in C_i \\ 0, & otherwise \end{cases} \tag{38.3}$$

nn_{hj} is the *jth* nearest neighbor of the object h, n is the all data points to be clustered, $|C_i|$ is the data points in C_i and L is used to determine how many neighbors that make contributions to connectivity.

38.2.4 AMOH-Cluster Algorithm

The steps involved in the automatic multi-objective clustering algorithm AMOH-Cluster are described in detail in Algorithm 38.2.

Algorithm 38.2 AMOH-Cluster Algorithm

Input: Dataset objects O $= \{O_1, \cdots, O_i, \cdots, O_n\}$
Output: Set of K clusters C $= \{C_1, C_2, \cdots, C_K\}$
1: Initialize the datasets ;
2: Encoding the expression string ;
3: Decoding the expression tree(ET) according to Algorithm 38.1 ;
4: Allocating data objects to the nearest cluster according to the distance of the data object to each cluster center;
5: Recalculate the mean of all data points in each cluster as the center of the cluster and search for a new center point to represent the cluster;
6: Repeat the above steps until no change;
7: Optimize the initial clustering result according to Algorithm 38.3;
8: Return the Set of K clusters C $= \{C_1, C_2, \cdots, C_K\}$.

38.2.5 Automatic Merge Process(AMP)

Based on the hierarchical clustering method, we propose a new automatic merge process method, called AMP. Then, AMP is used to merge the set of small clusters into larger ones. The automatic merge process is given in Algorithm 38.3:

Algorithm 38.3 Automatic Merge Process (AMP) (Merging the Original Clusters)

Input: The original clusters
Output: The final clustering results
1: Randomly choose two clusters, C_i and C_j;
2: Calculate the distance between the random two data points in C_i and C_j, find the nearest two data points x_i and x_j;
3: $D_{ij} \leftarrow$ Calculate the distance between x_i and x_j;
4: D_i, $D_j \leftarrow$ Calculate the mean of C_i and C_j respectively;
5: If ($D_{ij} \leq D_i$ and $D_{ij} \leq D_j$), $C_{ij} \leftarrow$ merge cluster C_i and C_j into a new cluster;
6: Remove C_i and C_j from the list;
7: Repeat the above steps until no clusters can be merged;
8: Output the cluster list.

38.3 Results and Discussion

We conducted experiments to compare the clustering results of AMOH-Cluster with MOCK and VAMOSA. As shown in Table 38.3, two aspects are considered: the results of Cluster Accuracy and the number of clusters. We use Java to implement all the experiments.

38.3.1 Data Sets

In the experiment, four artificial data sets from [12] are used to compare the performance of the three clustering algorithms. A brief description of the data sets is provided in Table 38.2, involving four aspects of the number of data points(n), the number of clusters(k), attributes(d) and the number of points in each cluster.

Table 38.2 Description of the data sets

Dataset	n	K	D	Points per cluster
Data_3_2	76	3	2	43,20,13
Data_5_2	250	5	2	50,50,50,50,50
Data_6_2	300	6	2	50,50,50,50,50,50
Data_4_3	400	4	3	100,100,100,100

38.3.2 Performance Measures

In addition to the number of clusters, we have also used the widely used measurement, Cluster Accuracy (CA), which measures the proportion of the data points that is correctly classified in the clustering results (clustering labels obtained by the algorithm) compared to pre-defined class labels to evaluate the performance of the three algorithms. The CA is defined below [13]:

$$\mathrm{CA(C)} = \sum_{i=1}^{K} \frac{max(C_i|L_i)}{|\,\Omega\,|} \qquad (38.4)$$

where *max* $(C_i| L_i)$ is the number of data points with the majority label (the label obtained by clustering are identical with the real ones) in the *ith* cluster; $|\Omega|$ is the data points in the *ith* cluster. The larger the CA value, the better the clustering effect.

38.3.3 Experimental Results

Since the starting point (initial center or initial division) of the three algorithms is randomly selected, for each data set, each algorithm runs independently 100 times for the clustering results to be comparable, and then the mean values of the optimal numbers of clusters (K)and Cluster Accuracy (CA) obtained by the three algorithms are calculated respectively. Table 38.3 provides the results of the comparison of K and CA for the three algorithms. The clustering results obtained by AMOH-Cluster on the four artificial data sets are shown in the Fig. 38.2.

As shown in Table 38.3, the mean values of the number of clusters obtained by the proposed AMOH-Cluster is closest to the actual number of clusters in each data set, so this illustrates the ability of the AMOH-Cluster for estimating the number of clusters outperforms MOCK and VAMOSA. In addition, it can be observed from Table 38.3 that the mean values of Cluster Accuracy (CA) produced by the proposed AMOH-Cluster is higher than that of MOCK and VAMOSA for all data sets. This illustrates that the AMOH-Cluster can find the better clustering partitioning for different datasets.

Table 38.3 The number of clusters (K) and Cluster Accuracy (CA) obtained by AMOH-Cluster, MOCK and VAMOSA

		AMOH-Cluster		MOCK		VAMOSA	
Dataset	Actual K	K	CA	K	CA	K	CA
Data_3_2	3	3.06	0.9832	3.91	0.8647	3.52	0.9001
Data_5_2	5	4.98	0.9755	3.86	0.7708	4.45	0.8724
Data_6_2	6	6.00	0.9938	6.68	0.9083	6.37	0.9465
Data_4_3	4	4.05	0.9849	4.51	0.8961	4.36	0.9244

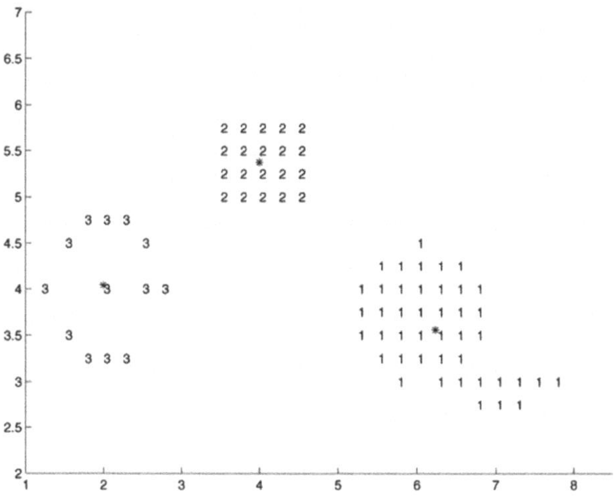

(a)

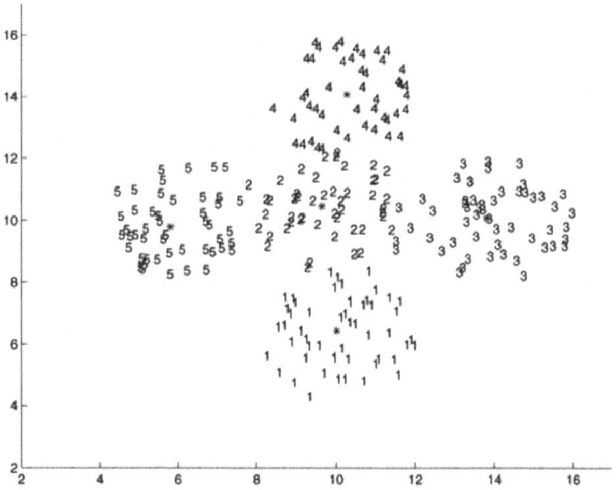

(b)

Fig. 38.2 Clustering results of artificial datasets (**a**) Data_3_2, (**b**) Data_5_2, (**c**) Data_6_2, (**d**) Data_4_3

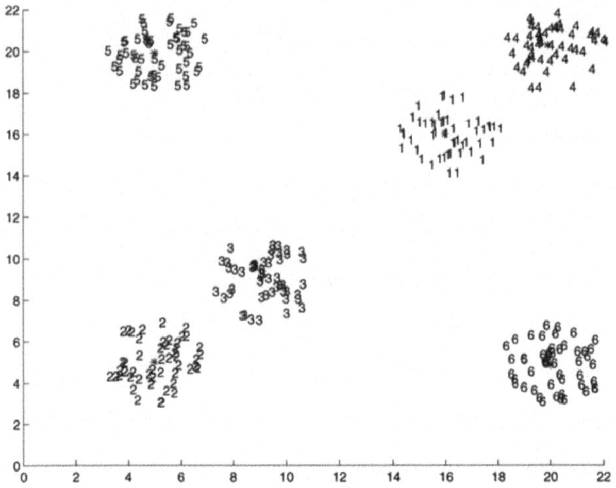

(c)

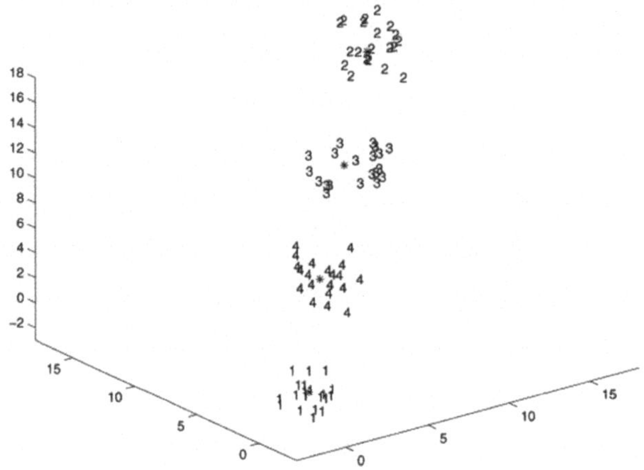

(d)

Fig. 38.2 (continued)

38.4 Conclusions

In the paper, we introduce the clustering algebraic operators and hierarchical method to solve the automatic multi-objective clustering problem of data sets and propose an automatic multi-objective clustering based on hierarchical method, called AMOH-Cluster. The proposed AMOH-Cluster could automatically calculate the optimal number of clusters and provide higher clustering accuracy by optimizing multiple conflicting objectives, intra-cluster data compactness and inter-cluster data connectivity. In addition, requiring only a few parameters is also a major highlight of the algorithm. We compared the clustering performance of the AMOH-Cluster, MOCK and VAMOSA in four artificial datasets. The experimental results show that the proposed algorithm is effective in calculating the optimal clustering number and the accuracy of the final clustering results.

Although the AMOH-Cluster has shown good performance in clustering a variety of data sets, it still has some shortcomings to be improved, such as high computation. In future work, we consider the use of parallel operations or natural heuristic evolutionary algorithms to minimize runtimes and improve the accuracy of the final clustering results.

Acknowledgments This work was supported by the Natural Science Foundation of China (No. 61472231). Natural Science Foundation of China (No. 61502283). Natural Science Foundation of China (No. 61640201).

References

1. Jose-Garcia, A., & Gomez-Flores, W. (2016). Automatic clustering using nature-inspired metaheuristics: A survey. *Applied Soft Computing, 41*, 192–213.
2. Heloulou, I., Radjef, M. S., & Kechadi, M. T. (2017). Automatic multi-objective clustering based on game theory. *Expert Systems with Applications, 67*, 32–48.
3. Handl, J., & Knowles, J. (2007). An evolutionary approach to multiobjective clustering. *IEEE Transactions on Evolutionary Computation, 11*(1), 56–76.
4. Dempster, A. P., Laird, N. M., & Rubin, D. B. (1977). Maximum likelihood from incomplete data via the EM algorithm. *Journal of the Royal Statistical Society: Series B: Methodological, 39*(1), 1–38.
5. Handl, J., & Knowles, J. (2004). *Evolutionary multiobjective clustering*. International conference on parallel problem solving from nature. Springer Berlin Heidelberg, London (pp. 1081–1091).
6. Saha, S., & Bandyopadhyay, S. (2010). A symmetry based multiobjective clustering technique for automatic evolution of clusters. *Pattern Recognition, 43*(3), 738–751.
7. Bandyopadhyay, S., & Saha, S. (2007). GAPS: A clustering method using a new point symmetry-based distance measure. *Pattern Recognition, 40*(12), 3430–3451.
8. Bandyopadhyay, S., Maulik, U., & Mukhopadhyay, A. (2007). Multiobjective genetic clustering for pixel classification in remote sensing imagery. *IEEE Transactions on Geoscience and Remote Sensing, 45*(5), 1506–1511.
9. Xia, H., Zhuang, J., & Yu, D. (2013). Novel soft subspace clustering with multi-objective evolutionary approach for high-dimensional data. *Pattern Recognition, 46*(9), 2562–2575.
10. Deb, K., Pratap, A., Agarwal, S., et al. (2002). A fast and elitist multiobjective genetic algorithm: NSGA-II. *IEEE Transactions on Evolutionary Computation, 6*(2), 182–197.
11. Chen, Y., Tang, C., Zhu, J., et al. (2007) Clustering without prior knowledge based on gene expression programming 2007. ICNC 2007. *Third International Conference on Natural Computation, IEEE, 3*, 451–455

12. Bandyopadhyay, S., & Pal, S. K. (2007). *Classification and learning using genetic algorithms: applications in bioinformatics and web intelligence* (Vol. 31). Springer Science & Business Media.
13. Fahad, A., Alshatri, N., Tari, Z., et al. (2014). A survey of clustering algorithms for big data: Taxonomy and empirical analysis. *IEEE transactions on emerging topics in computing, 2*(3), 267–279.

Chapter 39
Big Data Analytics for High Frequency Trading Volatility Estimation

Henry Han and Maxwell Li

Abstract High frequency trading has been dominating finance industry recently. It brings big data and new problems in finance. How to estimate security volatility in high frequency trading remains a challenge in business analytics. In this study, we propose a novel section volatility estimation model and implement it via a big data analytics approach. The proposed method conquers the weakness of the conventional realized volatility by demonstrating the capability to capture both global and local behavior of volatility in the whole trading period besides robustness to the fine time intervals. To the best of our knowledge, this work is the first volatility study in high frequency trading by using big data analytics. It not only provides a fast and more accurate volatility estimation in high frequency trading, but also has its significance in finance theory and trading practice.

Keywords Big data · High frequency trading · Volatility · Spark

39.1 Introduction

High frequency trading (HFT) has been dominating financial market in recent years. It counts about 55% of trading volumes in U.S. equity market and takes almost 80% of foreign exchange futures volumes in 2016 [1]. Unlike traditional trading, it is a computerized trading by using computer algorithm to conduct trading in a high frequency mode automatically, in which a transaction can be done in a few seconds or even milliseconds.

H. Han (✉)
The Laboratory of Big data and Analytics, Department of Computer and Information Sciences, Fordham University, New York, NY, USA
e-mail: xhan9@fordham.edu

M. Li
Business Analytics, Gabeli School of Business, Fordham University, New York, NY, USA

M. Tavana, S. Patnaik (eds.), *Recent Developments in Data Science and Business Analytics*, Springer Proceedings in Business and Economics,
https://doi.org/10.1007/978-3-319-72745-5_39

The rise of high frequency trading brings challenges in finance both in practice and theory. It increases market liquidity and prefers the most liquid, large cap stocks in trading [2]. As a result, there is a divergence between bid-ask spreads, which is the difference between bid price and ask price in trading, for large-caps and small-caps. To some degree, it greatly increases trading risks for its nature of 'high-frequency'. For example, it wiped 998.5 points (9.2%) off the DOW in 7 min on May 6, 2010 Flash Crash, besides wiping 3% off S&P 500 in 4 min on that day. It was even blamed as one 'important' reason for financial crisis happened in 2008.

HFT data are typical structured big data with a huge number of observations but a few features. In fact, it is a type of well-structured streaming data generated at ultra-high speed with huge volumes. For example, a HFT profile of five stocks in a month can have more than 20 million transaction records and asks more than 80 Gigabytes storage. Traditional data analytics methods actually can't apply to such streaming big data very well not to mention to dig trading knowledge from them.

Securities demonstrate special characteristics under HFT than traditional trading. Different stocks under HFT have totally different trading frequencies and even demonstrate periodic patterns or diurnal patterns [3]. Stock prices can be fixed or with low standard deviations in a transaction period with more than half million transactions. There are no mature theory and techniques available to analyze and predict the stock patterns though deep learning, social network analytics, and text mining are involved in current research [4].

More importantly, HFT challenges traditional financial theory because HFT data are essentially big data with milliseconds transaction records instead of daily-based traditional time series data. As such, traditional financial concepts and models may no longer work well for HFT data. For example, volatility is a fundamental concept to measure risk of a stock using historical data. It can be simply viewed as a special 'standard deviation' of a stock return under traditional daily based trading. Many volatility models have been developed in the past 30 years, which include ARCH, GRACH and their variants [5, 6].

However, these models assume trading data are traditional data collected from 252 business days annually. They cannot apply to HFT data that are collected in milliseconds not to mention its big data nature. A few volatility models have been proposed in recent years by using intraday returns [7–9]. These models extend the annualized volatility to monthly, daily or transaction unit-based volatility to handle HFT data. Realized volatility is a representative of these models and seems to gain more attention than other models [8, 9].

These proposed volatility models for HFT data make a good development of modern volatility theory under HFT. However, they are not data driven or at least not implemented in a data driven approach. Instead, they employ sampled data or even sparsely sampled data to estimate the standard deviation of stock log-returns in an extremely data-intensive situation. For example, authors in [9] picked four stocks, MMM, AIG, INTC, and MSFT and used only ten trading days of April 2004 in realized volatility calculation via a sparse sampling. Similarly, many researchers chosen to reduce HFT data into normal or low frequency data via sampling or transformation techniques, for the convenience of data analysis [8, 9]. These approaches would have a risk to change the data-intensive nature of HFT data volatility estimation into a plain variance estimation problem. As a result, it may inevitably lead to incomplete or even biased estimation that may not represent whole volatility dynamics in the market correctly, though it may look like theoretically correct according to traditional sampling theory.

On the other hand, it is urgent to use a big data analytics approach to handle high frequency trading data to take advantage of its data-intensive nature instead of seeking to reduce it to 'normal data'. In this study, we propose a novel volatility estimation method: section volatility for HFT data based on big data analytics. We compare it with realized volatility that is also implemented in a big data analytics approach. It demonstrates meaningful results in modelling the dynamics of high frequency trading well and provides a good way to capture the global risk tendency for HFT data. Moreover, we have evaluated the speed-up of our proposed method in comparison with its sequential implementations.

This paper is organized as follows. Section 39.2 proposes section volatility besides introducing realized volatility. Section 39.3 compares the performance of section volatility with that of realized volatility. Section 39.4 discusses the pros and cons of proposed section volatility model and concludes this paper.

39.2 Realized and Section Volatility for High Frequency Trading Data

We give the definition of realized volatility as follows, but skip most of its mathematical proof for the space limit.

Given a high frequency trading dataset of a security with K transactions in [0, T], it is partitioned as equally spaced transaction units I_1, I_2,... I_n, each of which is a time interval with a length say $\tau = 5$ minutes. Let r_{ti} be the ith log return in the tth interval, the realized volatility is defined as the sum of log prices from all intervals: $R_v = \sum_{t=1}^{n} \sum_{i=1}^{n_t} r_{ti}^2$, where n_t is the number of transactions in t^{th} interval that is often set as 1 in traditional sampling-based implementation [8, 9]. Such a definition assumes the price of the security s_t is subject to an *Ito* process $ds_t = s_t\mu_t dt + s_t\sigma_t dw_t$, where μ_t and σ_t are drift and diffusion of $\ln s_t$, and w_t is a standard Brownian motion [8]. R_v will finally converge to the corresponding integrated variance in probability in the interval [0,T], when the number of transaction units is large enough i.e. $\lim_{n\to\infty} \sum_{t=1}^{n} \sum_{i=1}^{n_t} r_{ti}^2 \to \int_0^T \sigma_t^2 dt$.

The realized volatility has some obvious weakness. It is hard for realized volatility to capture the global behavior of stock returns because a small increase in the next log-return can be amplified to its square value in the realized volatility. As a result, many local peaks appear in the realized volatility and somewhat prevent capturing the global volatility patterns. Furthermore, it is difficult to compare realized volatilities between different stocks from visualization because they may share many overlapped local peaks. Moreover, realized volatility can be quite sensitive to the length of time intervals and microstructure noise. The realized volatilities of a security can be quite different for different time intervals. On the other hand, microstructure noise such as ask-bid bounce and nonsynchronous trading would bring biases to realized volatility, especially for finer time intervals [10].

39.2.1 Section Volatility

Section volatility proposed in this work aims at conquering the weakness of realized volatility. It does not artificially separate transaction information into true signals, which are always unknown, and microstructure noise. Instead, it assumes all transaction information collected in trading are meaningful in volatility calculations. It then employs a novel way to construct log-returns by using the mean price in each interval (e.g. 5 min). The section volatility is derived from a sequence of log returns calculated from the mean price from each interval.

Given a high frequency trading dataset of a security with K transactions in a time interval [0, T], it is partitioned as equal time intervals $I_1, I_2, \ldots I_n$, t = 1,2...n. Let $r_t^{(m)}$ be the log return in the tth interval, $r_t^{(m)} = \ln \frac{s_t^{(m)}}{s_{t-1}^{(m)}}$, where $s_t^{(m)} = \frac{1}{n_t} \sum_{k=1}^{n_t} s_k$ and $s_{t-1}^{(m)} = \frac{1}{n_{t-1}} \sum_{k=1}^{n_{t-1}} s_k$ are the mean prices in the two adjacent intervals respectively. The section volatility is defined as $R_s = \left[\frac{1}{n-1} \sum_{t=1}^{n} \left(r_t^{(m)} - \overline{r_t^{(m)}} \right)^2 \right]^{1/2}$, in which $\overline{r_t^{(m)}}$ is the average of the log return $\overline{r_t^{(m)}}$.

Let $x_t = \ln s_t^{(m)}$ be the log price of the security, we have $dx_t = \mu_t^{(m)} dt + \sigma_t^{(m)} dw_t$ by following Ito's process, where drift $\mu_t^{(m)}$ is a zero or close to it, when the time interval is small. That is $dx_t = \sigma_t^{(m)} dw_t = \sigma_t^{(m)} \varepsilon_t \sqrt{dt}$, $\varepsilon_t \sim N(0, 1)$. By the definition of x_t, $dx_t = r_t^{(m)}$ in the interval, we have $\left(r_t^{(m)} \right)^2 = \left(\sigma_t^m \right)^2$. If we assume $\overline{r_t^{(m)}} = 0$ and replace n-1 by n in the section volatility equation, we will have the following relationships: $R_s^2 \approx \frac{1}{n} \sum_{t=1}^{n} \left(r_t^{(m)} \right)^2 = \frac{1}{n} \int_0^T \left(\sigma_t^m \right)^2 dt = \frac{\theta_n}{n} \int_0^T \sigma_t^2 dt$, where θ_n is a constant with respect to the number of time intervals. Such a result shows that section volatility has an approximately linear relationship with the integrated variance in [0, T]. Furthermore, it is noted that section volatility will be the square root of sampling based realized volatility when all prices are equal in each time interval [8].

Given a time interval I_n, there are actually two ways to implement section volatility. One is to use all intervals before it to calculate its section volatility. Such a 'look-back-all' approach is good at capturing the global behavior of volatility but miss some its local behavior in a small time interval. Another is to use a rolling-window that includes a set of previous intervals in a certain time period say 50 h, to calculate the section volatility. This 'rolling-window' approach can capture more local behavior of volatility than the previous way. Considering the short term nature of high frequency trading, we recommend the 'rolling window' implementation in practice, that is, $R_s = \left[\frac{1}{l-1}\sum_{t=1}^{l}\left(r_{n-t}^{(m)} - \overline{r_t^{(m)}}\right)^2\right]^{1/2}$. The 'look-back-all' approach only works well for the case where the whole trading period is short (e.g. 30 days).

39.3 Big Data Analytics for Section Volatility and Realized Volatility

39.3.1 Big HFT Data and Spark Analytics

Our HFT dataset consists of total 2004 stocks high frequency trading data from Dec 02, 2002 to July 30, 2003 collected from NYSE. As a typical big dataset, it has total 247,547,671 transaction records with five variables: symbol, data, time, price and size. Such a dataset asks total more than 8 Gigabytes storage.

We employ Spark to conduct big data analytics for its efficiency besides implementing a sequential version via python under an iMac with eight cores 24 Gigabytes RAM and 3.0 Ghz CPU. Spark is a state-of-the-art big data processing engine built on Hadoop and resilient distributed datasets (RDDs) [11, 12]. In addition to friendly programing model, Spark is faster than map-reduce and Hadoop in real-time big data management, and analytics [12].

It is interesting to point out that our Spark section volatility estimation can be easily run under any computer with 4.0 RAM and 2.26 Ghz CPU though it may need a little bit long time to calculate all section volatilities. However, it is almost impossible to run its corresponding sequential implementation under the same level computer for the size of input data and its computing recourse demand.

Unlike traditional sampling base realized volatility calculation, where last transaction price is counted as the price in a time interval and other transactions are dropped in sampling [9], we don't drop any transactions in our volatility calculations. Moreover, unlike traditional volatility is only calculated for one or few securities, we estimate volatility to a large set of stocks with more than 2000 entries. Such an approach can avoid possible biases brought by limited data and contribute to evaluating the modeling efficiency of different volatilities.

Figure 39.1 compares the realized volatility and section volatility for randomly selected 10 stocks under the 15 min intervals besides including corresponding stock prices and log-returns. The rolling window sizes of section volatilities are chosen as

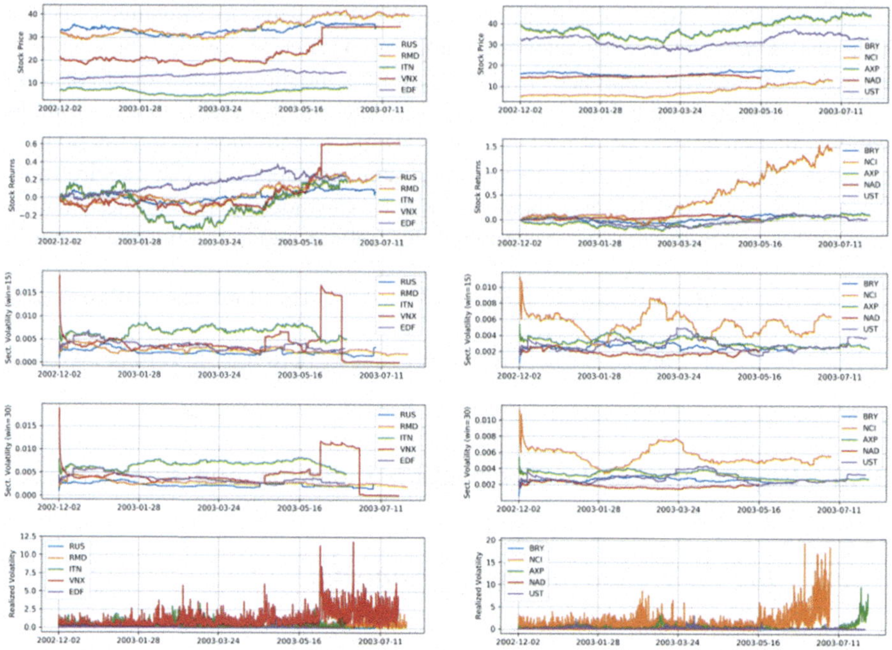

Fig. 39.1 The comparisons of realized volatility and section volatility for 10 randomly selected stocks on 15 min intervals from 12/02/2002 to 07/31/2003. The stock prices, stock log returns are also included for the sake of volatility comparisons

15 and 30 days i.e. 90 and 180 trading hours respectively. Those results from other time intervals (e.g. 5 or 30 min) demonstrate almost same or similar results as the 15 min intervals.

It is easy to see that realized volatility actually cannot model the variability of stock prices well for its too many peaks in each fixed time interval. The many local peaks may model volatility changes in few time intervals, but it cannot track volatility the behavior in a little bit long time period (e.g., 6 months). For the same reason, their plots are jammed together, which makes it hard to distinguish the variability behavior of stocks in the whole trading period. That is, it only shows variance behavior in each time interval locally for only modelling the impacts from previous intervals.

On the other hand, section volatility not only demonstrates the global tendency of stock variability in the whole trading period, but also captures local behavior of stock returns quite well. It is clear that the section volatility patterns 'match' the original stock-return price tendency better than the realized volatility in the whole trading period. Interestingly, the section volatility values by nature are much smaller than those of realized volatility. It implies that section volatility can model more subtle variability change in each time interval than realized volatility.

Moreover, section volatility is not as sensitive to the fine time intervals as realized volatility (data not shown). The section volatilities obtained from 5-min time intervals share similar patterns with those from 15-min and 30-min. This is because the mean log returns in each interval and rolling-window lessen the effects of the length of time intervals in section volatility. Instead, the log-return square in each transaction in a time interval contributes to the final realized volatility.

Rolling-window size. We have found that the rolling window size in section volatility can be chosen between 10–30 days for the sake of volatility modeling for almost all 2, 5, 15, and 30 min time intervals. Too small window size (e.g. 3 days) will lead to more trivial local patterns and weaken the global volatility pattern capturing. On the other hand, too large window size (e.g. 50 days) would miss some important local volatility patterns.

39.3.2 Big Data Analytics Speedup

We further compare the performance of big data section volatility implementation (5 min time interval) under Spark and its sequential version, both of which are implemented in python, under the iMac with 8 cores 24Gb RAM and 3.0Ghz CPU. Figure 39.2 illustrates the time usage of section volatility computing under the Spark

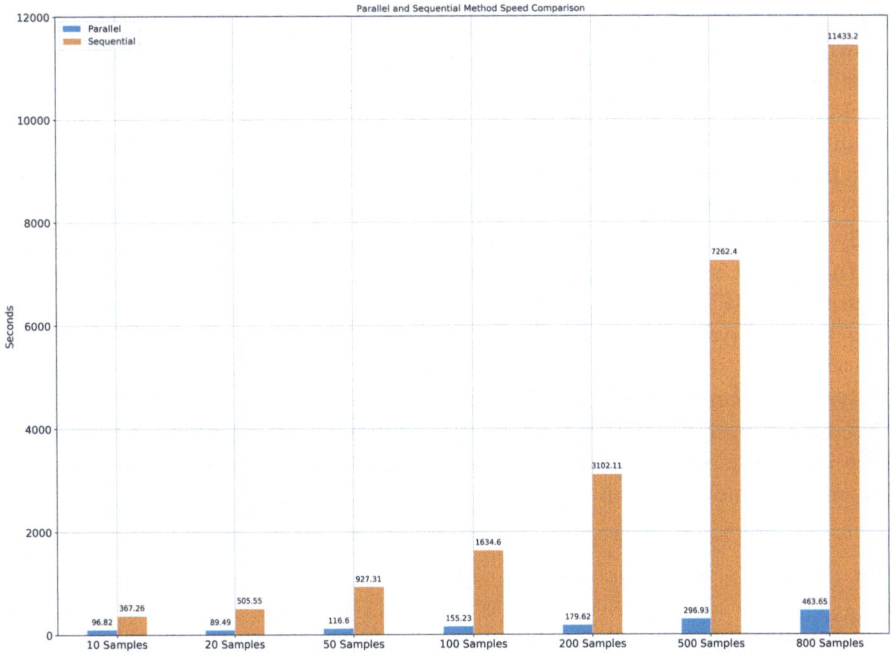

Fig. 39.2 The comparisons of Spark (parallel) and sequential implementations of section volatility under 5 min intervals. Parallel implementation demonstrate a sub-linear property with respect to input data. The sequential implementation seems to suffer from an exponential complexity with the increase of input data size

and sequential implementations for randomly selected 20 to 800 stocks under the whole 8 months trading period. It is easy to see that the big data analytics approach demonstrates a good speedup compared with its sequential implementation. The speedup is only 3 for a small input dataset such as 10 stocks, but it reaches 10 for 100 stocks and 24 for 800 stocks. In fact, our Spark implementation demonstrates a sublinear property in handling such big data analytics, in which time complexities demonstrate a linear or sublinear tendency with respect to the increase of input data. However, the sequential implementation demonstrates an obviously exponential complexity with respect to input data. Thus, our big data analytics implementation for the proposed section volatility also has time complexity efficiency compared with conventional sequential approaches.

39.4 Discussion and Conclusion

In this study, we propose a novel data-driven section volatility to estimate volatility for high frequency trading data and implement it via a big data analytics approach. The proposed section volatility demonstrates better modeling capabilities in capturing global and local patterns of volatilities for different stocks than conventional realized volatility. In addition, it conquers the weakness of realized volatility by showing robust patterns to finer time intervals. We also point out that section volatility has an approximately linear relationship with the integrated variance of stock returns in the trading period of interest mathematically. Unlike previous volatility models that rely on sampling to reduce high frequency trading data to low frequency trading, our method considers the whole dataset via a big data analytics. To the best of our knowledge, this study work is the first big data analytics study in high frequency trading volatility estimation [13].

The current section volatility is applied to all stocks irrespective of its type. Considering HFT prefers the most liquid and large-cap stocks in trading, we are interested in investigating the different section volatility patterns for the large-cap and small cap stocks in high frequency trading. Furthermore, we would like to analyze their corresponding liquidity dynamics via big data analytics to further unveil the impacts of HFT on liquidity. In addition, we are interested in employing deep learning models to predict unknown section volatility for different securities according to known section volatility in our future work [14].

References

1. Cespa, G., & Vives, X. (2017). *High frequency trading and fragility* (Working Papers Series). European Central Bank (2020).
2. Hendershott, et al. (2011). Does algorithmic trading improve liquidity? *Journal of Finance, 66*, 1–33.

3. Brownlees, T., Cipollini, F., & Gallo, M. (2011). Intra-daily volume modelling and prediction for algorithmic trading. *Journal of Financial Economics, 9*(3), 489–518.
4. Sirignano, J. (2017, June). Deep learning models in Finance. *SIAM News*.
5. Mitra, S. (2009). A review of volatility and option pricing: arxiv.org.
6. Spiegel, M. (1998). Stock price volatility in a multiple security overlapping generations model. *Review of Financial Studies, 11*(2), 419–447.
7. Ait-Sahalia, et al. (2005). *Ultra high frequency volatility estimation with dependent microstructure noise* (BER Working Paper No. 11380).
8. Andersen, et al. (2001). The distribution of realized stock return volatility. *Journal of Financial Economics, 61*(1), 43–76.
9. Zhang, et al. (2005). A tale of two time scales: Determining integrated volatility with noisy high-frequency data. *Journal of the American Statistical Association, 100*, 1391–1411.
10. Tsay, R., & Yeh, J. (2011). Random aggregation with applications in high-frequency finance. *Journal of Forecast, 30*, 72–103.
11. Spark. (2017). https://spark.apache.org/
12. Armbrust, et al. (2015). Spark SQL: Relational data processing in spark. *Proceedings of the ACM SIGMOD International Conference on Management of Data* (pp. 1383–1394).
13. Seddon, J., & Currie, W. (2017). A model for unpacking big data analytics in high-frequency trading. *Journal of Business Research, 70*, 300–307.
14. Schmidhuber, J. (2015). Deep learning in neural networks: An overview. *Neural Networks, 61*, 85–117.

Chapter 40
A Simulation of Sample Variance Calculation in the Teaching of Business Statistics to English Majors

Shili Ge, Rou Yang, and Xiaoxiao Chen

Abstract Variance is important for statistical description of a data set. Yet, the denominator of (*n*–1) in sample variance calculation confuses many Business English learners of statistics. In order to give learners an impressive instruction, a statistical simulation of population and sample variance calculation is designed with self-code Python program. The experimental simulation shows that the sample variances calculated with divisor of (*n*–1) are averagely closer to population variance than with *n*. The latter underestimates the population variance. The simulation offers an important explanation for statistics learners and can help them learn business statistics better.

Keywords Business statistics · Business English · Sample variance · Statistical simulation

---

S. Ge
School of English for International Business, Guangdong University of Foreign Studies, Guangzhou, China

Collaborative Innovation Center for Language Research and Service, Guangdong University of Foreign Studies, Guangzhou, China

R. Yang
School of Interpreting and Translation Studies, Guangdong University of Foreign Studies, Guangzhou, China

X. Chen (✉)
School of English for International Business, Guangdong University of Foreign Studies, Guangzhou, China
e-mail: gracekot@qq.com

M. Tavana, S. Patnaik (eds.), *Recent Developments in Data Science and Business Analytics*, Springer Proceedings in Business and Economics,
https://doi.org/10.1007/978-3-319-72745-5_40

40.1 Introduction

Business statistics is as an important course for Business English majors in China. Yet, due to their lack of adequate mathematical training, it is often difficult to explain some statistical notations to these students. The first difficulty is sample variance calculation after the introduction of central tendency in data description. There is an obvious difference between the calculation of population variance and sample variance, but few course books of statistics offer a simple and clear explanation for Business English major undergraduates, which leads to confusion among many learners. Beside accurate but complicated mathematical derivation, experimental simulation is usually an optimal choice for explaining statistical concept to beginners just like the simulation of sampling distribution by [1].

40.2 Variance as a Measure of Variation

The description of data using numerical measures includes central tendency and variation. Central tendency, including mean, median and mode, is usually easy for English major undergraduate students to understand. Variation covers mainly the notations of range, interquartile range (IQR), variance and standard deviation, and mean absolute deviation (MAD) of "limited utility" [2]. Among them, variance, especially the calculation of sample variance is hard for beginners to understand [3].

40.2.1 The Definition of Variance

"Variance and its related measure, the standard deviation, are arguably the most important statistics (other than the mean). They are used to measure the variability in a set of numerical data [2]." The standard deviation is the square root of the variance and "the population variance is the average of the squared distances of the data values from the mean" [4]. It can be shown very clearly by formula (40.1):

$$\sigma^2 = \frac{\sum_{i=1}^{N} (x_i - \mu)^2}{N} \tag{40.1}$$

Where: μ is population mean, N is population size, and σ^2 is population variance.

With the understanding of range, IQR, and MAD, it is not difficult for learners to understand population variance as an important characteristic of data variability or dispersion. Problem arises when sample variance comes into view as shown in formula (40.2):

$$s^2 = \frac{\sum_{i=1}^{n} (x_i - \bar{x})^2}{n - 1} \tag{40.2}$$

Where: n is sample size, $\bar{x}$ is sample mean, and s^2 is sample variance.

It is easy to notice that in formula (40.2), the denominator is $n-1$ (sample size minus 1) while in formula (40.1), the denominator is simply N, the population size. Clear explanation is needed for English major undergraduates to fully understand the meaning of sample variance calculation.

40.2.2 The Explanation of Sample Variance Calculation

In teaching the notation of sample variance, most course books of statistics offer the explanation as follows: "When we compute a sample variance, we are often interested in using it to estimate the population variance σ^2" [5]. "It turns out that the sample variance s^2 with $(n-1)$ in the denominator provides better estimates of σ^2 than would an estimator calculated with n in the denominator" [6]. For this reason, the sample variance s^2 is always computed with the denominator of $(n-1)$. Some course books name this better estimator as "unbiased estimator", and s^2 with the denominator of $(n-1)$ would have resulted in a biased estimator of σ^2; "one that produced an average s^2 that was smaller than the true value of σ^2" [2, 4].

As for the reason, some claims "a detailed explanation is beyond the scope of this text" [5], and some tries to explain from the perspective of degrees of freedom [7], which results in more confusion for Business English major statistics learners.

A course book claims that "if we were to select all possible samples of size n from a given population and for each sample we computed the sample variance using [formula (40.2)], the average of all the sample variances would equal σ^2 (the population variance), provided we used $n-1$ as the divisor" [4]. This is worth experimenting. No matter what the result is, it can give the learners a straightforward impression that there is a difference, more or less, for the calculation of s^2 with the denominator of n and $n-1$ and which one is a better estimator for population variance σ^2.

40.3 Simulation Design

A statistical simulation contains mainly three parts: the generation of random numbers, the simulation of discrete events and the statistical analysis of the simulation data [8]. For the simulation of variance calculation, the 3 steps are as following:

First, to generate N random numbers as a population of uniform distribution, and the mean, μ, and variance, σ^2, of the population are calculated.

Then, to sample from the population exhaustively with sample size n, and calculate the sample variance with the denominator of n and $(n-1)$, respectively, for each sample.

At last, to analyze the two populations of sample variances resulted by the denominator of n and $(n-1)$ and their relationship with the original population variance σ^2.

40.4 Data Analysis and Discussion

To fulfill the simulation, a Python program is written to perform the task. The algorithm of the program includes 4 steps:

1. Setting up a population of N individuals and calculate population variance;
2. Setting sample size n;
3. Drawing all possible samples of size n with replacement;
4. Calculating and outputting the variances of all these samples with n and $(n-1)$ as denominators.

The program is run 7 times with different N and n as shown in Table 40.1. All population items are random integer numbers in the range of 0 to 100 following uniform distribution. The means of sample variances with denominators of n and $(n-1)$ are analyzed in comparison with population variances.

As for n and $(n-1)$ as the denominators in the calculation of sample variances, Table 40.1 shows clearly that the means of all sets of sample variances with the divisor of $n(s^2)$ are much smaller than the corresponding population variances in comparison with the larger values of sample variances with the divisor of $(n-1)$ (s_1^2). That means, $(n-1)$ as denominator in the calculation of sample variance is closer to the true population variance, which is the reason that it is referred to as "unbiased estimator" of population variance. This can be seen in Fig. 40.1 that the population variances of the seven experiments forming the blue line in between, the sample variances calculated with the denominator of n forming the red line below the population variances, and the sample variances with $(n-1)$ forming the green line

Table 40.1 Variances of different populations and samples

No.	N	σ^2	n	Mean of s^2	Mean of s_1^2	No. of Samples
1	5	311.44	2	194.65	389.30	10
2	10	992.61	3	735.27	1102.90	120
3	10	327.81	4	273.18	364.23	210
4	20	436.19	5	367.32	459.15	15,504
5	20	816.35	8	751.90	859.31	125,970
6	30	643.77	6	554.98	665.97	593,775
7	30	655.85	10	610.62	678.46	30,045,015

Note: s^2 means sample variance is calculated with the denominator of n and s_1^2 means $(n-1)$

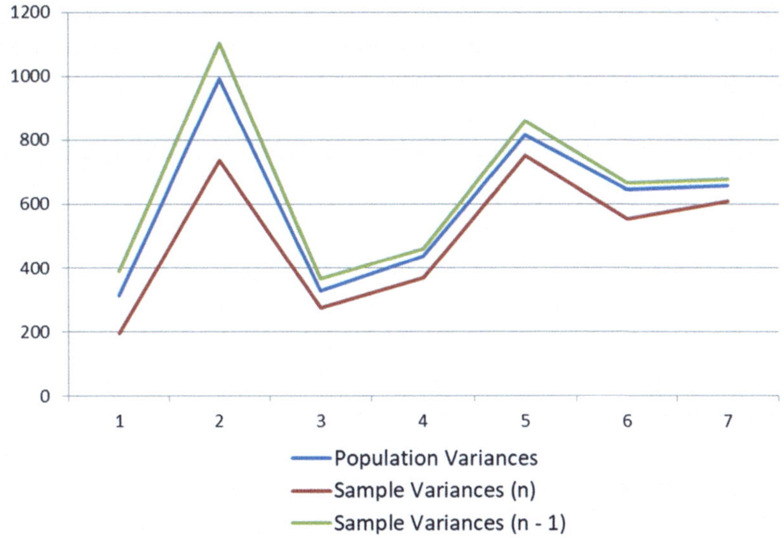

Fig. 40.1 Variance comparison of populations and samples

above the population variances. It is obvious that the green line is closer to the blue line all the way to the end, which graphically shows that the sample variances calculated with $(n-1)$ is better than with n.

There is also a trend from both Table 40.1 and Fig. 40.1 that along with the increase of population size N and sample size n, the means of sample variances with the divisor of $(n-1)$ (s_1^2) are closer to population variances, though there is no mean that equals to the corresponding population variance, yet. This experiment result proves that, although sample variance with the divisor of $(n-1)$ (s_1^2) can be a better estimator of population variance, σ^2, than that with the divisor of n (s^2), it cannot be expected that the mean of s_1^2 exactly equals to σ^2 as stated in some course books as [4].

40.5 Conclusion

With the help of a self-coded Python program, the statistical simulation of sample variances calculation is conducted with the denominator of n and $(n-1)$. The experiment shows that n as the denominator in calculation leads to smaller values of sample variance results and $(n-1)$ leads to larger values, but these larger ones are closer to population variances averagely. From Table 40.1, it also can be seen that, with the increase of N and n, the exhaustive total sample numbers grow exponentially, which leads to the difficulty of calculation of sample variances. Therefore, the largest N is limited to 30 and n to 10, which certainly have some limitation. In the future, larger populations and samples have to be simulated with more efficient algorithm and the statistical simulation results need more complicated analysis including graphic and numerical presentation of the results.

Acknowledgments This research was supported by Graduate Education Innovation Plan of Guangdong Province (2015JGXM-MS22) and the Science and Technology Innovation Project of Guangdong Province (2013KJCX0070).

References

1. Ge, S., Fang, F., & Chen, X. (2015). The simulation of sampling distribution in the teaching of business statistics to English majors. *Journal of Simulation, 3*(2), 43–46.
2. Selvanathan, E. A., Selvanathan, S., & Keller, G. (2014). *Business statistics: Australia and New Zealand* (6th ed.). Sydney: Cengage Learning Australia Pty Limited.
3. Zhao, J., & Guo, L. (2016). An analysis on the definition of sample variance. *Journal of Science of Teachers' College and University, 36*(7), 61–62+.
4. Groebner, D. F., Shannon, P. W., Fry, P. C., & Smith, K. D. (2011). *Business statistics: A decision-making approach* (8th ed.). Upper Saddle River: Pearson Education, Inc.
5. Anderson, D. R., Sweeney, D. J., & Williams, T. A. (2011). *Statistics for business and economics* (11th ed.). Mason: South-Western Cengage Learning.
6. Mendenhall, W., Beaver, R. J., & Beaver, B. M. (2013). *Introduction to probability and statistics* (14th ed.). Boston: Brooks/Cole, Cengage Learning.
7. Walpole, R. E., Myers, R. H., Myers, S. L., & Ye, K. (2012). *Probability & statistics for engineers & scientists* (9th ed.). Boston: Pearson Education.
8. Ross, S. M. (2013). *Simulation*. San Diego: Elsevier.

Chapter 41
Application of Clinical Diagnosis and Treatment Data of Coronary Heart Disease Based on Association Rules

Kun Zhang, Xiaoyan Chen, Haifeng Wang, and Yufei Wang

Abstract According to the characteristics of Chinese medicine diagnosis and treatment of coronary heart disease and the need for mining, this paper will introduce the association rules mining, from the medical treatment of patients with all aspects of the disease and the basis of coronary heart disease diagnosis and treatment of Chinese medicine between the basis of digging out the law of Chinese medicine, the application of association rules algorithm, Get a series of rules of coronary heart disease syndrome, for the diagnosis and prevention of coronary heart disease provides an important basis for decision-making.

Keywords Association rules · Data mining · Data medical treatment · Intelligent data analysis

---

K. Zhang
College of Ocean Information Engineering, Hainan Tropical Ocean University, Sanya, Hainan, China

State Key Laboratory of Marine Resources Utilization in South China Sea, Hainan University, Haikou, Hainan, China

College of Information Science and Technology, Hainan University, Haikou, Hainan, China

X. Chen
College of Ocean Information Engineering, Hainan Tropical Ocean University, Sanya, Hainan, China

H. Wang (✉) · Y. Wang
College of Ocean Information Engineering, Hainan Tropical Ocean University, Sanya, Hainan, China

State Key Laboratory of Marine Resources Utilization in South China Sea, Hainan University, Haikou, Hainan, China
e-mail: wyfxxz@163.com

M. Tavana, S. Patnaik (eds.), *Recent Developments in Data Science and Business Analytics*, Springer Proceedings in Business and Economics,
https://doi.org/10.1007/978-3-319-72745-5_41

41.1 Introduction

Coronary heart disease is a serious harm to human health, common diseases and frequently-occurring disease, has become one of the major diseases leading to human death. It is of great theoretical and practical significance to study how to quickly and effectively extract the association rules from the clinical diagnosis and treatment data of traditional Chinese medicine and to apply the relevant association rules to the clinical diagnosis and treatment of coronary heart disease.

41.2 Data Mining Association Rules

Coronary heart disease diagnosis and treatment data are often multi-valued attributes [1], multi-class standard data, the traditional association rules mining algorithm is usually not a good combination of domain knowledge, direct mining of the rules, the efficiency is low. In this paper, combined with the characteristics of TCM diagnosis and treatment data of coronary heart disease[2–4], the decision attribute and the non-decision attribute are block coded, and an association rule mining algorithm is proposed to eliminate the rule of traditional Chinese medicine in treating coronary heart disease. Improve the efficiency of the algorithm. This algorithm has some theoretical and practical value for finding the data mining problem of association rules between decision attributes and non – decision attributes [5].

Traditional association rules [6] often use a unified support threshold to mine frequent item sets, resulting in low support items with low support levels that can not be tapped. If you reduce the support threshold will produce a large number of redundant short project set, and will lead to greatly reduce the efficiency of the algorithm. In this paper, an association rule mining method is proposed, which can effectively exploit more long patterns and reduce useless short patterns, so that the generated rules are more theoretical and practical. Effective diagnosis and treatment of coronary heart disease.

At present, many scholars [7, 8] have made a lot of research on the correlation between the pre-order and the back of the rules, and put forward a number of correlation measures, but the most of the measures are based on the fact that the correlation in the process of development is not the premise. However, this kind of global correlation research has great chance, if the correlation between the itemsets change, then there will be the current mining association rules support and confidence is higher, and is positive correlation, but With the passage of time or the occurrence of certain special events will appear the current rules of support and confidence changes, or even irrelevant. Aiming at the lack of global correlation in association rules, this paper proposes a method of segmented nonlinear regression and reverse verification, and verifies the relevance of association rules. This method is more accurate to analyze the relevance of association rules, and the number of rules is greatly reduced, and more meaningful association rules can be excavated.

Experiments on TCM diagnosis and treatment data show that this method is more practical.

41.3 The Basics of Associating Rule Data Sets

Association rule mining is an important branch of data mining research [9]. It is the first time that people have proposed. It is a description of the interdependence and association of a thing and other things. It is used in the market shopping analysis for the discovery of supermarket shopping. The database between the different goods in the link to find out the customer buying habits and found that often buy goods together to help businesses develop targeted marketing strategies, commodity shelf arrangements and according to the purchase model of the user classification, etc., more typical Examples are the story of beer and diapers. Some young parents in the United States after get off work often go to the supermarket to buy baby diapers, supermarkets through the customer's shopping information mining and found that the probability of buying diapers customers will buy beer, then immediately adjust the supermarket shelves placed, the diaper And beer together, the results of diapers, beer sales increased significantly. Association rules mining has been widely concerned by experts in the field of data mining in the past 10 years, and has applied to many other areas of the database, have achieved good mining effect [10, 11].

Most of the mining algorithms of association rules are based on the support confidence framework theory, which is not very effective if the time factor of the data and the correlation between the rule front and the back are taken into account. In this paper, we propose a new association rule framework: aging support degree matching. The new matching method is used to replace the traditional confidence, which can effectively solve the problem of the correlation between the preamble and the latter. The time entropy factor is added on the basis of the support degree matching framework, so that the generated rules can reflect the significance of the data. The feasibility and superiority of the framework are illustrated by an example, and the association rule maintenance algorithm and implementation idea of adding the database are presented. The effectiveness of the algorithm is proved by experiments.

Definition 41.1 [12] Suppose $\mathrm{I} = \{i_1, i_2, \cdots, i_m\}$ is a collection of all items, that is, all the fields in the database; is a collection of all transactions, that is, the database $\mathrm{D} = <\mathrm{T}_1, \mathrm{T}_2, \cdots, \mathrm{T}_n>$, where each transaction is a collection of items that makes $\mathrm{T} \subseteq \mathrm{I}$. So that each transaction has an identifier, called TID. Suppose X is an item set, affair T include A if and only when $\mathrm{X} \subseteq \mathrm{T}$. The association rule is $\mathrm{X} =>$ Yimplication, among them $\mathrm{X} \subset \mathrm{Y}$, $\mathrm{Y} \subset \mathrm{I}$, and $\mathrm{X} \cap \mathrm{Y} = \varnothing$.

Definition 41.2 k- item set: a commodity or an attribute is called a project. a collection of multiple items is called an item set. Suppose I be a collection of all the items in the D database, set $\mathrm{x} = \{i_1, i_2, \cdots, i_k\}$($\mathrm{x} \subseteq \mathrm{I}$ and $|\mathrm{x}| = \mathrm{k}$) called k-item set.

Definition 41.3 Support: The degree of support for rule A = > B in transaction database D is the percentage of transactions that contain D in transaction A ∪ B, that is, probability P(X ∪ Y), is sup(X = > Y).

The degree of support reflects the frequency to which the rule applies. The rules of low support generally do not make sense, and people are less interested in the rules that are less likely to apply [13].

Definition 41.4 Confidence: c is a transaction containing X in D and also contains a percentage of Y, denoted by conf(X = > Y).

The goal of mining association rules is to find the association rules that support and trust certain conditions, that is, strong association rules.

Definition 41.5 Frequent item sets: if the number of transactions containing X data item is greater than the minimum support, it is called {X} frequent item sets, ie |X|/|D| ≥ min _sup. Where min_sup is the minimum support.

Definition 41.6 Strong association rules: For each frequent item set S, find all nonempty subsets s of S, if sup(S)/ sup (s) ≥ minconf, generate strong association rules s = > S-s. Where S-s is the confidence level for rule s = > S-s. Where s is the rule of the front, the rule of the back S-s.

The data preprocesses the existing EMR data before entering the data warehouse. The presence of incomplete, inconvenient and inconsistent data is the biggest reason why raw data can not be used directly for mining. Some attributes may be of interest to us but are not available; some useful attributes may be considered unimportant and missing; some data with personal idioms and lack of uniformity; and some data before the format conversion is not Used for a variety of mining algorithms. Into the garbage, out of garbage, in order to dig out the correct knowledge, must be the original data for cleaning, conversion, to ensure accurate and reliable data. Data preprocessing is the basis for data analysis and mining. The data quality of preprocessing directly affects the mining results. The purpose of data preprocessing is to prepare data sources for data mining to meet the actual mining requirements of the application domain, so that the rules found are more effective, as shown in Fig. 41.1.

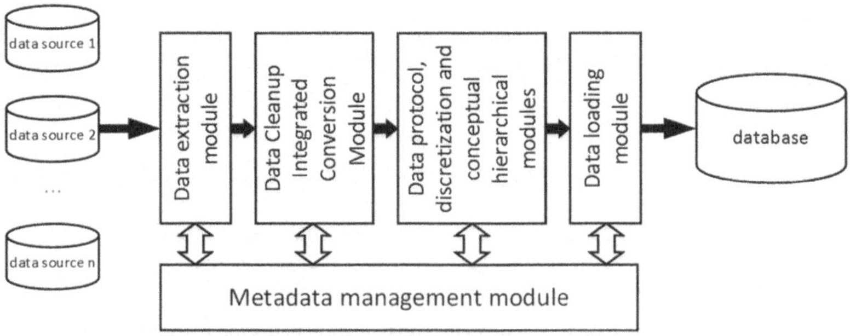

Fig. 41.1 Data preprocessing module

Table 41.1 Patient diagnosis and treatment analysis results

Patient ID	Check the time	Field name	Value
10011	2016-09-13	Whether coronary heart disease	Y
10012	2016-09-13	Whether coronary heart disease	Y
10013	2016-09-15	Whether coronary heart disease	Y
10014	2016-09-17	Whether coronary heart disease	Y
...	...	...	...

Table 41.2 Partial data of individual patient basic fact sheets

Patient ID	Check the time	Field name	Value
10011	2016-09-13	Tongue color	Dark red
10011	2016-09-13	Pulse	Chord
10011	2016-09-13	Moss color	White
10011	2016-09-13	Texture of the moss	Greasy
...	...	...	...

The data in this article comes from a hospital clinical acquisition system in Sanya City, the original data after the confidentiality of the data obtained after the table, as shown in Tables 41.1 and 41.2.

In this hospital clinical data collection, and data mining related to the table has more than 10, this article cited two tables as an example, for data mining. In order to facilitate data mining, will be related to the theme of the various attributes into a data warehouse. This model builds the data warehouse with the theme of TCM syndrome differentiation of coronary heart disease.

41.4 Summary

This paper is based on data mining association rules mining. By finding the frequency set directly, the generation of candidate sets is reduced. The data were pretreated according to the characteristics and mining requirements of coronary heart disease in a hospital in Sanya. The TCM syndrome differentiation rule was excavated from 18 aspects of pulse, tongue and morbidity, and 8 TCM syndromes of coronary heart disease. The feasibility of the algorithm is confirmed by experiments, and a series of syndrome differentiation rules of coronary heart disease are obtained, which provide important decision basis for the diagnosis and prevention of coronary heart disease. The future work will combine the knowledge of experts in the field of Chinese medicine to further improve and improve the algorithm to dig out more practical application of the rules of the value.

Acknowledgments This research was financially supported by the Key Project of Scientific Research of Hainan Province(No.Hnky2015ZD-14); the Scientific and Technological Cooperative Project for College and Region of Sanya(No.2015YD16).

References

1. Yuan, F., & Chen, S. (2011). Model construction on efficient mMining association rules in clinical data of hypertension. *Computer Engineering and Applications, 47*(36), 226–229.
2. Yan, G. (2009). Application of association rule algorithm in TCM data mining, silicon valley, No. 24: 65.
3. Wang, J. (2013). Realization of diagnosis and treatment data Dmining system based on asocciation rules. *Modern Electronics Technique, 36*(19), 124–126.
4. Liu, G., & Liu, P. (2011). Prospects of data mining technology in TCM diagnosis and treatment. *Medical Information, 24*(09), 5616–5617.
5. Jin, L., Wang, Z. D., Kan, H., & Han, R. (2013). Research on the application of data mining in the extraction of TCM diagnosis and treatment rules. *Lishizhen Medicine and Materia Medica Research, 24*(04), 1015–1017.
6. Liu, G., & Wu, L. (2011). Study on the mining of medication and prescription of traditional Chinese medicine based on association rules. *Jiangsu Journal of Traditional Chinese Medicine, 43*(11), 72–74.
7. Fengxia, O., & Wang, Z. (2011). Application of association rules in data mining technology to TCM diagnosis. *Journal of Henan Institute of Engineering(Natural Science Edition), 23*(02), 53–58.
8. Liu, Z. (2012). *Research of association rule mining algorithms and their applications of coronary heart disease diagnosis and treatment with traditional Chinese medicine* (Master Dissertation). Dalian Maritime University, Dalian.
9. Shi, L. (2008). *Data warehousing and data mining on the clinical database of coronary treatment with Chinese traditional medicine* (Master Dissertation). Dalian Maritime University, Dalian.
10. Zhang, K., Wang, H. X., Wang, H. F., & Li, Z. (2012). New exploration on definition of similarity measures between vague sets. *Journal of Natural Science of Heilongjiang University, 29*(03), 412–415.
11. Zhang, K., Wang, H. X., Wang, H. F., & Li, Z. (2014). Vague spatial decision method and its an application for the location of tailings dam. *Computer Science, 41*(04), 260–262.
12. Zhang, K., Wang, H.-X., Wang, H.-F., & Li, Z. (2015). The time series prediction algorithm of enrollment based on the hesitancy degree correction rule of vague sets. *ICIC Express Letters, 9* (5), 1311–1318.
13. Zhang, K., Wang, H., Wang, H., & Li, Z. (2015). Fuzzy time series prediction model and application based on fuzzy inverse, processing. *International Journal of Signal, Image Processing and Pattern Recognition, 8*(10), 121–128.

Chapter 42
Bibliometric Analysis of Spatial Econometrics

Jianhua Liu and Wei Li

Abstract This paper collects 5865 spatial econometrics academic papers published in China National Knowledge Infrastructure(CNKI) in 2002 to 2017 as the research object, using the method of bibliometric analysis to analyze and sort out the trends and characteristics of domestic space econometrics research, and has developed out the hotspots of space econometrics research, hot institutions and popular authors in China. And draws the keyword co-occurrence network of space econometrics research, which has profound guiding significance for the further research and development of China's space econometrics.

Keywords Space econometrics · Bibliometric analysis · CNKI · Keyword co-occurrence network

42.1 Introduction

Space econometrics is a new interdisciplinary, which based on econometrics, spatial statistics, geographic information systems and other disciplines to explore the establishment of space economic theory model as the main task, and use economic theory, mathematical models, spatial statistics and professional software and other tools to study the phenomenon of space economy [1]. Paelinck and Klaassen [2] first proposed the concept of "space econometrics" and a series of methods, and systematically discussed the object of spatial econometrics research, research content and basic model. Then, Anselin and Hudak [3], Haini ng [4] and Cressie [5] further developed the space econometrics on the basis of the previous research, and established a more effective spatial measurement model, which makes the spatial measurement theory gradually improved. Anselin and Hudak [3], Anselin et al. [6]

J. Liu (✉) · W. Li
School of Management Engineering, Zhengzhou University, Zhengzhou, China
e-mail: 277894576@qq.com

M. Tavana, S. Patnaik (eds.), *Recent Developments in Data Science and Business Analytics*, Springer Proceedings in Business and Economics,
https://doi.org/10.1007/978-3-319-72745-5_42

and LeSage [7] used computational methods to study spatial econometrics, which strongly promoted the application of spatial econometrics in regional economics.

Throughout the development of China's space econometrics research, the earliest began in the late 1980s and in the early twenty-first century began a rapid development. This paper, from the perspective of bibliometrics, summarizes and analyzes the development, research status and characteristics of space econometrics in China for more than 10 years, and provides some reference and guidance for the further study of spatial econometrics in China.[1]

42.2 Data Source and Processing

42.2.1 Data Collection and Selection

Based on the consideration of time and representative factors, this paper chooses the periodical articles related to spatial econometrics contained in the CNKI Chinese Academic Literature Publishing Network as the data source. CNKI China Academic Literature Publishing Network as the most well-known and comprehensive knowledge reserve database in China, it covers a wide range of literature types, including academic journals, doctoral dissertations, excellent master's degree papers, important conference papers, yearbooks, newspapers, patents, standards and scientific and technological achievements [8]. The related research of space econometrics usually needs to establish the relevant spatial measurement model, and accordingly analyze this model. Therefore, the search term of this paper is "space econometrics" or "space econometric model" or "space econometric study" or "space econometric analysis", the time range is January 1, 2001 – August 14, 2017, then, searching the full-text field, we finally obtained 5907 related literature.

42.2.2 Data Processing

The systematic review of the literature on spatial economics retrieved and the deletion of articles that are not related to space econometrics, such as "Understanding Monetary Policy: New Keynesian View" published in Journal of Nanjing University of Finance and Economics. After careful verification, a total of 42 unrelated documents were deleted, 5865 articles were retained after screening.

[1]Foundation project: the key project of science and technology of Henan: "Research on the Innovation and Development Strategy of the Central Plains Urban Agglomeration Based on Spatial Econometrics";"Research on model system and key technology of medium and long term development plan of science and technology and economy in Henan" (No.142102310141)

42.3 Distribution of Domestic Space Econometrics Papers

42.3.1 Time Distribution

The number of papers published in a particular field in a given period can largely reflect the level of research and development of the subject. Therefore, the retrieval of 5865 articles in the annual analysis of the content, to reflect the status of China's space econometrics research. According to the original data sorting available space econometric annual report volume statistics table, see Table 42.1.

According to the data in Table 42.1, we will draw the corresponding trend graph of the annual measurement of space econometrics. As shown in Fig. 42.1, we can see the change of relevant research situation more clearly. From the figure we can see that from 2000 to 2005, China's domestic space econometrics related research basically stay in the blank stage, mainly because this stage involved in the study of very few scholars; from 2006 onwards, the number of research papers published of China's space econometric began to increase year by year, this can be attributed to macroeconomic policy on the needs of space econometrics increased year by year.

Table 42.1 Space econometrics annual issuance

Years	Quantities of articles	Years	Quantities of articles
2000	1	2009	214
2001	0	2010	330
2002	1	2011	450
2003	1	2012	612
2004	3	2013	715
2005	5	2014	850
2006	37	2015	925
2007	98	2016	1049
2008	150	2017	411

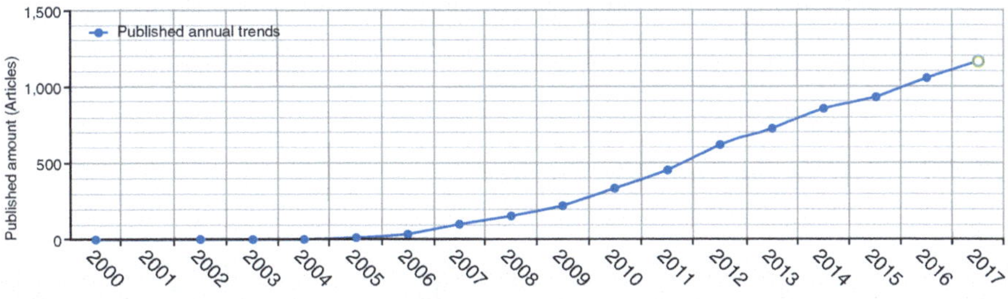

Fig. 42.1 Spatial econometrics annual issuance trend char

Table 42.2 Spatial econometrics important author statistics

Name of author	Quantities of articles	Name of author	Quantities of articles
Gu Guofeng	13	Liu Huajun	9
Wu Yuming	13	Su Fanglin	9
Deng Ming	12	Lin Guangping	8
Wang Liping	11	Huang Sen	8
Long Zhihe	11	Qian Zhengming	8
Pu Yongjian	10	Xie E	7
Wang Jiankang	10	Zhou Huimin	7
Shen Tiyan	9	Zhu Pingfang	7
Wu yuming	9	Jiang Lei	7
Tao Changqi	9	Zhao Guoqing	7

42.3.2 Author and Research Organization Distribution

An author's ability of research in a given field can usually be determined by the number of papers published and the number of citations cited. Likewise, the research level of a research institution is usually determined by the cumulative value of all scholarly research levels within the institution. And the level of research in a region in this area is determined by the number of highly qualified scholars and the high level of research.

Through the screening and analysis of the original data, we get the distribution of important authors (see Table 42.2), and the distribution of the research institutions (as shown in Table 42.3). As the number of authors in the field and the number of research institutions are too large, so this is important to define that the important authors are those published no less than 7 articles; the important research agencies are those among the top 20 institutions issued a document number for this field.

It can be seen from Tables 42.2 and 42.3 that in the field of spatial econometrics, Gu Guofeng of Northeast Normal University, Wu Yuming of Guangxi Normal University and Deng Ming of Xiamen University and other scholars have been awarded the forefront of this research field; Hunan University, Chongqing University, Chongqing University, East China Normal University, Zhejiang University, Jinan University, Jilin University, Northeast University of Finance and Economics, Xiamen University and other schools through in-depth study of space econometrics received a large number of research results, embodies a large number of experts in this field, therefore have a strongly research level in this field.

42.3.3 Article Source Distribution

According to the original data from 2000 to 2017 to draw the space econometrics literature published source statistics table, as shown in Table 42.4. It can be seen from Table 42.4 that the number of Hunan University applied econometric

Table 42.3 Major research institutions statistics of spatial econometrics

Name of organization	Quantities of articles	Name of organization	Quantities of articles
Hunan University	208	Nanjing University	85
Chongqing University	137	Southwest University of Finance and Economics	82
East China Normal University	120	South China University of Technology	82
Zhejiang University	108	Peking University	81
Jinan University	104	Zhejiang Business University	78
Jilin University	99	Anhui University of Finance and Economics	76
Northeast University of Finance and Economics	93	Nanjing Normal University	75
China People's University	92	Jiangxi University of Finance and Economics	75
Xiamen University	91	Wuhan University	71
Nankai University	85	Huazhong University of Science and Technology	69

Table 42.4 Major article sources statistics of spatial econometrics

Article source	Quantities of articles	Article source	Quantities of articles
Hunan University	130	Inquiry Into Economic Issues	40
Statistics and Decision	95	Zhejiang Business University	40
Economic geography	87	Areal Research and Development	36
East China Normal University	64	Scientia Geographica Sinica	36
Jilin University	64	Science Research Management	36
Chongqing University	56	China Population Resources and Environment	35
Northeast University of Finance and Economics	56	East China Economic Management	35
Statistics & Information Forum	53	Journal of Business Times	34
Zhejiang University	47	Journal of Industrial Technology and Economics	34
Science and Technology Management Research	40	Science & Technology Progress and Policy	32

economics reached 130, ranking the first in the statistical results. "Statistics and Decision" published in the number of 95, ranked second; "Economic geography" 87, ranked third; East China Normal University, Jilin University to 64 tied for fourth; Chongqing University and Northeast University of Finance and Economics ranks sixth with 56. It can be seen from the literature that most of the literature on spatial econometrics is from universities and core journals, which corresponds to the theory that the British philologist Bradford's "scientific literature is unevenly distributed in related journals".

42.4 Space Econometrics Research Hotspot Analysis

Key words as the identification of the paper search, to a certain extent, shows the main research content of the paper. Therefore, it is helpful to reveal the development history, research status, research hotspot and future development trend of the field by analyze a large number of keywords in the field of research. Therefore, the statistical analysis of the keywords can be an effective method for the analysis of the content of the paper [9].

By sorting out the keywords in the original data we get the higher frequency key words in the space econometrics literature,In CNKI we use the clustering analysis to draw the keywords with higher frequency as the keyword co-occurrence Network view as shown in Fig. 42.2.

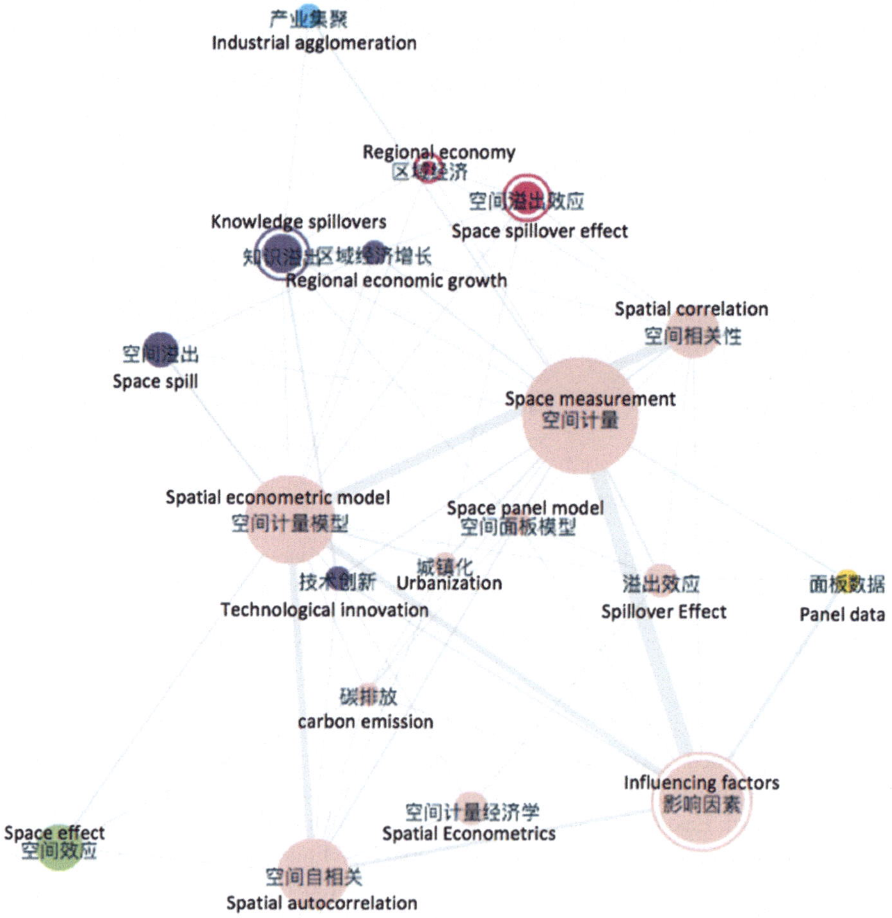

Fig. 42.2 Keyword co-occurrence network view

The larger the number of keywords in the graph, the higher the frequency of the occurrence of the two nodes, the thickness of the connection between the two nodes represents the co-occurrence of the two keywords, the more the number of co-occurrence, the thicker lines. From the figure we can see that the future development trend of space econometrics is mainly in the field of regional economic development, industrial structure adjustment, knowledge transfer, overflow and technological innovation and other fields.

42.5 Conclusion

This paper collects 5865 spatial econometrics academic papers published in China National Knowledge Infrastructure(CNKI) in 2002 to 2017 as the research object, using the method of bibliometric analysis to clarify the temporal distribution of space econometrics in China, the distribution of authors and institutions, the distribution of literature sources, the distribution of research levels, etc. And explores the future development trend of China's space econometrics through the co-occurrence of the key words in order to provide some help for the study of space econometrics scholars. As the number of documents in the original data is too large, it is difficult to manually sort out the sources data without subjective mind in the screening process, which may have an impact on the research results. The author will improve it in the follow-up study.

References

1. Qiong, Y., & Cong-kun, H. (2015). Study on the latest development of domestic space econometrics. *Technology Innovation and Application*, (11), 260–261.
2. Paelinck, J. H. P., & Klaassen, L. H. (1979). *Spatial econometrics*. Farnborough: Saxon House.
3. Anselin, L., & Hudak, S. (1992). Spatial econometrics in practice: A review of software options. *Regional Science and Urban Economics, 22*(3), 509–536.
4. Haining, R. P. (1990). *Spatial data analysis in the social and environmental sciences*. Hurtado MA, Moscatelli GN, Godag-none RE.
5. Cressie, N. (1991). *Statistics for spatial data, revised edition*. New York: Wiley.
6. Anselin, L., Bera, A. K., Florax, R., et al. (1996). Simple diagnostic tests for spatial dependence. *Regional Science and Urban Economics, 26*(1), 77–104.
7. LeSage, P. (1999). *The theory and practice of spatial econometrics*. A manual to accompany the spatial econometrics tool-box. Available at: www.spatial-econometrics.com
8. Wei, C., Runxi, Z., & Wang, G. (2011). Chinese network public opinion research bibliometric analysis. *Information Science, 29*(01), 131–135.
9. Yongmiao, H. (2007). The status, function and limitations of econometrics. *Economic Research*, (05), 139–153.

Chapter 43
Using Diagnostic Analysis to Discover Offensive Patterns in a Football Game

Tianbiao Liu, Philippe Fournier-Viger, and Andreas Hohmann

Abstract Football is a popular team sport, for which analyzing a team strategies can reveal useful information for understanding and improving a team's performance. For this purpose, a promising approach is to analyze data collected about a match using data mining algorithms. However, designing such approach is not trivial as a football match involves both the time dimension and the spatial dimension. In this paper, a diagnostic analysis based approach is proposed, which consists of preparing data from a match by considering the spatial dimension and then extracting sequential rules from the data. The proposed approach is illustrated in a case study to analyze the match between Germany and Italy at the 2012 European Championship. Results of this study show that threatening offensive patterns from the Germany team are identified, illustrating complex interactions between players for performance analysis.

Keywords Football · Performance analysis data mining · Sequential rules · European Championship

T. Liu (✉)
College of Sports and PE, Beijing Normal University, Beijing, China
e-mail: LTB@bnu.edu.cn

P. Fournier-Viger
School of Humanities and Social Sciences, Harbin Institute of Technology (Shenzhen), Shenzhen, China

A. Hohmann
Institute of Sports Science, University of Bayreuth, Bayreuth, Germany

M. Tavana, S. Patnaik (eds.), *Recent Developments in Data Science and Business Analytics*, Springer Proceedings in Business and Economics,
https://doi.org/10.1007/978-3-319-72745-5_43

43.1 Introduction

Football is an interesting team sport because even in a single football match, there is a massive amount of technical and tactical data to be analyzed. Thus, coaches are in great need of scientific support to analyze match data, to take appropriate strategic decisions.

Because of the complexity of a football game, and to capture the various aspects of the game, it is necessary to design new algorithms and adapt them to the actual game. Football is a game with a single direction time flow, that is match events occur in chronological order, and they constitute the match database. However, most studies [5–7] handle each controlling sequence of a match database as a transaction, where information about the sequential ordering (time) of events is discarded or ignored. Thus, important patterns involving the time dimension cannot be discovered.

To address this issue, a promising approach is to discover sequential patterns or rules in sequences of events. Thus, this paper designs a diagnostic based approach based on this idea and shows its application in a case study where the European Championship 2012 semi-final match between Germany vs. Italy is studied. Results show that the proposed method can provide important insights about the interactions between players in a football for performance analysis.

The rest of this paper is organized as follows. Section 43.2 presents the proposed method. Section 43.3 presents the results. Finally, Sect. 43.4 provides a discussion and draw a conclusion.

43.2 Proposed Method

The proposed method is composed of the following main steps. First data is collected from game footage to encode interactions between players as sequences. Then, a sequential rule mining algorithm is applied to extract sequential patterns from these sequences. These rules represent effective attacking tactics (defined to have a close relationship to the chance of scoring a goal) in a match. The details about all these steps are given in the following paragraphs.

43.2.1 Data Collection

To analyze a football match and compare tactical offensive behaviors, a match is firstly video recorded so that all game elements during the ball possession phases can then be analyzed.

43.2.2 Reliability of Data

The next step is to analyze all match events in the game analysis with two observers and an inter-rating consistency (inter-observer agreement) quantified by Cohen's Kappa. The Cohen's kappa value (K) of the model is found to be K = 0.93 for the "player" in our studied match, a value that indicates a very good agreement between the observers [2, 4].

43.2.3 Data Definition

From this annotated data, sequences are created in the match database. A sequence begins when a team takes possession of the ball (kick-off, free kick, throw-in, corner kick, goal kick, goalkeeper throw of the ball after the opponent's attack, possession of the ball from a tackle or a one-on-one situation). Then, the sequence ends when the team loses possession of the ball (goal, shot on the goal, foul or any loss of control of the ball).

A key challenge to analyze of football matches is to also consider the spatial dimension. For this purpose, the football field is divided into 30 zones, as illustrated in Fig. 43.1. Consider that the attacking direction is from left to right. Then, zones #1–10 are the backfield, zones #11–20 are the midfield and zones #21–30 are the attacking third. Each third covers the same length of 35 m out of a total length of 105 m.

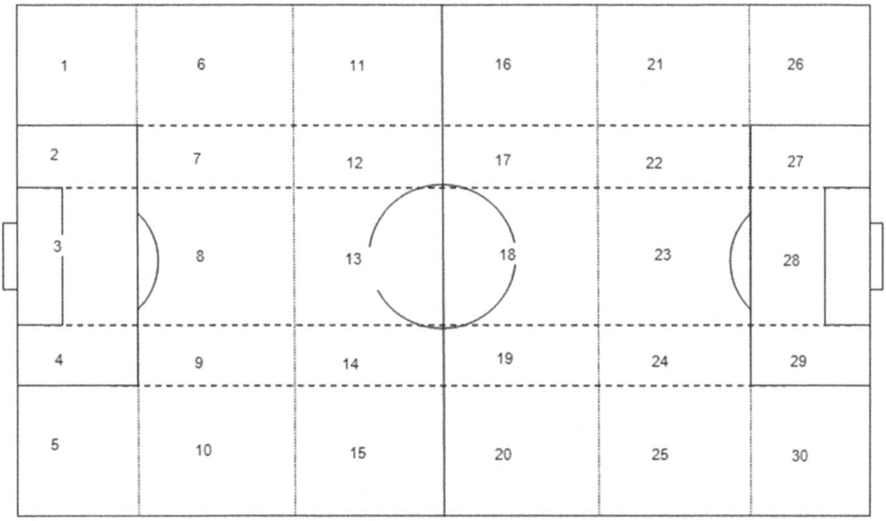

Fig. 43.1 Spatial division of the soccer field areas

43.2.4 Data Pre-processing

After all the data is recorded, to obtain useful diagnostic results, data need to be pre-processed in an appropriate way given the analysis methods. Thus, the last five passes of each sequence are kept in accordance with the five passes theory [3, 9].

43.2.5 Attacking Index

To better express a team's offensive quality, the attacking index defined by Liu [7] is employed to determine the weight (importance) of each attacking sequence. It is defined as a function

$$AI = f(x, n, t)$$

where, x is tactics and techniques applied in a sequence, n represents the number of controlling chains, and t denotes the time. In this paper, time is measured using a measurement unit of 5 min. If a sequence is indexed 3, then it is given a weight of 3 in the database.

43.2.6 Algorithm

Then, to analyze the collected data, an algorithm is applied. In this work, we chose to mine sequential patterns [8], as we are interested in finding patterns common to several attacking sequences. Sequential pattern mining consists of finding subsequences appearing frequently in a set of sequences. However, knowing that a sequence appears frequently does not provide a measure of the certainty that a pattern will occur. An alternative that adds the concept of confidence is sequential rule mining [1]. A sequential rule indicates that if some event(s) occur in a sequence, some other event(s) are likely to occur afterward with a given confidence or probability.

This study used the RuleGrowth algorithm, which is a very efficient algorithm developed for discovering sequential rules that appear in sequence databases [1]. The RuleGrowth algorithm outputs all sequential rules having a support and confidence higher than user-specified thresholds min_sup and min_conf, respectively. A sequential rule $X \Rightarrow Y$ is a sequential relationship between two sets of items X and Y such that X and Y are disjoint and unordered. The support of a rule $X \Rightarrow Y$ is the number of sequences that contain $X \cup Y$ divided by the number of sequences in the database, that is support($X \Rightarrow Y$) = P($X \cup Y$)/Number_of_sequences. The confidence of a rule is the number of sequences that contain $X \cup Y$ divided by the number of sequences that contain X, that is confidence ($X \Rightarrow Y$) = P ($X \cup Y$)/P(X).

43.3 Results

The proposed approach was applied to a match that took place on June 29th, 2012 and was video recorded. For the case study presented in this section diagnostic analyses of this game were conducted and passing trends were identified for sequence leading to goals. The diagnostic analysis aimed to find effective playing patterns that had were likely to lead to a goal. Using the proposed method, results could be provided to a coach, and the team could accordingly adjust their strategy and optimize their play.

43.3.1 First Half for Germany

Table 43.1 shows the results for Germany's chance of goal playing patterns in the 1st half (min_sup = 0.1, min_conf = 0.45), which created difficulties for their opponents. The direction No. 7 (Schweinsteiger) to No. 8 (Özil) and No. 20 (Boateng) to No. 23 (Gomez), as well as No. 7 (Schweinsteiger) or No. 20 (Boateng) to No. 23 (Gomez) had a good chance of scoring a goal. The chance of goal direction was greater on the right side. That means, a German attack from the right side created a much more threatening situation for Italy's defense.

43.3.2 Second Half for Germany

Table 43.2 displays the mining results for the effective playing patterns for Germany in the second half (diagnostic results, min_sup = 0.1; min_conf = 0.24 according to the domain expert). They are from No. 7 (Schweinsteiger) to No. 18 (Kroos) on the left, No. 8 (Özil) in the middle and No. 21 (Reus) on the right side.

Table 43.1 Diagnostic sequential rules for Germany in the 1st half

Series	Rules			Support (%)	Confidence (%)
1	7	==>	8	0.181818182	0.52
2	20	==>	23	0.125874126	0.461538462
3	7, 20	==>	23	0.104895105	0.833333333

Table 43.2 Diagnostic sequential rules for Germany in the 2nd half

Series	Rules			Support (%)	Confidence (%)
1	7	==>	21	0.105769231	0.255813953
2	7	==>	8	0.105769231	0.255813953
3	7	==>	18	0.100961538	0.244186047

43.4 Summary and Discussion

In this article, using a sequential pattern mining method, we analyzed Germany attacking pattern in the football match Germany vs. Italy in the 2012 European championship. The goal of this study is to find effective attacking patterns of the Germany national team. It was found that for the Germany national team, the key players in this game are No.7 (Schweinsteiger), No.20 (Boateng), No. 23 (Gomez), No.18 (Özil) who created a high chance of scoring. Using the Rulegrowth algorithm, key player combinations have been identified with respect to scoring opportunity. Comparing previous studies [5, 7], the proposed approach has the benefit of taking the passing order into account, which better describes a football game and is thus more realistic. Moreover, incorporating the attacking index helped researchers to understand the weight of each offensive chain, which assisted to find key players. Future work should continue seeking algorithms that can best simulate the football game which best describe and diagnose the game, stream data might be a good direction.

Acknowledgements The authors would like to thank all the participants involved in this work. Markus Kraus and Simon Scholz are acknowledged for their important data collection and comments. The Project is sponsored by the Scientific Research Foundation for the Returned Overseas Chinese Scholars, State Education Ministry and also supported by "the Fundamental Research Funds for the Central Universities (Youth Scholars Program of Beijing Normal University)".

References

1. Fournier-Viger, P., Wu, C.-W., Tseng, V. S., Cao, L., & Nkambou, R. (2015). Mining partially-ordered sequential rules common to multiple sequences. *IEEE Transactions on Knowledge and Data Engineering, 27*(8), 2203–2216.
2. Greve, W., & Wentura, D. (1997). *Wissenschaftliche Beobachtung: Eine Einführung*. PVU/ Beltz: Weinheim.
3. Hughes, C. F. (1990). *The winning formula*. London: Collins.
4. Landis, J. R., & Koch, G. G. (1977). The measurement of observer agreement for categorical data. *Biometrics, 33*, 159–174.
5. Liu, T., Hohmann, A., Chen, Q., Lei, T., & Xue, J. (2017). Apriori-based performance analysis on offense models of Elite Women's Football Games: A case study of Algarve Cup 2012. *Journal of Shanghai University of Sport, 41*(1), 77–82.
6. Liu, T., & Hohmann, A. (2013). Applying data mining to analyze the different styles of offense between Manchester United and FC Barcelona in the European Champions League Final. *International Journal of Sports Science and Engineering, 7*(02), 067–078.
7. Liu, T. (2014). Systematische spielbeobachtung im internationalen leistungsfußball.
8. Mabroukeh, N. R., & Ezeife, C. I. (2010). A taxonomy of sequential pattern mining algorithms. *ACM Computing Surveys, 43*(1), 1–41.
9. Xue, J., Li, Y., & Guo, C. (2007). Study on general pattern of attacking for goal character of each team in the final in 18th World Soccer Cup. *China Sport Science and Technology, 43*(1), 36–40.

Chapter 44
Fuzzy Game Based on Fuzzy Comparison Operator

Cunlin Li, Lin Zhang, and Zhifu Jia

Abstract A fuzzy comparison operator was proposed to explore the fuzzy game. By comparing the fuzzy income of different strategy, the fuzzy equilibrium strategy was established based on a fuzzy comparison operator. At last the necessary and sufficient condition of the fuzzy equilibrium was discussed.

Keywords Fuzzy games · Extension principle · Fuzzy comparison operator

44.1 Introduction

With the complexity of the social and economic system, it is very difficult to the decision maker to exactly predict the outcomes of an event. More and more experts focus on the uncertain phenomena in their research. Several tools to deal with the uncertain phenomena have been established, such as fuzzy set theory, intuitionistic fuzzy number, rough set theory and interval number. Among these tools, fuzzy theory is widely used in game theory. Sakawa and Nishizaki [1, 2] proposed a kind of muti-objective model to investigate the bi-matrix game, whose payoff was a fuzzy variable. They used fuzzy goals for payoffs to model the uncertainty of a player's judgement and maximized reachability of the fuzzy goals to establish the equilibrium strategies. Bector and Chandra [3, 4] introduced a dual fuzzy linear programming model, whose parameters were fuzzy variable. And they transformed the fuzzy game into a dual fuzzy linear mathematical programming. Takashi Maeda [5, 6] considered four kinds of fuzzy order operators, based on these fuzzy order operator, several equilibrium strategies were established for the fuzzy game. Yang and Gao [7] proposed the optimistic equilibrium strategy and the maximum chance equilibrium

C. Li
School of Management, North Minzu University, Yinchuan, China

L. Zhang (✉) · Z. Jia
College of Mathematics and Information Science, North Minzu University, Yinchuan, China
e-mail: bitlcl@163.com

M. Tavana, S. Patnaik (eds.), *Recent Developments in Data Science and Business Analytics*, Springer Proceedings in Business and Economics,
https://doi.org/10.1007/978-3-319-72745-5_44

strategy according to the confidence levels and payoff levels. Li [8] established an asymmetric model based on Maeda's symmetric model. By the crisp parametric bi-matrix games and the pseudo-inverse of a monotone function, three kinds of equilibrium strategies were established.

In Sect. 44.2, some commonly used notations and definitions about the fuzzy set theory were introduced. The fuzzy comparison operator was given. In Sect. 44.3, the fuzzy game was presented, the existence of equilibrium strategies was discussed.

44.2 Prelininaries

Definition 44.1 If $\mu_{\tilde{y}}(t)$ is a membership function of a fuzzy variable $\tilde{y}$, it satisfies the followings

1. $\mu_{\tilde{y}}(t)$ is a continuous function from $\mathcal{R}$ to [0, 1],
2. $\mu_{\tilde{y}}(t) = 0$ for all $t \in (-\infty, a]$.
3. $\mu_{\tilde{y}}(t)$ is strictly increasing on (a, b) with t.
4. $\mu_{\tilde{y}}(t)$ is strictly decreasing on (b, c) with t.
5. $\mu_{\tilde{y}}(t) = 0$, $t \in [c, +\infty)$.
6. $\mu_{\tilde{y}}(t) = 1$, $t = b$.

Here $\tilde{y}_\alpha = \{t \in \mathcal{R} | \mu_{\tilde{y}}(t) \geq \alpha, \alpha \in (0, 1]\}$, let $y_\alpha^L = \inf \tilde{y}_\alpha, y_\alpha^R = \sup \tilde{y}_\alpha$, $\tilde{y}_\alpha = [y_\alpha^L, y_\alpha^R]$.

Definition 44.2 Zadeh [9] Let $\tilde{x}, \tilde{y}$ be fuzzy variables, $\forall \alpha \in [0, 1]$ then,

(i) $\tilde{x} \prec \tilde{y}$ if and only if $(y_\alpha^L, y_\alpha^R) \leq (x_\alpha^L, x_\alpha^R)$ holds.
(ii) $\tilde{x} \cong \tilde{y}$ if and only if $(y_\alpha^L, y_\alpha^R) = (x_\alpha^L, x_\alpha^R)$ holds.

Lemma 44.1 *Sakawa and Yano [10] If* g(t_1, t_2, ···, t_m) *is a continuous function with respect to* t_i, *let* $\tilde{t}_i$ *be fuzzy variables,* $\tilde{t}_{i\alpha} = [t_{i\alpha}^L, t_{i\alpha}^R]$, i = *1*, 2, ···, m. *Then,*

$$\left(\tilde{g}\left(\tilde{t}_1, \tilde{t}_2, \cdots, \tilde{t}_m\right)\right)_\alpha = \{g(t_1, t_2, \cdots, t_m) \in \mathcal{R} | \forall t_i \in [t_{i\alpha}^L, t_{i\alpha}^R], i = 1, 2, \cdots, m\}.$$

Lemma 44.2 *Sakawa and Yano [10]* $\tilde{y}$ *is a fuzzy variable on* $\mathcal{R}$, $\tilde{g}$ *is a fuzzy function derived from crisp functiong, it holds*

(I) $\tilde{g}(\tilde{y}) = \bigcup_{\alpha \in [0,1]} \alpha g(\tilde{y}_\alpha)$,

(II) $\tilde{g}(\tilde{y})_\alpha = \bigcap_{\lambda < \alpha} g(\tilde{y}_\lambda)$.

44.3 The Equilibrium of Game with Fuzzy Payoffs

The game theory that Nash presented is based on an assumption that the payoff of each player was a real number in his paper. In the real world, the decision maker always make his decision in an uncertain environments. In other words, the payoff of the player is not a real number, but an uncertain variable. In this article, for characterizing the real complex world, the payoffs of the players is modeled by an fuzzy variable.

Next, several commonly notations were given. The sets of mixed strategy for Player 1 and player 2 are as the followings

$$S_1 \equiv \left\{ t^T = (t_1, t_2, \cdots, t_m) | t_i \geq 0 \ \sum_{i=1}^{m} t_i = 1, t_i \in \mathcal{R} \ , i = 1, 2, \cdots, m \right\},$$

$$S_2 \equiv \left\{ s^T = (s_1, s_2, \cdots, s_n) \big| s_j \geq 0 \ \sum_{j=1}^{n} s_j = 1, s_j \in \mathcal{R} \ , j = 1, 2, \cdots, n \right\},$$

where $\mathcal{R}$ is real number set.

As our assumption, the payoff of the player is a fuzzy variable like Definition 44.1 giving, it is obvious that the payoff of strategy pair $(i,j) \in S_1 \times S_2$ still is a fuzzy variable.

Definition 44.3 Let $t \in S_1, s \in S_2, \tilde{X}$ and $\tilde{Y}$ are fuzzy matrices, $\tilde{G}_2 = \left(S_1, S_2, \tilde{X}, \tilde{Y}\right)$ is a fuzzy bi-matrix game.

Here, $\tilde{X}$ and $\tilde{Y}$ are the payoff matrices whose element is fuzzy variable.

$$\tilde{X} = \begin{pmatrix} \tilde{x}_{11} & \cdots & \tilde{x}_{1n} \\ \vdots & \ddots & \vdots \\ \tilde{x}_{m1} & \cdots & \tilde{x}_{mn} \end{pmatrix}, \tilde{Y} = \begin{pmatrix} \tilde{y}_{11} & \cdots & \tilde{y}_{1n} \\ \vdots & \ddots & \vdots \\ \tilde{y}_{m1} & \cdots & \tilde{y}_{mn} \end{pmatrix}.$$

Definition 44.4 Let $\tilde{G}_2 = \left(S_1, S_2, \tilde{X}, \tilde{Y}\right)$ be a fuzzy bi-matrix game. If a pair (t^*, s^*) $\in S_1 \times S_2$ satisfies (I) and (II),

$$\begin{aligned} &(\mathrm{I})\ t^T \tilde{X} s^* \prec t^{*T} \tilde{X} s^*, \quad \forall t \in S_1, \\ &(\mathrm{II})\ t^{*T} \tilde{Y} s \prec t^{*T} \tilde{Y} s^*, \quad \forall s \in S_2. \end{aligned}$$

The pair (t^*, s^*) is a fuzzy equilibrium strategy of $\tilde{G}_2$.

Theorem 44.1 *Let* $\tilde{G}_2 = \left(S_1, S_2, \tilde{X}, \tilde{Y}\right)$ *be a fuzzy game, if* $t \in S_I$, $s \in S_J$, $\alpha \in [0, 1]$, *then*

$$\begin{aligned} \left(t^T \tilde{X} s\right)_\alpha &\triangleq \left[t^T X_\alpha^L s, t^T X_\alpha^R s\right], \\ \left(t^T \tilde{Y} s\right)_\alpha &\triangleq \left[t^T Y_\alpha^L s, t^T Y_\alpha^R s\right]. \end{aligned}$$

Proof We just prove one of above equations, the other can be proved by same way.

From Lemma 44.1, it is obvious that

$$\left(t^T\tilde{Y}s\right)_\alpha = \left(\sum_{i=1}^{m}\sum_{j=1}^{n} t_i\tilde{y}_{ij}s_j\right)_\alpha = \left\{\sum_{i=1}^{m}\sum_{j=1}^{n} t_i u_{ij}s_j \middle| u_{ij} \in \left(\tilde{y}_{ij}\right)_\alpha\right\}$$
$$= \left[\sum_{i=1}^{m}\sum_{j=1}^{n} t_i\left(y_{ij}\right)_\alpha^L s_j, \sum_{i=1}^{m}\sum_{j=1}^{n} t_i\left(y_{ij}\right)_\alpha^R s_j\right] = \left[t^T Y_\alpha^L s, t^T Y_\alpha^R s\right]. \qquad \square$$

Theorem 44.2 $(t^*, s^*) \in S_1 \times S_2$ *is the fuzzy equilibrium strategy of* $\tilde{G}_2 = \left(S_1, S_2, \tilde{X}, \tilde{Y}\right)$ *if and only if (44.1) and (44.2) hold, for* $i = 1, 2, \cdots, m$, $j = 1, 2, \cdots, n$.

$$e_i^T\tilde{X}s^* \prec t^{*T}\tilde{X}s^*, \tag{44.1}$$
$$t^{*T}\tilde{Y}d_j \prec t^{*T}\tilde{Y}s^*, \tag{44.2}$$

where e_i and d_j are as followings

$$e_i = (\overbrace{\underbrace{0,\cdots,0}_{i-1},1,0,\cdots,0}^{m})^T, d_j = (\overbrace{\underbrace{0,\cdots,0}_{j-1},1,0,\cdots,0}^{n})^T.$$

Proof We only prove (44.2), (44.1) can be proved by same way.

If (t^*, s^*) is a fuzzy equilibrium strategy of $\tilde{G}_2$, next inequality holds

$$t^{*T}\tilde{Y}s \prec t^{*T}\tilde{Y}s^*, \quad \forall s \in S_2.$$

From Theorem 44.1, it indicates

$$\left(t^{*T}Y_\alpha^L s, t^{*T}Y_\alpha^R s\right) \le \left(t^{*T}Y_\alpha^L s^*, t^{*T}Y_\alpha^R s^*\right), \forall\alpha \in (0, 1].$$

For, $j = 1, 2, \cdots, n$, by the randomness of s, let $s = d_j$, we have

$$\left(t^{*T}Y_\alpha^L d_j, t^{*T}Y_\alpha^R d_j\right) \le \left(t^{*T}Y_\alpha^L s^*, t^{*T}Y_\alpha^R s^*\right).$$

From the arbitrariness of α and Theorem 44.1, (44.2) holds.

Otherwise, we assume that (44.1) and (44.2) hold. Then, From Theorem 44.1, for $\alpha \in [0, 1]$ we obtain

$$t^{*T}Y_\alpha^L d_j \le t^{*T}Y_\alpha^L s^*, \tag{44.3}$$
$$t^{*T}Y_\alpha^R d_j \le t^{*T}Y_\alpha^R s^*, \tag{44.4}$$

where $j = 1, 2, \cdots, n$.

If y_j is the j-th element of y, the followings hold

$$\begin{aligned}
\left(t^{*T} Y_\alpha^L d_j\right) s_j &\leq \left(t^{*T} Y_\alpha^L s^*\right) s_j, \\
\left(t^{*T} Y_\alpha^R d_j\right) s_j &\leq \left(t^{*T} Y_\alpha^R s^*\right) s_j.
\end{aligned}$$

Because of

$$\left(s^{*T} Y_\alpha^L d_j\right) s_j = \left[\sum_{i=1}^{m} t_i \left(y_{ij}\right)_\alpha^L\right] s_j = \sum_{i=1}^{m} t_i \left(y_{ij}\right)_\alpha^L s_j \tag{44.5}$$

and

$$\left(t^{*T} Y_\alpha^L s^*\right) s_j = \left[\sum_{i=1}^{m} \sum_{j=1}^{n} t_i \left(y_{ij}\right)_\alpha^L s_j\right] s_j \tag{44.6}$$

To sum (44.5) and (44.6) from $j = 1$ to $j = n$, it indicates that

$$\begin{aligned}
\sum_{j=1}^{n} \left(t^{*T} Y_\alpha^L d_j\right) s_j &= \sum_{j=1}^{n} \sum_{i=1}^{m} t_i \left(y_{ij}\right)_\alpha^L s_j = t^{*T} Y_\alpha^L s, \\
\sum_{j=1}^{n} \left(t^{*T} Y_\alpha^L s^*\right) s_j &= t^{*T} Y_\alpha^L s^* \sum_{j=1}^{n} s_j = t^{*T} Y_\alpha^L s^*.
\end{aligned}$$

It means the following inequality holds

$$t^{*T} Y_\alpha^L s \leq t^{*T} Y_\alpha^L s^*,$$

in a similar manner, we have

$$t^{*T} Y_\alpha^R s \leq t^{*T} Y_\alpha^R s^*,$$

namely,

$$\left(t^{*T} Y_\alpha^L s, t^{*T} Y_\alpha^R s\right) \leq \left(t^{*T} Y_\alpha^L s^*, t^{*T} Y_\alpha^R s^*\right).$$

According to Definition 44.2, it holds

$$t^{*T} \tilde{Y} s \prec t^{*T} \tilde{Y} s^*, \quad \forall s \in S_J.$$

Then, the Theorem is proved.

In next of this paper, we will consider a fuzzy matrix game $\tilde{G}_1 = \left(S_1, S_2, \tilde{Y}\right)$.

Definition 44.5 The function $\underset{t\in S_1}{\text{Fmax}}\, t^T\tilde{Y}s$ and $\underset{t\in S_1}{\text{Fmin}}\, t^T\tilde{y}s$ are defined as follow

$$\text{(i) } \underset{t\in S_1}{\text{Fmax}}\ t^T\tilde{Y}s = \underset{\alpha\in[0,1]}{\cup} \alpha\left\{\underset{t\in S_1}{\max} t^T Bs \middle| t^T Bs \in \left[t^T Y_\alpha^L s, t^T Y_\alpha^R s\right]\right\},$$

$$\text{(ii) } \underset{t\in S_1}{\text{Fmin}}\ t^T\tilde{Y}s = \underset{\alpha\in[0,1]}{\cup} \alpha\left\{\underset{t\in S_1}{\min} t^T Bs \middle| t^T Bs \in \left[tY_\alpha^L s, t^T Y_\alpha^R s\right]\right\}.$$

$$\text{here } B = (b_{ij})_{m\,\times\,n},\ b_{ij} \in \left[\left(y_{ij}\right)_\alpha^L, \left(y_{ij}\right)_\alpha^R\right].$$

Theorem 44.3 $\tilde{G}_1 = \left(S_1, S_2, \tilde{Y}\right)$ *is a fuzzy matrix game, if* t $\in$ S_I, s $\in$ S_2, *the following inequalities hold*

(i) $t^T\tilde{Y}s \prec \underset{t\in S_1}{\text{Fmax}}\, t^T\tilde{Y}s,$

(ii) $\underset{s\in S_2}{\text{Fmin}}\, t^T\tilde{Y}s \prec t^T\tilde{Y}s.$

Proof From Lemma 44.2, it implies

$$\left(\underset{t\in S_1}{\text{Fmax}}\, t^T\tilde{Y}s\right)_\alpha = \underset{\lambda<\alpha}{\cap} \{\underset{t\in S_1}{\max} t^T Bs | t^T Bs \in \left[t^T Y_\lambda^L s, t^T Y_\lambda^R s\right]\}.$$

From above we have

$$\underset{t\in S_1}{\max} t^T Y_\lambda^L s \le \underset{t\in S_I}{\max} t^T Bs \le \underset{t\in S_1}{\max} t^T Y_\lambda^R s,$$

$$\text{Then,}\quad \left(\underset{t\in S_1}{\text{Fmax}}\, t^T\tilde{Y}s\right)_\alpha = \underset{\lambda<\alpha}{\cap} \left[\underset{t\in S_1}{\max} t^T Y_\lambda^L s, \underset{t\in S_1}{\max} t^T Y_\lambda^R s\right]$$

Since $t^T Y_\lambda^L s$ monotonically increases about λ and $t^T Y_\lambda^R s$ monotonically decreases about λ. Then

$$\left(\underset{t\in S_1}{\text{Fmax}}\, t^T\tilde{Y}s\right)_\alpha = \underset{\lambda<\alpha}{\cap} \left[\underset{t\in S_1}{\max} t^T Y_\lambda^L s, \underset{t\in S_1}{\max} t^T Y_\lambda^R s\right] = \left[\underset{t\in S_1}{\max} t^T Y_\alpha^L s, \underset{t\in S_1}{\max} t^T Y_\alpha^R s\right].$$

Thus, it is obvious that $\left(t^T Y_\alpha^L s, t^T Y_\alpha^R s\right) \le \left(\underset{t\in S_1}{\max} t^T Y_\alpha^L s,\ \underset{t\in S_1}{\max} t^T Y_\alpha^R s\right).$

Based on the arbitrariness of α, it implies that $t^T\tilde{Y}_\alpha^L s \prec \underset{t\in S_1}{\text{Fmax}}\, t^T\tilde{Y}_\alpha^R s.$

As similar way, we obtain the proof of (II). □

Theorem 44.4 $(t^*, s^*) \in S_I \times S_J$ *is a fuzzy equilibrium strategy of* $\tilde{G}_1 = \left(S_1, S_2, \tilde{Y}\right)$ *if and only if for* t $\in$ S_I, s $\in$ S_2, *(44.7) holds that*

$$t^{*T}\tilde{Y}s^* \cong \underset{t\in S_1}{\text{Fmax}}\ \underset{s\in S_2}{\text{Fmin}}\, t^T\tilde{Y}s \cong \underset{t\in S_1}{\text{Fmin}}\ \underset{s\in S_2}{\text{Fmax}}\, t^T\tilde{Y}s. \tag{44.7}$$

Proof Let $t^* \in S_1$ and $s^* \in S_2$ such that (44.7) hold. It implies

$$\underset{s\in S_2}{\text{Fmin}}\, t^{*T}\tilde{Y}s \cong \underset{t\in S_1}{\text{Fmax}}\ \underset{s\in S_2}{\text{Fmin}}\, t^T\tilde{Y}s,\ \underset{t\in S_1}{\text{Fmax}}\, t^T\tilde{Y}s^* \cong \underset{s\in S_2}{\text{Fmin}}\ \underset{t\in S_1}{\text{Fmax}}\, t^T\tilde{Y}s.$$

From (44.7), it implies that $\underset{t\in S_1}{\text{Fmax}}\, t^T\tilde{Y}s^* \cong \underset{s\in S_2}{\text{Fmin}}\, t^{*T}\tilde{Y}s \cong t^{*T}\tilde{Y}s^*$.
According to Theorem 44.2, it holds that

$$t^T\tilde{Y}s^* \prec \text{F}\underset{t\in S_1}{\max}\, t^T\tilde{Y}s^* \cong t^{*T}\tilde{Y}s^* \cong \text{F}\underset{s\in S_2}{\min}\, t^{*T}\tilde{Y}s \prec t^{*T}\tilde{Y}s, t\in S_1, s\in S_2.$$

if the point (t^*, s^*) is the fuzzy equilibrium strategy. It implies that

$$t^T\tilde{Y}s^* \prec t^{*T}\tilde{Y}s^* \prec t^{*T}\tilde{Y}s, t\in S_1, s\in S_2$$

From Definition 44.3, for $\alpha\in[0,1]$, we have

$$\begin{aligned}\left(t^TY_\alpha^Ls^*, t^TY_\alpha^Rs^*\right) &\le \left(t^{*T}Y_\alpha^Ls^*, t^{*T}Y_\alpha^Rs^*\right),\\ \left(t^{*T}Y_\alpha^Ls^*, t^{*T}Y_\alpha^Rs^*\right) &\le \left(t^{*T}Y_\alpha^Ls, t^{*T}Y_\alpha^Rs\right).\end{aligned}$$

Then, from the arbitrariness of t and s, it is obvious that

$$\begin{aligned}\left(\max_{t\in S_1} t^TY_\alpha^Ls^*, \max_{t\in S_1} t^TY_\alpha^Rs^*\right) &\le \left(t^{*T}Y_\alpha^Ls^*, t^{*T}Y_\alpha^Rs^*\right),\\ \left(t^{*T}Y_\alpha^Ls^*, t^{*T}Y_\alpha^Rs^*\right) &\le \left(\min_{s\in S_1} t^{*T}Y_\alpha^Ls, \min_{s\in S_2} t^{*T}Y_\alpha^Rs\right).\end{aligned}$$

Based on the arbitrariness of α and Definition 44.2, it holds that

$$\underset{t\in S_1}{\text{Fmax}}\, t^T\tilde{Y}s^* \prec t^{*T}\tilde{Y}s^* \prec \underset{y\in S_2}{\text{Fmin}}\, t^{*T}\tilde{Y}s.$$

By Theorem 44.3, it holds

$$\underset{s\in S_2}{\text{Fmin}}\ \underset{t\in S_1}{\text{Fmax}}\, t^T\tilde{Y}s \prec \underset{t\in S_1}{\text{Fmax}}\, t^T\tilde{Y}s^* \prec \underset{s\in S_2}{\text{Fmin}}\, t^{*T}\tilde{Y}s \prec \underset{t\in S_1}{\text{Fmax}}\ \underset{s\in S_2}{\text{Fmin}}\, t^T\tilde{Y}s.$$

As the similar way, it indicates that $\underset{s\in S_2}{\text{Fmin}}\ \underset{t\in S_1}{\text{Fmax}}\, t^T\tilde{Y}s \prec \text{m}\underset{t\in S_1}{\text{Fmax}}\ \underset{s\in S_2}{\text{Fmin}}\, t^T\tilde{Y}s$.
Otherwise, from Theorem 44.3, it indicates that

$$\underset{s\in S_2}{\text{Fmin}}\, t^{*T}\tilde{Y}s \prec t^{*T}\tilde{Y}s^* \prec \underset{t\in S_1}{\text{Fmax}}\, t^T\tilde{Y}s^*.$$

By Definition 44.3, if $\alpha\in[0,1]$, it means

$$\left(\min_{t\in S_2} t^{*T}Y_\alpha^Ls, \min_{s\in S_2} t^{*T}Y_\alpha^Rs\right) \leqq \left(t^{*T}Y_\alpha^Ls^*, t^{*T}Y_\alpha^Rs^*\right), \left(t^{*T}Y_\alpha^Ls^*, t^{*T}Y_\alpha^Rs^*\right)$$
$$\le \left(\max_{t\in S_1} t^TY_\alpha^Ls^*, \max_{t\in S_1} t^TY_\alpha^Rs^*\right).$$

It implies that

$$\left(\min_{s\in S_2} t^{*T} Y_\alpha^L s,\ \min_{s\in S_2} t^{*T} Y_\alpha^R s\right) \le \left(\max_{t\in S_1} t^T Y_\alpha^L s^*,\ \max_{t\in S_1} t^T Y_\alpha^R s^*\right)$$

From $\left(\min_{s\in S_2}\max_{t\in S_1} t^T Y_\alpha^L s,\ \min_{s\in S_2}\max_{t\in S_1} t^T Y_\alpha^R s = \left(\max_{t\in S_1} t^T Y_\alpha^L s^*,\ \max_{t\in S_1} t^T Y_\alpha^R s^*\right)\right.$, then

$$\left(\min_{s\in S_2} t^{*T} Y_\alpha^L s,\ \min_{s\in S_2} t^{*T} Y_\alpha^R s\right) \le \left(\min_{s\in S_2}\max_{t\in S_1} t^T Y_\alpha^L s,\ \min_{s\in S_2}\max_{t\in S_1} t^T Y_\alpha^R s\right).$$

From $\left(\max_{t\in S_1}\min_{s\in S_2} t^T Y_\alpha^L s,\ \max_{t\in S_1}\min_{s\in S_2} t^T Y_\alpha^R s\right) = \left(\min_{s\in S_2} t^{*T} Y_\alpha^L s,\ \min_{s\in S_2} t^{*T} Y_\alpha^R s\right)$, then

$$\left(\max_{t\in S_1}\min_{s\in S_2} t^T Y_\alpha^L s,\ \max_{t\in S_1}\min_{s\in S_2} t^T Y_\alpha^R s\right) \le \left(\min_{s\in S_2}\max_{t\in S_1} t^T Y_\alpha^L s,\ \min_{s\in S_2}\max_{t\in S_1} t^T Y_\alpha^R s\right).$$

From the arbitrariness of α and Definition 44.2,

$$\operatorname*{Fmax}_{t\in S_1}\operatorname*{Fmin}_{s\in S_2} t^T \tilde{Y} s \prec \mathrm{m} \operatorname*{Fmin}_{s\in S_2}\operatorname*{Fmax}_{t\in S_1} t^T \tilde{Y} s.$$

So, it holds the Eq. (44.7) . □

Acknowledgments This paper is supported by the National Natural Science Foundation of China (No. 71561001); the Key Social Science Research Base of State Ethnic Affairs Commission of China-Governance and social management research center of Northwest Ethnic regions (No. MWJD 201612); the Foundation of North Minzu University (GLXY201608).

References

1. Nishizaki, I., & Sakawa, M. (2000). Uilibrium solutions in multi-objective bi-matrix games with fuzzy payoffs and fuzzy goals. *Fuzzy Sets and Systems, 111*, 99–116.
2. Nishizaki, I., & Sakawa, M. (2001). *Fuzzy and multi-objective games for conflict resolution*. New York: Physica-Verlag.
3. Bector, C. R., & Chandra, S. (2005). *Fuzzy mathematical programming and fuzzy matrix games*. Berlin: Springer.
4. Bector, C. R., & Chandra, S. (2004). Duality in linear programming with fuzzy parameters and matrix games with fuzzy pay-offs. *Fuzzy Sets and Systems, 146*, 253–269.
5. Takashi, M. (2003). On characterization of equilibrium strategy of two-person zero-sum games with fuzzy payoffs. *Fuzzy Sets and Systems, 139*, 283–296.
6. Takashi, M. (2000). Characterization of the equilibrium strategy of the bimatrix game with fuzzy payoff. *Journal of Mathematical Analysis and Applications, 251*, 885–896.
7. Xiangfeng, Y., & Jinwu, G. (2014). Bayesian equilibria for uncertain bimatrix game with asymmetric information. *Journal of Intelligent Manufacturing, 12*(1), 65–78.
8. Cunlin, L. (2012). Characterization of the equilibrium strategy of fuzzy bimatrix games based on L-R fuzzy variables. *Journal of Applied Mathematics, 2012*, 1–15.
9. Zadeh, L. A. (1975). The concept of a linguistic variable and its applications in approximate reasoning. *Information Sciences, 8*, 199–252.
10. Sakawa, M., & Yano, H. (1991). Feasibility and Pareto optimality for multi-objective nonlinear programming problems with fuzzy parameters. *Fuzzy Sets and Systems, 43*, 1–15.

Chapter 45
An Analysis of Business English Translation Research in China

Xue Yu and Shili Ge

Abstract CiteSpace II is used to draw the scientific knowledge map of China's business translation research. The literature of business translation studies from 2012 to 2016 in CNKI database are used as data source. Research on hotspots and fronts is realized by analyzing keywords frequency and burst terms. It is found that the research hotspots include Business English, cultural differences, translation techniques, and so on. Translation teaching is receiving more and more attention. The research fronts include Skopos theory, foreign business correspondence, metaphor, Business English translation teaching, higher vocational education and foreign trade English.

Keywords Business English translation · Research hotspots · Research fronts · CiteSpace II

45.1 Introduction

Business English is regarded as a branch of English for Specific Purposes (ESP). The vocabulary, syntax and discourse of Business English have specific features. Due to the features, Business English translation is different from general English. In 1990s, the study of Business English translation followed up [1], and has gradually evolved into a hot research area. The diversity of needs, size and content forms is undergoing tremendous changes, and Business English translation is becoming the main content of translation activities [2]. The analysis of the research results of Business English

X. Yu
School of English for International Business, Guangdong University of Foreign Studies, Guangzhou, China

S. Ge (✉)
Collaborative Innovation Center for Language Research and Service, Guangdong University of Foreign Studies, Guangzhou, China
e-mail: garyfirst@qq.com

M. Tavana, S. Patnaik (eds.), *Recent Developments in Data Science and Business Analytics*, Springer Proceedings in Business and Economics,
https://doi.org/10.1007/978-3-319-72745-5_45

translation can not only reveal its development status, but also assist translators in grasping the research trends and development trends in the field.

45.2 Data Sources and Analysis Tools

45.2.1 Data Source

CNKI (China National Knowledge Infrastructure) is the origin of relevant literatures. Referring to the data retrieval condition in [2], and choosing philosophy and humanities, social sciences, information technology as well as economics and management science as disciplinary fields, we use advanced search and chose "journal". The search condition is set to "abstract", and words such as "business" "translation" "business trade" are input. As a result, 1806 articles are found. By deleting conference notification, advertisement and other irrelevant documents, we finally selected 1500 articles. The retrieval date is May 24, 2017.

45.2.2 Analysis Tools

CiteSpace is a freely available application for visualizing and analyzing trends and patterns in literature. The data containing information such as author, title, keywords, source publication, abstract and so on from CNKI are saved as Refworks form and converted by the inbuilt function in CiteSpace. We select "Title" "Abstract" "Author Keyword" and "Keyword Plus" as "Term Source", "Noun Phrases" as "Term Type", "Terms" as "Node Type" and keep other parameters unchanged. "Time Slicing" is set as "2012–2016".

45.3 Data Analysis and Discussion

45.3.1 Keywords Analysis

As keywords are generally the summarization of the key content of the literature, the statistics and analysis of keywords can be used to determine the research hotspots and important themes in a certain field of study [3]. The threshold is set as 20, which means the frequency of keywords more than 20 will be counted and displayed in CiteSpace. In Table 45.1, the high-frequency keywords in Business English translation research are presented.

"Business English", "Translation", and "Business English translation" are the first three keywords with the highest frequency. "Cultural Difference", the fourth high frequency keywords, is the key factor in influencing Business English

Table 45.1 High-frequency keywords of domestic Business English translation research

Keywords	Counts	Keywords	Counts
Business English	439	Skopos theory	36
Translation	322	Linguistic features	36
Business English translation	181	Problems	35
Cultural difference	71	English translation	29
Translation strategy	69	Translation principles	29
Business translation	61	Translation methods	29
Translation techniques	60	Techniques	28
Translation teaching	56	Application	27
Features	41	Methods	26
Strategies	40	Teaching mode	26
Functional equivalence	37	Functional translation theory	21
Business contracts	36		

translation. Business English translation is cross-culture and cross-language communication activities [4]. It is not only the conversion between language symbols but also the result of the consideration of the existence of cultural difference between countries. Translation theories such as "Functional Equivalence" "Skopos Theory" "functional translation theory" are the theoretical basis of Business English translation. "Translation Strategies" "Techniques" "Methods" occur many times. Under the guidance of translation theories, specific translation principles and strategies are put forward to realize the guiding role of translation theory in the process of translation. Translation teaching should be connected with market needs. Ref. [5] points out several problems in translation teaching: translation teaching lacks the market-oriented guiding ideology; the preparation of teaching material fails to keep up with the pace of the times; the curriculum lacks the consideration of the combination of language teaching and knowledge teaching; and teaching methods are outdated. It is imperative to reform the Business English translation teaching mode. Business English is interdisciplinary and the study of the linguistic features in vocabulary, sentence and discourse is beneficial to translation work.

45.3.2 Analysis of Burst Terms

Burst refers to the change of frequency. The burst strength in the key literature and key words can reflect the changes in research interests in the field and is an important basis to judge the frontier of the study [6]. In Table 45.2, burst terms of domestic Business English translation are presented.

Burst terms in 2012 are "Skopos theory", "foreign business correspondence", "metaphor", and "Business English translation teaching". Burst terms in 2013 are "business letters", "higher vocational education" and "foreign trade English". The research fronts mainly focus on two topics: "translation theory" and "Business

Table 45.2 Burst terms of domestic Business English translation

Keywords	Year	Strength	Begin	End
Skopos theory	2012	2.1245	2012	2013
Foreign business correspondence	2012	2.7345	2012	2013
Metaphor	2012	1.82	2012	2013
Business English translation teaching	2012	2.7345	2012	2013
Business letters	2012	1.9374	2013	2014
Higher vocational education	2012	1.9281	2013	2016
Foreign trade English	2012	2.6537	2013	2016

English teaching". Skopos theory, providing an insight into the nature of translation as a purposeful activity, is also the hotspot of domestic translation study. Metaphor, as a way of thinking and a means of cognition, serves as an important component of English language. The understanding of metaphor will assist in Business English translation. Other five keywords belong to Business English translation teaching. Foreign business correspondence and business letters are widely used in international trade. But there are still problems in the teaching. Great efforts should be made to improve the teaching effect.

45.4 Conclusion

With the help of CiteSpace, research hotspots and research fronts of domestic Business English translation studies from 2012 to 2016 are presented by keywords analysis and burst terms. As the data we obtained are not complete, the analysis result of this paper is not comprehensive. Further studies can focus on the cited references analysis to find out the most influential articles and authors in domestic Business English translation studies and comparative study of both domestic and foreign studies of Business English translation can be conducted to generate greater reference value.

Acknowledgments This research was supported by Graduate Education Innovation Plan of Guangdong Province (2015JGXM-MS22), the Science and Technology Project of Guangdong Province, China (2017A020220002) and the fund of Center for Translation Studies, Guangdong University of Foreign Studies (CTS2014-13).

References

1. Fang, M. (2012). Thirty years of application translation studies (1980–2010). *Shanghai Journal of Translators*, (2), 22–27.
2. Xu, J., & Xia, R. (2013). Analyzing the ten-year developments of business translation research in China (2002–2011). *Foreign Languages in China, 10*(4), 11–18.

3. Li, H. (2014). Visualization and analysis of hotspots and fronts in international translation studies. *Chinese Translators Journal, 35*(02), 21–26.
4. Zhao, X. (2016). Cultural differences in Business English translation and countermeasures. *Ability and Wisdom*, (27), 255.
5. Hu, D., & Wang, J. (2009). Understanding of translation teaching based on market demand and its countermeasures. *Foreign Language World*, (04), 43–47.
6. Zhu, J., & Cai, J. (2013). The evolution path and forefront hot issues of e-educology of foreign languages: Knowledge map analysis based on CSSCI data. *Technology Enhanced Foreign Language Education*, (02), 13–18.

Part V
Communications and Information Systems Analytics

Chapter 46
Optimum Design and Function Improvement of Mobile Nursing System

Gang Tong, Huan Liu, and Luxing Wang

Abstract To improve the convenience and efficiency of nursing work, in order to better promote the nurses groups and the use of nursing care to break the limitations of the PDA. An objective and fair performance evaluation method was used to improve the working statistics module in mobile nursing system and to increase the information storage and transmission function of handheld mobile terminal (PDA), and to improve and optimize the mobile nursing system. After the optimization of the mobile care system, you can better promote the enthusiasm of the nursing work to improve, but also can improve the convenience of PDA use. The improvement of mobile nursing system improves the enthusiasm of nurses' nursing work, enlarges the application scope of PDA, and promotes the process of digitization and informationization in medical field.

Keywords Mobile nursing · PDA · Performance appraisa · Temporary storage of information

46.1 Introduction

The clinical application of mobile care system is an important step in the transformation of information technology, which promotes the modernization, networking and informationization of medical business. After the actual use of the hospital department, although the existed system has brought great convenience in the nursing work greatly, but in the process of latter maintenance and use of the feedback, there are some problems to be solved, After handling these problrms, It can Optimize and improve the practicality of the system greatly.

G. Tong (✉) · H. Liu · L. Wang
School of Information Science and Technology, Qingdao University of Science and Technology, Qingdao, China
e-mail: tg.gg@163.com

M. Tavana, S. Patnaik (eds.), *Recent Developments in Data Science and Business Analytics*, Springer Proceedings in Business and Economics,
https://doi.org/10.1007/978-3-319-72745-5_46

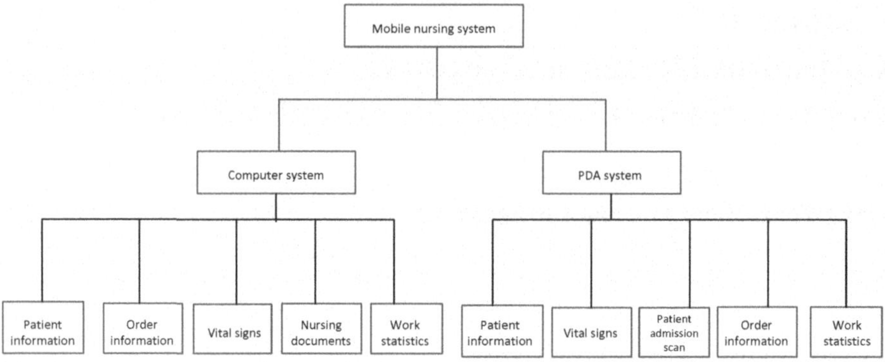

Fig. 46.1 System architecture

46.2 Overview of Mobile Nursing System

Mobile care system can be divided into two parts, including PC-side system and PDA-side system. PC system operator interface can provide convenient and efficient information processing, PDA-side system is able to achieve the collection of nursing information and network convenient transmission. Figure 46.1 is the schematic diagram of the system function.

46.3 Research Objectives

Although the application of mobile nursing system has realized the digital transformation of nursing work to a great extent, it still have some problems to be solved in practical application. Through the actual maintenance of the system and the feedback of the nursing staff, the following problems are summarized and need to improve.

1. The work of the system statistics module for the care of the work is too simple, It can't objectively reflect the merits of the work of nursing staff. In order to promote the enthusiasm of nurses, the Ministry of Nursing suggest to use the method of performance appraisal to implement the incentive policy, in order to reflect the principle of fair and reasonable, more labor and more benefits, To carry out the digital processing of performance evaluation, it is suggested that the performance evaluation should be added into the mobile nursing system, and the work statistics should be carried out in an objective and scientific way.
2. PDA in the actual use of the process has its limitations. It is too dependent on the hospital LAN, if the local area network connection is abnormal, the network signal is weak, the network signal coverage and database server is failure and so on, Then the PDA can't be put to use normally.

3. PDA scan head alignment and are not sensitive enough, based on the scanning operation is not smooth; the wireless network can't achieve the desired roaming, communication quality is not stable, sometimes there will be some abnormal conditions, such as data submission exception [1].

46.4 The Specific Method of Mobile Nursing System Improvement

46.4.1 Work Statistics Module Improvement

Throughout the domestic and foreign research on nursing evaluation criteria, there is no uniform measure of nursing performance evaluation. In order to establish an objective and fair management method of performance evaluation, the following methods are summarized.

The assessment criteria were divided into five modules: nursing staff qualifications, nursing workload, nursing quality, nursing research and teaching, and work attitude. The performance of wages into a score of 100 points, the performance of wages = performance of the total wage/100 * performance wage scores, performance payroll score = seniority score + care workload score + care quality score + nursing research and teaching score + work attitude score.

According to the work experience qualifications, titles (primary, intermediate, high), education (secondary school, college, undergraduate, graduate students), According to the special skills of nursing care to score the nursing staff, the higher the score, the more senior. Qualification score can account for 10% of the total score.

Nursing workload should be calculated respectively by the number of nursing posts, nursing care, nursing level, day shift and night shift, working hours. Nursing workload scores can account for 20% of the total performance score.

Nursing quality can be divided into nursing difficulty, work completion, etc.. The difficulty of nursing is the most difficult to take effective indicators to measure. Many scholars at home and abroad have put forward the evaluation criteria, including Yamase [2], and so on, using Delphi method to construct a comprehensive nursing intervention scoring system; Bjork et al. [3] proposed that the coordination, flexibility, decision-making ability, patient coordination and communication skills of nurses affect the nursing behavior of nurses; Ma Jingjing et al. [4] and Zhang Yan et al. [5] established the evaluation index system of nursing difficulty by using Delphi method respectively, and the relevant difficulty coefficient and weight index are obtained, which makes the difficulty of nursing can be scientifically measured [6]. Through the research of domestic and international measurement standards.

The difficulty of nursing evaluation can be divided into two aspects: the difficulty of nursing operation and the evaluation of the difficulty of nursing. According to each index to give the score, nursing quality score accounted for the proportion of total performance score is 45%.

Nursing research and teaching can be refined into the hospital teaching situation inside and outside, nursing skills training, professional skills training, to help guide the nurse, published papers (municipal, provincial, national), nursing research and nursing new technology development. For each of the detailed reference, nursing research and teaching accounted for 15% of the total performance.

Work attitude is directly reflected in the patient's service experience and personal honor award. According to the patient's visit to the satisfaction of the survey, comprehensive consideration of the patient's praise and complaints, etc., personal honor can be divided into provincial, municipal, hospital level and the Department of reward assessment. The score of nurses' work attitude can account for 10% of the total score.

The system is implemented in the follow way: Each person's system login account bind their personal information together to achieve qualifications statistics, By the system automatically read the qualification information to score; Nursing workload can be calculated through personal account login statistics, including PC-side system login and PDA-side system login usage; According to the classification of the factors affecting the difficulty of nursing to automatically judge the quality of work; By the individual to fill in nursing research and teaching and work attitude and score it, the head nurse to review.

46.4.2 PDA Data Temporary Design

When the PDA collects the data, the data is stored in the storage format required by the hospital database. There are two ways for PDA to upload the information, wired uploads and wireless uploads. Cable upload is to use USB data cable to connected the PDA device with the computer, through the computer upload to the server. In this way, the data can be stored in the database when the wireless network fails. Wireless upload mainly make use of the function of the traditional PDA WLAN connection. When the PDA access to the hospital LAN, PDA can start the data upload function. As shown in Fig. 46.2 PDA temporary – upload function diagram.

PDA temporary storage – upload function is the realization of the specific system of nurses patrol, drug delivery instructions (pendulum check, dispensing scan, doctor order execution), oral medication doctor instructions (pendulum check, dispensing scan, doctor's orders), blood samples Collection of operations, vital signs/access to the collection of information need to collect the information module

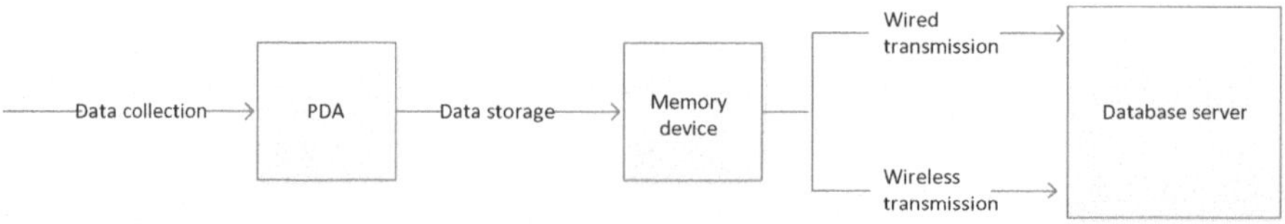

Fig. 46.2 PDA temporary – upload schematic

to add the corresponding data storage table, in the form of a table to write data storage, and thus with the doctor or nurses workstation data exchange and reprocessing, can further improve the patient diagnosis information and greatly improve the quality of history collection. When the PDA is unable to exchange data with the database in real time, the data is stored in the form of tables. When the PDA is connected to the database, the data in the table is output directly to the database. The table that has been transferred is automatically cleared.

When PDA is used in these places that the hospitals Lan can't cover, the data storage function of PDA can be used to record the information of the pre-clinical stage. This will help to record the information of patients in time, and provide more valuable information for clinical treatment. And the concept of mobile medical information is realized in deed.

46.4.3 Future Improvement

The future development trend of mobile nursing system should be the integration of hospital information resources. Today's medical data industry is growing mature, Gao Han Song [7] combined with the cloud computing and proposed a mining platform architecture of medical cloud data based on hadoop. In this context, the mobile nursing system not only needs the continuous optimization and improvement of its own functions, but also the realization of the integration with various medical systems to build information medical platform, information sharing.

46.5 Conclusion

Through the continuous improvement and optimization of the mobile care system, it is possible to improve the working efficiency of the nurses group and provide the high quality medical service for patients. It is believed that the synergistic development of information technology and medical care will bring a new era of development for medical care.

Acknowledgment My deepest gratitude goes first and foremost to Professor Tong, my supervisor, for her constant encouragement and guidance. He has walked me through all the stages of the writing of this thesis. Without his constant and illuminating instruction, this thesis could not have reached its present form. Last my thanks would go to my beloved family for their loving consideration and great confidence in me all through these years.

References

1. Wang, H., Hu, C., Ji, Y., et al. (2015). Design and implementation of a new generation of mobile nursing system [J]. *Chinese Journal of Digital Medicine, 10*(10), 43–45.
2. Hiroaki, Y. (2003). Development of a comprehensive scoring system to measure multifaceted nursing workloads in ICU. *Nursing and Health Sciences, 5*(4), 299–308.
3. Bjørk, I. T., & Kirkevold, M. Issues in Nurses' practical skill development in the clinical setting. *Journal of Nursing Care Quality, 14*(1), 72–84.
4. Ma, J., Song, J., & Feng, X. (2008). Construction of evaluation index system of technical difficulty in nursing operation [J]. *Zhonghua Nursing, 43*(7), 625–628.
5. Yan, Z., et al. (2014). A preliminary study on the degree of difficulty of nursing programs. *Journal of Nursing Management, 14*(7), 457–459.
6. Bishan, H., et al. (2015). Advances in the evaluation of nursing difficulty. *Journal of Nursing, 50* (11), 1361–1364.
7. Han Song, G., et al. (2013). Medical data mining platform based on cloud computing. *Journal of Medical Informatics*.

Chapter 47
Technology Development and Networking Application of a Mobile Passive Optical Access Network System

Yin Shuhua

Abstract There are various emergencies including disaster relief, military actions, anti-terrorist and peace maintaining operation or occasional real-time information transmission in poor geological environments. Aiming at these, a mobile emergency optical access network system is proposed to achieve the seamless connection between mobile and fixed communications. After integrating various emergency demands, system technology indices, optical line termination (OLT) and optical network unit (ONU) technology indices are formulated. For systematic design, a PAS5001-NM3 chip is applied at OLT while a PAS6201-NM3 chip, multi-function interface technology fusing various device interfaces of links and dynamic bandwidth allocation (DBA) control technology are employed at ONU end. The test results show that the throughputs from the initiator to the target are 75.47 (transmission control protocol (TCP)) and 20.56 Mbps (user datagram protocol (UDP)), respectively. At different geographical locations in different areas, multi-service fulfillment trials such as voice, data, image and video are carried out separately at the transmission distances of 8, 10 and 18 km. The technology indices of the system all meet the design requirements with a stable network operation and normal signal transmission.

Keywords Optical access system · PAS5001-NM3 chip · Optical line termination (OLT) · Dynamic bandwidth allocation (DBA) · Multi-function interface

47.1 Introduction

When emergent events occur, mobile equipments are connected to nearby core networks while short-distance, multi-service and rates information is integrated to core networks to achieve the seamless connection between mobile and fixed

Y. Shuhua (✉)
Xi'an Fanyi University, Xi'an, China
e-mail: Shuhua_y@126.com

M. Tavana, S. Patnaik (eds.), *Recent Developments in Data Science and Business Analytics*, Springer Proceedings in Business and Economics,
https://doi.org/10.1007/978-3-319-72745-5_47

communications. In this way, broadband terminals can realize broadband communication with various mobile equipments.

The mobile emergency optical access system is proposed for irresistible natural disasters. The proposal of the system is to implement the mass-flow real-time data interaction, achieve large capacity information transmission in real time under mobile conditions and promote the planning, design and technology upgrade of mobile optical access networks. The study analyzes the mobile emergency passive optical access network and focuses on the application on mobile conditions while considering the access application to fixed transport networks to guarantee the stable performances and powerful functions of equipments. The networks need to be absolutely real-time, reliable, stable, valid and dynamic under protection switching mechanism so as to be easily connected with different interfaces of light-ware terminal equipments (LTEs) and computer networks. Additionally, the system has flexible layouts and can be conveniently utilized to meet the mobile emergency demands under various environments [1–7].

Aiming at problems of traditional access modes, the designed mobile emergency optical access system plays its role in information protection of emergent events and highlights flexible organization and application under mobile environments considering the practical and future service demands. This is based on Ethernet passive optical network (EPON) technology according to area communication equipment of field communication networks, fixed fiber communication (G703) and computer network technology standards (IEEE802.3). The mobile and fixed services are synchronously accessed to build the integrated access transport platform of mobile optical image, data and voice services and the flexible multi-function interface is applied to promote the command and control system to the field. The primary technologies include the key technologies relating the design of optical line termination (OLT) and optical network unit (ONU) technology programs, peripheral circuit design of core chips, fusion design of link interfaces and embedded technology design of coordinated software [8–10].

47.2 Key Technology Indices of the System

G652 fiber is used under emergency conditions to achieve that the up- and down-line rates of system capacities are both 1.25 Gb/s. The largest optical splitting ratio is 1:16 with the largest grade of 3 and the largest transmission distance of 20 km. The requirements for the system indices include the primary technology indices of the system, OLT technology indices (shown in Table 47.1) and ONU technology indices. In terms of service interfaces, STM-1 optical interface of ATM satisfies the I.432 specification of ITU-T and 10BaseT Ethernet interface meets the 802.3 of IEEE while E1 interface satisfies G.703 of ITU-T [1–13].

Table 47.1 OLT technology indices

OLT fiber interface technology index		
Emitter type	Average optical power	≥−4 dBm
	Working wavelength	1310 nm
Burst mode optical receivers	Receive sensitivity	−30 dBm
	Saturated receiving optical power	−9 dBm
	Working wavelength	1510 nm
Optical connector	Connection mode	SC
	Insert loss	0.5 dB (max)

47.3 Key Technology of Systematic Design

47.3.1 Hardware Design Technology at Olt Terminal

OLT hardware is mainly composed of OLT and HOST subsystems. The OLT subsystem includes a PAS5001-NM3 chip and the peripheral devices while the HOST subsystem includes an AT91RM9200 chip and the peripheral devices.

The PAS5001-NM3 chip is employed to fully support IEEE802.3ah protocol, VLAN service and extended DBA algorithm. The ARM9 processor produced by ATMEL Company is utilized as the AT91RM9200 and it has application peripherals and standard interface conditions.

As the control center of the whole system, the HOST subsystem can complete multiple tasks (configuration of the OLT subsystem, fault detection and failure processing) and manage EPON network. The AT91RM9200 is connected with remote management clients through the RJ45 interface to manage the whole EPON system while it is connected with the PAS5001-NM3 through 16-Local Bus to transmit the management information to the OLT and the ONU.

The ARM9-processor is connected with the PAS5001-NM3 by using the 16-Local Bus and one ARM9 chip can be used to manage eight OLTs. The Boot Loader of the AT9200 embedded platform is stored in a serial FLASH chip of 128 KB. The embedded platform requires running LINUX operating system and constructing various contents such as databases in OLT and the other FLASH and SDRAM connected with the AT9200 are 32 and 64 MB, respectively. The 4 KB EEPROM is mainly employed to store parts of parameters of the network management system and embedded system. The AT9200 embedded platform provides a RJ45 interface of 100 MB to manage networks while the RS232 interface is mainly used to debug the embedded system and software.

The interfaces of the system on PAS5001 chip of the OLT subsystem include GE-PON, core network, HOST processor and External Memory/Flash interfaces. The GE-PON is a staticizer (SERDES) interface, accompanying with the control line for burst-mode receiver and burst clock data recovery (CDR). The core network interface is divided into active and passive modes. In active mode, PAS5001 is connected with PHY as a multi-access computer (MAC) dielectric layer device while on the contrary, it is taken as a physical layer equipment to be connected with the MAC equipment in passive mode.

47.3.2 Multi-function Interface Technologies

Fusion of link interfaces is to move the information access to the scene and achieve the integration of fiber transport, mobile communication and computer network. Thus, the seamless connection between optical transport equipment and mobile spread-spectrum connecting equipment or mobile program-controlled equipment can be achieved to construct the transport platform for multiple services of links. According to the emergency node setting or flexible networking requirements, setup of the multi-function interface is convenient for rapid networking and forward and backward compatibility of equipments. The small peripheral mobile equipments are connected with the fixed optical transport network and for interfaces, to integrate the conversion functions of group A, LTE E1 and computer Ethernet ET interfaces. As a result, the conversion of group A interface of spread-spectrum connecting equipment, fundamental group rate E1 interface of optical transport network and Ethernet ET interface of computer network communication can be realized.

Interface function module is composed of various circuits including those for code conversion of data flow, level adjustment, impedance matching, clock extraction, synchronous and alarm information processing. The main circuits of three interface module, namely, the core chips of code conversion, are all formed by using large-scale integration chip (CD22103) based HDB3 codec.

47.4 Test Results of Systematic Performances

According to the built platform of the test system, the downward waveband, and the optical power, bandwidth as well as performance of three main points are separately tested. The test system is displayed in Fig. 47.1.

The instruments for test include the AQ6317B spectrum analyzer and the 8153A multi-wavelength power meter, and the test software is Qcheck v1.3. The downward waveband data of J-EPON measured at the test point 1 are shown in Fig. 47.2.

The optical powers of the system are tested and the data measured at various test points are displayed as follows: OLT transmission power P_1, the entrance power P_2 of the optical splitter and ONU received power P_3 are 0.4, −3 and −12.4 dBm, respectively. The bandwidth and performance are tested by using Qcheck v1.3 and the data are shown in Table 47.2.

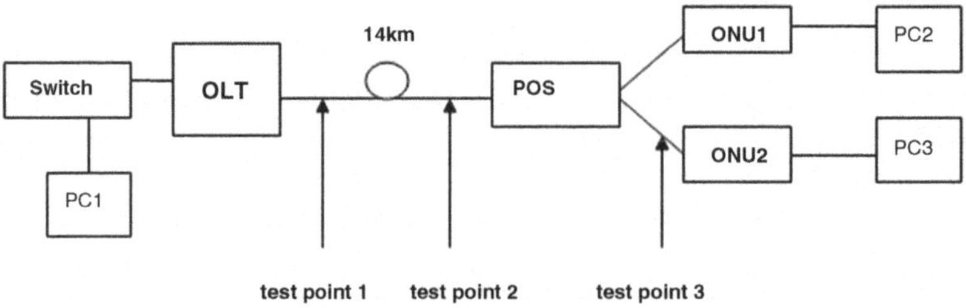

Fig. 47.1 Test system block diagram

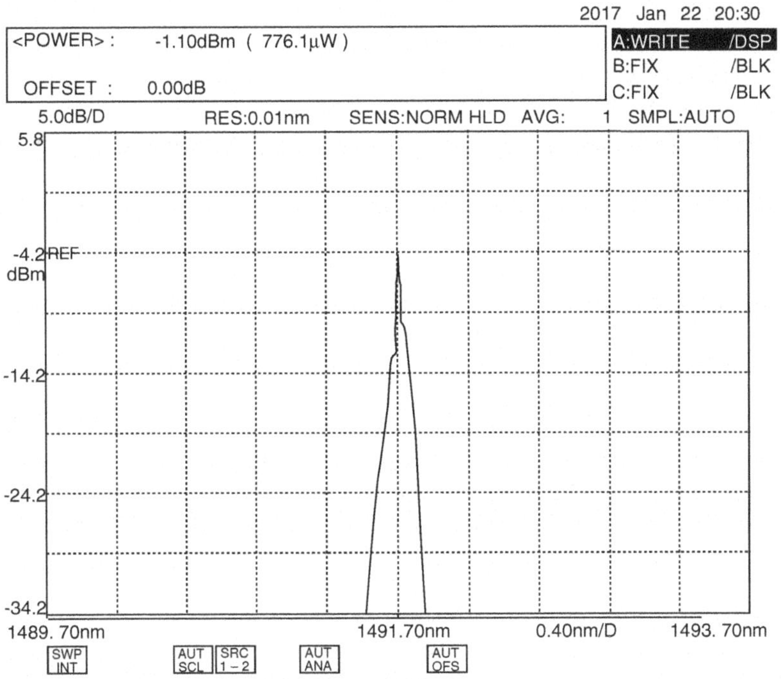

Fig. 47.2 Measured downward waveband

Table 47.2 Measured data

Test protocol	Test project	Test result	
		Source end PC1 → destinatio PC2	Source end PC1 → destinatio PC3
TCP	Answering time (ms)	Max: 3 Min: 3 Average: 3	Max: 3 Min: 1 Average: 2
	Handling capacity (Mbps)	75.472	21.622
UDP	Answering time (ms)	Max: 3 Min: 2 Average: 3	Max: 3 Min: 1 Average: 2
	Handling capacity (Mbps)	20.566	14.815

47.5 The Tryout and Effect of Link Networking

The emergency link networking was conduct in a steep geological position to carry out emergency settings in the areas. Six command posts are constructed including the overall command post (SJZ), safeguard command post, support group (SPBQ), team 1 (1 T), team 2 (2 T) and team 3 (3 T). In general, the distance and depth of the emergency link are 10~15 km and 20 km, respectively. In organization operation program, OLT is set in the preset head-end computer room for national defense. Leveled command posts are connected to nearby optical access points by using the optical splitter (POS) in the optical splice box to transfer information with the mobile backbone network using the preset optical cables.

As for the performance standard of OC-1 × 4-U POS chosen by allocation factors, the splitting ratio of POS is 1:4. By distributing optical fibers, the command post users are connected, and the whole system can cover 20 ONU users in the early stage. With the development and requirement of situations, relevant configurations can be added to complete the access of ONU at any moment. The groups (Y) only offer the data solution and the LAN service using switchers within areas of all command posts is connected to the core network through FE interface of ONU. Additionally, ONU is directly connected to the command posts, so data and voice services can be simultaneously provided here. According to various service demands in emergency sites, the construction methods with different channels are applied and the interface functions are separately tested by using different connection modes on trial. After performing multiple trials on multiple services separately at three distances of 8, 10 and 18 km in different areas, the various functions and technology indices of the system all meet the design requirement, companying with stable system operation and normal signal transmission.

Acknowledgments The work was supported by the Special Natural and Science Foundation in Education Department of Shaanxi Provincial Government in China (No. 13BZ69).

References

1. Xiaoling, G., Wei, S., Liangdong, G., et al. (2013). Application research on maneuvering communication network in the testing field based on clustering algorithm of classification structure. *Journal of Academy of Equipment, 24*(05), 107–111.
2. Yan-jin, W., Yu-feng, S., & Nan, C. (2013). Design and demonstration of a novel self-coherent PON scheme based on high-order modulation. *Journal of Fudan University (Natural Science), 52*(03), 380–385.
3. Yiduo, L., Weiwei, Q., Kai, L., et al. (2015). Development of universal type emergency device for electric power optical fiber communication system. *Guangxi Electric Power, 38*(06), 52–55.
4. Gang, W., Jian-hua, L., & Zhong-hui, G. (2010). Modeling and optimization of maneuver communication strength dispatching. *Computer Simulation, 27*(03), 23–26.
5. Jing, Y., & Hong-de, D. (2013). Research on some key techniques of emergency communication based on satellite. *Computer & Networks*, (07), 60–63.
6. Han, L., & Yi, H. (2016). Technical plan analysis of PON fiber multiplexing and reach extending. *Study on Optical Communications, 197*(05), 32–35.
7. Hong, D., & Hai-qing, W. (2014). QoS guaranteed method for mobile communication. *Journal of Military Communications Technology, 35*(01), 49–53.
8. Wei, C., Jian, Y., Zuo-wei, H., et al. (2014). Study on the low-loss single-mode optical fibers and their transmission experiment. *Journal of Optoelectronics·Laser, 25*(12), 2300–2304.
9. Hai-tao, Y., Wei, L., Ji-long, H., et al. (2014). Upstream data transmission in passive optical networks based on orthogonal waveform multiplexing technology. *Chinese Journal of Lasers, 41*(08), 61–68.

10. Feng-liang, S., & Yu, Y. (2016). Research on and implementation of voice service simulation in passive optical network. *Study on Optical Communications, 197*(05), 32–35.
11. Zhao, Z., Hai-tao, N., & Yao-jun, W. (2016). Long reach DWDM-PON with 12.5 GHz channel spacing based on comb source seeding. *Optoelectronics Letters, 12*(4), 304–307.
12. Jian-guo, Y., Meng-qi, L., Yong, W., et al. (2016). A novel construction scheme of QC-LDPC codes based on the RU algorithm for optical transmission systems. *Optoelectronics Letters, 12*(2), 132–135.
13. Ming-yu, J. (2015). A peak signal denoising method for mobile communication network. *Computer Simulation, 32*(10), 310–313.

Chapter 48
Improvement of E-MIMLSVM+ Algorithm Based on Semi-Supervised Learning

Wenqing Huang, Hui You, Li Mei, Yinlong Chen, and Mingzhu Huang

Abstract The MIMLSVM algorithm is to transform the MIML learning problem into a single-instance multi-label learning problem, which is used as a bridge to degenerate into a single-instance single-label learning. However, this degradation algorithm is relatively easy to understand, but in the degradation process will lose some information, affecting the classification effect. By using multi-tasking learning, E-MIMLSVM+ is used to combine tag relevance to improve the algorithm MIMLSVM+. In order to make full use of the unlabeled samples to improve the classification accuracy, the paper improves MIMLSVM algorithm by using the semi-supervised learning method. Experimental results show that the proposed method can achieve higher classification accuracy.

Keywords SVM · MIML · Semi-supervised learning · Multitask learning

48.1 Introduction

As time goes on, more and more attention has been paid to multi-instance multi-label [1] learning. In our reality, many problems can be considered as MIML problems, such as picture classification, Web classification and so on. SVM [2] have excellent performance in two categories, so they are applied to a variety of classification problems, including machine learning, binary, multi-class [3], and semi-supervised learning [4]. Supervised learning only use labeled data training, but in real life, many real data was very few, and unlabeled data was substantial and relatively easy to obtain, for example a large web exists, these unlabeled samples often can provide more accurate information of sample distribution therefore, semi-supervised learning is in fact based on supervised learning, combined with the unlabeled samples to train the classifier function, in order to obtain better classification results.

W. Huang · H. You (✉) · L. Mei · Y. Chen · M. Huang
School of Information, Zhejiang Sci-Tech University Hangzhou, Hangzhou Shi, China
e-mail: 405852890@qq.com; patternrecog@163.com

M. Tavana, S. Patnaik (eds.), *Recent Developments in Data Science and Business Analytics*, Springer Proceedings in Business and Economics,
https://doi.org/10.1007/978-3-319-72745-5_48

48.2 Relate Work

Before the proposed MIML learning framework, traditional single-sample single-label supervised learning framework, multi-instance learning [5] framework and multi-label [6] learning framework have been proposed. Multiple sample learning represents an object with an example set (sample package) that is associated with a class marker, which is designed to predict suitable class tags that do not see the sample package. MLL is an example that represents an object that is associated with multiple class tags. The purpose of learning is to predict the proper set of class tags that have not been shown. The MIML is to get a function$f : 2^x \rightarrow 2^y$, the MIL is $f: 2^x \rightarrow \{-1, +1\}$, and the MLL is $f: x \rightarrow 2^y$. The MIMLSVM [7] method was brought up by Zhi-Hua Zhou and Min-Ling Zhang after in-depth research and learning. The degeneration strategy is applied to the two algorithms that these are typically MIML algorithms as the core of the algorithm. But, this degradation method was relatively understanded, but in the degradation process will lose some information, affecting the classification effect. The M3MIML [8] algorithm are applied to solve the MIML learning problems that they are based on regularization. Although these methods can solve the MIML learning problem, but the effect of classification is not ideal. Zhang Minling et al. based on neural network MIMLRBF algorithm is proposed. The neurons in the first layer is composed of the center, and the center point is obtained through calling the k-MEDOIDS algorithm in MIML samples, of which two packets between the distance measurement using the Hausdorff distance. The hidden layer and the output layer between the weight matrix using the singular calculate the value and function of this decomposition is to minimize the error level.

Due to MIML learning problems, Ying-Xin Li et al. proposed a new method of MIML named MIMLSVM+, and the E-MIMLSVM+ [9] was extended method from it. In the algorithm, the kernel is a bag instance rather than a single feature vector. In contrast, the traditional SVM is different. Because MIMLSVM+ may lose some tag-related information during the degradation process, E-MIMLSVM+ extends MIMLSVM+ by introducing multitasking learning technology.

The MIMLSVM algorithm uses an example based Gauss kernel when degraded into traditional supervised learning problems, and the MIMLSVM + algorithm uses a package based multiple sample kernel in the degradation process. SVM and semi supervised clustering assumption based semi supervised learning, trying to find two types of samples can be separated through the data division and the low density region hyperplane, so you can also use labeled samples and unlabeled samples.

E-MIMLSVM+ algorithm has achieved better results than other multi-sample multi-tag classification algorithms, but requires a lot of training time and storage space, because at the same time to learn more, because the application of multi-tasking technology, the link between the information added to the SVM. Tasks will lead to more sample packages involved in the optimization process, the efficiency is not high enough.

48.3 Improved Algorithm

48.3.1 The MIMLSVM Algorithm and the E-MIMLSVM+ Algorithm

Starting with MIMLBoost, by adding a pseudo-label to each instance, the MIML problem is reduced to the multi-instance single-label problem, where we take a direct approach. That is, we each train a tag of the classifier; we collect all the bags, the label is positive for the bag, and the unlabeled bag is negative. Therefore, compared with other machine learning frameworks, the multi instance multi mark framework is more powerful for real world objects.

But the objects in the real world often have rich meanings, such as web page classification, a news report may also belong to “science and technology”, “travel” or “financial” categories, then only one sample to represent an object is too simplified, in order that section will lose a lot of useful information.

First, the sample number of training samples is assumed to be n. X_i is the i-th bag in the training set, y is a label. For each label y, $\varphi(X_i, y)$ be the index function named: $\varphi(X_i, y) = +1$. Therefore, the resulting SVM classification model is:

$$
\begin{aligned}
&\min_{w_y, b_y, \xi_{iy}} \frac{1}{2}\|W_y\|^2 + C\sum_{i=1}^{n} \xi_{iy}\tau_{iy} \\
&s.t: \varphi(X_i, y)\left(\langle w_y, \phi(X_i)\rangle + b_y\right) \geq 1 - \xi_{iy} \\
&\xi_{iy} \geq 0 (i = 1, 2, \ldots, n)
\end{aligned}
\tag{48.1}
$$

Among them, w_y and ξ_{iy} are the parameters to be optimized. w_y represents the normal direction of the hyperplane. The function $\varphi(X)$ can map the package X_i, ξ_{iy} is the amount of the relaxation variable, the amount of the function interval that corresponds to the allowable deviation of the sample point. The parameter C is the weight used to balance the model complexity and training sample errors. If the difference in the number of training samples of different categories will affect the classification effect, and even make the classifier no value. τ_{iy} is defined as follows

$$
\tau_{iy} = \frac{1 + \varphi(X_i, y)}{2} R_y + \frac{1 - \varphi(X_i, y)}{2} \tag{48.2}
$$

The R_y is the unbalanced level of y, it is estimated based on the number of positive samples in the training set and the number of negative samples.

The use of kernel functions makes SVM avoid complex computations directly in high dimensional space, and the use of different kernel functions can produce different classification effects. The kernel is defined as follows:

$$
K_{MI}(X_i, X_j) = \frac{1}{n_i n_j} \sum_{(x_{is\_0}, x_{is\_1}) \in X_i} \times \sum_{(x_{jz\_0}, x_{jz\_1}) \in X_j} e^{-\gamma_1 \|x_{is\_0} - x_{jz\_0}\|^2 - \gamma_2 \|x_{is\_1} - x_{jz\_1}\|^2} \tag{48.3}
$$

Where $\|x_{is\_0} - x_{jz\_0}\|^2$ is used to measure the visual feature similarity between the two instances. $\|x_{is\_1} - x_{jz\_1}\|^2$ measures the spatial distance of two instances. γ_1 and γ_2 represent the weight of visual information and spatial information, respectively.

In the degradation, the contact information between tags is not taken into account. In summary, the final classification model is defined as follows:

$$f_y(X) = \langle w_y, \phi(X) \rangle + b_y = \sum_{i=1}^{n} a_{iy}\varphi(X_i, y)K_{MI}(X_i, X) + b_y \tag{48.4}$$

Where w_y is the commonality between the different tasks. By using multi-tasking learning, E-MIMLSVM+ is used to combine tag relevance to improve the algorithm MIMLSVM+. Multi-task learning to optimize its problem as follows:

$$\begin{aligned} &\min_{w_0, v_y, b_y, \xi_{iy}} \frac{1}{2}\left(\sum_{y \in Y} \|V_y\|^2 + \mu\|W_0\|^2\right) + C\sum_{y \in Y}\sum_{i=1}^{n} \xi_{iy}\tau \\ &s.t: \varphi(X_i, y)(\langle(w_0 + v_y), \phi(X_i)\rangle + b_y) \geq 1 - \xi_{iy} \\ &\xi_{iy} \geq 0 (i = 1, 2, \ldots, n) \end{aligned} \tag{48.5}$$

Among them, the parameter μ is used to adjust the relationship between parameter w_0 and parameter v_y, that is, to balance the similarity of each task. If the number of clusters in the labeled space K is equal to the number of $|Y|$, i.e., $K = |Y|$, the E-MIMLSVM+ algorithm will degenerate into MIMLSVM+.

48.3.2 The Improved Method

Semi-supervised learning [10] is used to improve the learning performance by automatically using unlabeled samples to improve learning performance. The demand of semi- supervised learning is very strong, because it is easy to collect a number of unlabeled samples in the practical application, but it is a waste of manpower and material resources. For example, in computer-aided medical image analysis, you can get a number of medical images from the hospital, but if you want medical experts to identify all the lesions in the image is unrealistic. "There are fewer labeled data and unlabeled data", which is more obvious in Internet applications, for example, you need to ask the user to mark the page of interest, but few users are willing to spend a lot of time to provide a marker. As a result, there are few labeled web pages, but there are countless pages on the Internet that can be used as unlabeled samples. Semi-supervised learning is a way to make use of "cheap" unlabeled samples. To make full use of the spatial distribution information of the sample, we use the semi supervised learning method to improve the MIMLSVM algorithm, it can improve the classification accuracy.

The improved method is described as follows:

$Y = SLMIMLSVM^{+}(S, U, T)$

Input: S - the training set, labeled data

U - the unlabeled data

T - the testing set

Output: Y - the set predicted labels of T

1) According to the mark indicating matrix $Y = [\phi(X_i, y)]_{n\times|y|}$, the tag in the tag space is divided into k subset $L = L_1 \cup L_2 \cup \cdots \cup L_k$.Ranking the labels in label set randomly for K times, then we can get K different sets $L_k(k = 1,2, \ldots, K)$.

2) For training set $S_k = \{(X_i, Y_i \cap L_k)|i = 1,2, \ldots, N\}$, calculate the kernel matrix $[K_{ts}(X_i, X_j)]\ (i, j = 1, \ldots, N; s)$.

3) Train an SVM $f_y = SVMTrain(S_k), y \in L_k$ based on $[K_{st}(X_i, X_j)]$, using the formulation(7).Get mark assignment $y^* = (\hat{y}_{1+1}, \hat{y}_{1+2}, \ldots, \hat{y}_{1+u})$.

4) $y_i \in U(i = 1,2, \ldots, L)$

do $y_i \leftarrow (y_i, y_i, \ldots, y_i)_d^T$

$S_i \leftarrow \{\}$

for each $(X_j, Y_j) \in S$

do $X_j \leftarrow [X_j, y_1^*, y_2^*, \ldots, y_i^*]$

$S_i \leftarrow S_i \cup (X_j, \phi(X_j, y_i^*))$

end for

end for

Train an SVM $f_{ki} = SVMTrain(S_i)$ based on $[K_{MI}(X_i, X_j)]$ using the formulation(5) end for

---

48.4 Experiment

We use scene dataset and text data set to evaluate the classification effect of the improved algorithm. The scene data set has 2000 natural scene pictures. The picture on the scene sample set is divided into five categories, they are sunset, trees, sea, desert and mountains. There are more than 22% of the images in the sample set with multiple categories. In real, each picture is associated with 1.24 markers labels. Here, each picture is represented by a packet, and each packet contains nine examples, each of which is represented by a 15-dimensional feature vector by the SBN method.

The Reuters-21578 is a widely used text data set. Remove the text without marking and no text, and then randomly remove some of the text only one category mark, and ultimately get 2000 text samples, divided into seven categories. Of which about 15% of the text samples have multiple category markers, and an average of each text sample is associated with 1.15 ± 0.37 category markers labels. By using

Table 48.1 The scene data set

	Improved method	E-MIMLSVM+	MIMLSVM
Hamming loss	**0.3442±0.0001**	0.3445±0.0001	0.3476±0.0007
Ranking loss	**0.3955±0.0002**	0.3989±0.0009	0.5047±0.0002
One error	**0.6623±0.0001**	0.6749±0.0001	0.7508±0.0009
Coverage	**1.8186±0.0001**	1.8260±0.0001	1.9827±0.0008
Average precision	**0.5642±0.0001**	0.5586±0.0004	0.4891±0.0004

Table 48.2 The Reuters data set

	Improved method	E-MIMLSVM+	MIMLSVM
Hamming loss	**0.2132±0.0001**	0.2138±0.0001	0.2437±0.0003
Ranking loss	**0.2595±0.0001**	0.2658±0.0001	0.5027±0.0009
One error	**0.6714±0.0001**	0.6734±0.0001	0.7762±0.0001
Coverage	**1.7378±0.0001**	1.7896±0.0015	1.9706±0.0008
Average precision	**0.5776±0.0001**	0.5774±0.0001	0.4260±0.0001

the sliding window technique, each containing a set of 243 feature vectors, each representing a part of the document.

In order to evaluate our improved algorithm, the improved method is compared with each other. The Gaussian kernel parameters is $\gamma = 0.2^2$. The k is set to 20% . The kernel parameters with $\gamma_1 = 10^{-5}$ and $\gamma_2 = 10^{-2}$ is used to implement E-MIMLSVM+ and the q is 0.5.The samples is selected as small sample sizes, two for each class. For the "classification" task, we use a sample-based evaluation index. In simple terms, the smaller the value of one error, coverage, hamming loss, and ranking loss, the better the algorithm effect; the better the average precision value. Experiments were performed using 10 fold cross validation.

Tables 48.1 and 48.2 shows the experimental results of the five metrics, both on scene data set and Reuters data set. It can be seen that, on the whole, the effect of classification on the Reuters data is better than that on scene data, in each sample set, to improve the performance of our method is better than the other MIML classification algorithms.

In order to solve the small sample problem, improvement of E-MIMLSVM+ algorithm based on semi-supervised learning. Experiments on scene classification and text categorization show that our method is more efficient and can produce better performance than other MIML methods.

Acknowledgment The work was sponsored by the Institute of computer vision, image processing and pattern recognition, Zhejiang Sci-Tech University.

References

1. Zhou, Z. H., Zhang, M. L., Huang, S. J., Li, Y. F. (2008). MIML: A framework for learning with ambiguous objects. CORR abs/0808.3231.
2. Vapnik, V. (1995). *The nature of statistical learning theory*. New York: Springer-Verlag.
3. Qian, H. M., Mao, Y. B., Xiang, W. B., & Wang, Z. Q. (2010). Recognition of human activities using SVM multi-class classifier. *Pattern Recognition Letters, 31*, 100–111.
4. Chen, W. J., Shao, Y. H., & Xu, D. K. (2014). Manifold proximal support vector machine for semi-supervised classification. *Applied Intelligence, 40*(4), 623–638.
5. Dietterich, T. G., Lathrop, R. H., & Lozano-Pérez, T. (1997). Solving the multiple-instance problem with axis-parallel rectangles. *Artificial Intelligence, 89*(1–2), 31–71.
6. Tsoumakas, G., & Katakis, I. (2007). Multi-label classification: An overview. *International Journal of Data Warehousing and Mining, 3*(3), 1–13.
7. Zhou, Z. H., & Zhang, M. L. (2006). Multi-instance multi-label learning with application to scene classification. *Advances in Neural Information Processing Systems, 176*(1), 1609–1616.
8. Zhang, M. L., Zhou, Z. H. (2008). *M3MIML: A maximum margin method for multi-instance multi-label learning*. Proceedings of the 8th IEEE International Conference on Data Mining (ICDM'08), Pisa, Italy, 208: 688–697.
9. Li, Y. X., Ji, S. W., Kumar, S., Ye, J. P., & Zhou, Z. H. (2012). Drosophila gene expression pattern annotation through multi-instance multi-label learning. *Transactions on Computational Biology and Bioinformatics, 9*(1), 1445–1450.
10. Blum, A., & Chawla, S. (2001). *Learning from labeled and unlabeled data using graph mincuts*. In Proceedings of the 18th International Conference on Machine Learning (ICML), 19–26, Williamston, MA.

Chapter 49
The Method of Improving Learning Ability of "C Language Programming Design" Course

Chunxiu Xiong, Xinhua You, and Peng Yu

Abstract According to the students cannot do anything without mobile phone nowadays, the question of learning the "C language program design" course is increasingly serious, this paper has presented a series of methods of attracting students' interest to improve the students' ability of learning it. Firstly, arousing students' attention to the C language, secondly, changing the "cramming" teaching method of the past by easy course, and reviewing the curriculum appropriately, finally, choosing the appropriate code examples, drawing inferences, creating templates, and solving question by multi-answer to consolidate the teaching, and through the students' homework and the method of fast compiling and debugging program to grasp the students' learning interest of the course.

Keywords Teaching methods · Reviewing the curriculum · Code examples · Create a template

49.1 Introduction

Nowadays, people's lifestyles are inseparable from the mobile phone, watching the phone when opened eyes, walked, by bus, eating, and even sitting in the classroom as well.

For the various definitions are quite abstract, sounds boring, so students are attracted to the mobile phone. Lots of teachers will sigh the students who take less seriously to the class than past students, and sigh the various harm to the students when mention the mobile phone Therefore, in order to change this bad habit, it needs to take the students' eyes back from the cell phone, this is a difficult problem.

In order to let students give up their cell phones, our courses should be more attractive than the mobile phones. For attracting students' interests better and

C. Xiong (✉) · X. You · P. Yu
Computer Science of ZhiXing College of HuBei University, Wuhan, Hubei, China
e-mail: xiongchunxiu@126.com

M. Tavana, S. Patnaik (eds.), *Recent Developments in Data Science and Business Analytics*, Springer Proceedings in Business and Economics,
https://doi.org/10.1007/978-3-319-72745-5_49

improve the programming ability of students' C language, this article mainly studies from the following seven aspects.

49.2 Seven Methods to Attract Students' Interests

49.2.1 Arouse the Students' Attention to the C Language Course

In order to attract students' interests better, we should first arouse the students' attention to the C language course [1], and let the students be perfectly happy to learn the C language course. This requires students to understand the importance of learning C language course, C language and other languages whose basic part is basically interlinked, such as C++, Java, C#,.Net, VB......, and after learning C language, you would learn and even master a variety of other computer language, so it is still meaningful to learn C language.

Secondly, C language course can improve students' logical thinking ability and have profound and lasting influence on their study and life in the future.

Thirdly, C language has strong practicability, not a purely theoretical course, which can improve the students' hands-on practical ability and capacity.

In this way, the students have to learn the C language.

The students have the motivation to learn C language now, but it does not mean that students can learn the C language well, it needs the teachers well guide students to learn, a good teacher, just like the light in the dark, illuminates the direction of students.Interest is the best teacher, so how to arouse the interest of the students is the most important.

49.2.2 Explain Definition, Popular and Easy to Understand

Various definitions should be explained in a straightforward way; examples should be appropriate to the actual life. For example, global variables and local variables: you can simply understand global variables as public facilities that can be used by everyone, and understand local variables as family inside facilities that can only be used by the family.

The understanding of the variable range [2]: public facilities can be seen by all from the installation, so they can be used by all, until they were scrapped by the public, so the scope of global variables is from the start position where they are defined to end of the source file; internal facilities can only be seen by the family from installing, so they can only be used by the family, until they were scrapped by the family, so the scope of a local variable is from the starting position where it is defined to the end of the function.

In order to improve the interest of the C language course, we should not only to choose a variety of examples from practical life, but also to explain through the code examples when teach all kinds of definitions, which not only can attract the attention of students, but also can consolidate on the memory of the definitions, so as to let students happy to learn so that like to learn.

49.2.3 Speak Less and Practice More

Completely eradicate the cramming education, according to my teaching experience of many years, students will take decreasingly time to listen in the C language or other programming language courses [3], so in order to make full use of the only two classroom hours, the teachers should lengthen the interaction time with the students, and let the students listen with a task of the curriculum, through a task to learn, which can extend more interesting time of learning than aimlessly listening. Therefore, the choice of this task is very important, which should neither too simple, no challenging, nor too difficult, so that students retire without a fight, it must be adjusted according to student learning progress.

49.2.4 Course Review

Never expect the students to spend a lot of time to learn C language after class, so for the purpose of gaining new knowledge by reviewing old, it needs the teachers to review the last lesson before the new lesson together with the students, so as to link up the new curriculum.

49.2.5 Select the Appropriate Example Code and Draw Inferences from Others

It is especially important to select example code in the process of increasing interest.

With regard to the program of the C language, it is gradually raised from simple to difficult, and the code numbers are increased line by line, which can not be increased too fast. The difficulty of the control should be accurate; it should not be too difficult or too easy.

Practical ability is very important for the students to learn C language, so it needs to guide students to fill in the code according to the template, therefore the template is very important, and it needs to locate according to the specific chapter.

1. Detailed template

For example, the templates should be more detailed and simpler in the previous chapters, and the code should be as specific as possible. At the beginning the students are unfamiliar with the code, but we do not require the students to master the meaning of each line of codes in details and the usage method, it only needs to teach the students the function of each line of codes, which greatly reduces the difficulty of learning, so that the students accept the C language more easily. For example, the Fig. 49.1 is the simplest greeting code of C language.

We can draw inferences about other routine greeting sentences from the instance, take the Fig. 49.2 for example. In that way, the students understand the templates formed automatically in the Fig. 49.3.

As long as the students change the ellipsis in this template, they can print all kinds of sentences. That is, thinking the main function as a simple template is applied to the students to complete a variety of related tasks. At the same time, the teacher can guide the students to print some interesting statements autonomously, which makes the students think about the problem actively, attract the students' ideas, and then arouse the students' interest.

When the students master the simplest C language program, you can add an input statement to this simple template, which can acquire the ever-changing printing result. Such as talk about a light topic. How is the weather today? Take the Fig. 49.4 for example.

Through learning by analogy, let students study freely, and rewrite the code which they interest in, or the teachers put forward a series of topics to guide the students to choose. Therefore, the students will have a sense of achievement and interest in learning C language spontaneously as long as they write a similar simple procedure.

Fig. 49.1 Hello everyone

```
main ( )
{
  printf ("Hello everyone!");
}
```

Fig. 49.2 How are you

```
main ( )
{
  printf("How are you?");
}
```

Fig. 49.3 A simple template

```
main ( )
{
  printf("......");
}
```

Fig. 49.4 Weather

```
main( )
{
  string weather;
  scanf("%s", &weather);
  printf("Today is %s", weather);
}
```

Fig. 49.5 Sum

```
main( )
{
 int n=1000,sum=0;
 for(int i = 1; i<=n; i++)        sum = sum + i;
 printf("1+2+3+......+%d = %d", n, sum );
}
```

Fig. 49.6 For statement

```
int i = 1;
for(; i<=n;)
{      sum = sum + i ;
       i++;
}
```

Fig. 49.7 While statement

```
int i = 1;
while (i<=n)
{      sum = sum + i;
       i++;
}
```

Take the cycling for example, the teacher puts forward a question to attract the students, what is the sum from 1 to 100? All of the students know the answer is 5050, however, if the question changes from 1 to 1000, the students will calculate for a long time, if there is a way to acquire the results quickly (through programming, it needs only a few seconds), the students is certainly interested in. Simplification of complex problems is also a way to attract students, and the teachers can explain and write code both together, take the Fig. 49.5 for example.

With true examples, how to calculate the sum from 1 to 10,000, it only needs to change n to 10,000, and if the question is changed, what is the sum of odd numbers or even numbers between 1 and 100? Or other questions, then you can ask the students to solve one of the questions, or to think out a similar topic for themselves. In addition, you can also rewrite for statement into while statement and do... while statement, and this cannot be done overnight, firstly, you can move the assignment expression: int i = 1 above, secondly, you can move the expression: i++ to the body. As following Fig. 49.6:

In this way, you can directly rewrite for statement into while statement and do... while statement. Take the Figs. 49.7 and 49.8 for example. And then emphasize the

Fig. 49.8 Do while statement

```
int i = 1;
do{
        sum = sum + i;
        i++;
} while (i<=n);
```

Preprocessing command	Variables definition
Function definition	Data input
Global variable definition	Data processing
main ( )	Data output
{	}

Fig. 49.9 An abstract template

attention of some points on the program, which is simpler and more intuitive. The students will not only understand the use of the cycle statement, but also master a series of related algorithms as long as they rewrite it successfully.

2. Abstract template

Take the following sections for example; the basic knowledge has been mastered, so the template is slightly more abstract, such as Fig. 49.9.

49.2.6 Multi-answer Question

Typically, a program can be implemented by a variety of methods, the students should try to use in the learning process, not only can grasp different knowledge, but also improve their skills in different ways. Take the selective structure programming for example, you can use if... else as well as switch... case to solve problems, and take the cyclic structure programming for example, you can use one of three cycles, for loop, while loop, and do... while loop, and take the sorting algorithm for example, the students should try to use varied sorting algorithms, such as bubble sort, selection sort, quick sort, merge sort, two binary tree sort, radix sort, Hill sort, heap sort, etc., and then compare the advantages and disadvantages of each algorithm.

49.2.7 Pay Attention to Students' Homework

Paying attention to students' homework and correcting each student's homework carefully, you can find problems you can't imagine, and it will be a great help to teaching. In addition, the teachers' attitude to the students' homework will have great influence on the students' work, if the teachers ignore the students' homework, which is unimportant to the students who feel, and then spend less time on homework. However, if the teacher checks the students' homework carefully, corrects the programming ideas and the details seriously, the students will spend more time on it, this is an interactive process.

At last, students must conclude and classify various errors in the programming process, and list out all kinds of errors. For a long time, it will not only improve debugging speed, but also improve the efficiency of programming.

49.3 Experimental Results and Summary

This thesis puts forward seven ways to improve the students' learning interest in improving the students' ability on learning the C language. Some are suitable for classroom, some are suitable for the computer room, no matter which method needs the teacher's full guidance.

According to my comprehensive application of the seven methods, through the practice of teaching reform on the freshmen of three classes who majored in the E-commerce and two classes who majored in computer application technology in the department of Computer Science of ZhiXing College of HuBei University, and the "C Language Programming Design"course is their required course, which has a good teaching effect, and most students are able to write a program according to the template which has been set up, and complete the task of teaching.

Through the teaching practice, it is time to summarize the experimental data as follow Table 49.1.

It can be seen from Table 49.1:

1. It has a great relationship with the class management on improving the ability of the students (the instructor of the E-commerce class 1 and 2 has counted the numbers before the class weekly);
2. The learning basis and the learning consciousness of the undergraduate are stronger than the specialized students obviously, so the effect on curriculum reform of the undergraduate (computer application class 1 and 2) is superior to the specialized students (E-commerce class 1, 2 and 3);
3. The learning effect in the computer room is better than in the classroom, it indicates that the students prefer hands-on ability which is more practical.
4. As a whole, using the seven methods is better than using any part of seven methods independently.

Therefore, the comprehensive application of the seven methods can improve students' interest in learning obviously, it does not worry about students who cannot learn "C language programming"course well as long as the students all stay interested.

Table 49.1 Experimental result of using different methods in different classes

Methods	Class	E-commerce class 1	E-commerce class 2	E-commerce class 3	Application class 1	Application class 2
Total numbers		31	29	30	34	36
Classroom	Effectiveness	Excellent	Good	Common	Excellent	Excellent
	Raised number	22	17	14	28	27
	Ratio (%)	70.97	58.62	46.67	82.35	75
Computer room	Effectiveness	Excellent	Excellent	Excellent	Excellent	Excellent
	Raised number	28	27	25	33	34
	Ratio (%)	90.32	93.1	83.33	97.06	94.44
Seven methods are integrated	Effectiveness	Excellent	Excellent	Good	Excellent	Excellent
	Raised number	29	28	26	33	34
	Ratio (%)	93.54	96.55	86.67	97.06	94.44

References

1. He, X. (2012). Exploring the practice of “C language programming design” course for Computer Science Department of Independent Colleges. *The Science Education Article Collects, 04*, 2–3.
2. Xin-yue, L., & Wei, N. (2012). The new enlightenment C language program design course teaching reform. *Computer Knowledge and Technology, 33*, 2–3.
3. Min-lei, X., & Min-lian, X. (2014). A preliminary study on the C program design course teaching reform. *Computer Knowledge and Technology, 27*, 2–3.

Chapter 50
Implementation of Koch Curves Based on Html5 <canvas>

Zhengzhi Xu and Youhui Su

Abstract With the development of modern science, fractal theory has been become as a significant branch of Mathematics. Fractal theory mainly reveals that macroscopic and microscopic phenomenon has a startling similarity to one another. Koch curve looking like snowflake is one of a typical and strict fractal pattern. For good performance, the report uses a recursive algorithm to implement Koch curve on web platform with <canvas> element.

Keywords Koch curve · Fractal · Recursive algorithm · Html5 · <canvas> element

50.1 Introduction

In tradition, developers tend to use graphics library in C, C++, Java even Python to design or draw a graph. This report aims to open up a convenient way to draw graph like Koch curve vividly.

In Euclidean geometry, our world is constructed by beeline, ellipse, triangle, parabola and other curves. The geometry objects what they describes are regular or smooth, but lots of complex objects in nature such as winding coastline, strange shape of clouds are unable to be made up of the simple curve. To research mathematical properties of complex objects, fractal geometry has been raised up with the publishment of B.B.Mandelbrot's famous work [1].

In fractal theory, Koch curve is a classic sample to introduce fractal and its recursive algorithm. Koch curve has lots of mathematical properties [2] and "infinite length of border encloses finite area" is one of them. To get more mathematical properties, see some works about fractal theory.

Z. Xu (✉) · Y. Su
School of Mathematics and Physics, Xuzhou University of Technology,
Xuzhou, Jiangsu, China
e-mail: 793988650@qq.com

M. Tavana, S. Patnaik (eds.), *Recent Developments in Data Science and Business Analytics*, Springer Proceedings in Business and Economics,
https://doi.org/10.1007/978-3-319-72745-5_50

In web front-end area, <canvas> is a new element in Html5. <canvas> element has a great capacity in rendering image and Big Data drawing. Safari 1.3 Web browser leads <canvas> element in html. In fact, <canvas > element is an object, containing drawing context. The context not only provides series functions to draw lines, circles, Bezier curves, but also supports developers crafting animation, clipping images, operating pixel, etc.

With the accomplish of Html5 standard on October 29th, 2014, <canvas> element formally becomes a part of Html document. Currently <canvas> element is gradually replacing Flash in kinds of browser.

Firstly, to draw series of Koch curves with various dimensions, it is better to use a strong interactivity platform such as web. Secondly, none of programs uses <canvas> API to implement Koch curve in existing literatures. In summary, it is important to design a program on web page.

This report mainly research how Koch curves are drawn by <canvas> API instead of other graphic library because of its good performance in drawing graph with tremendous vertices.

50.2 Research Objectives

This report implements Koch curves using <canvas> API. This author tries to use simplest codes and most basic <canvas> API to draw n-dimensional Koch curves on browser platform.

The first goal of this report is using an effective algorithm to construct and draw Koch curves with <canvas> API.

Using <canvas> to drawing Koch curves has advantages in drawing big-data graph, interactively showing and convenience over other graphic libraries.

So, the more meaningful objective of the report is analyzing how these advantages of <canvas> element lie in drawing such fractal graph.

50.3 Detailed Finding

50.3.1 Algorithm

In fact, Koch curve originates in an equilateral triangle called 0-dimensional Koch curve which means 0th iteration. Suppose that an equilateral triangle, take 1/3 and 2/3 points of each edge. Then, take the points as a new edge and make an new equilateral triangle. Now we have a hexagram called 1-dimensional Koch curve. Repeat these steps until infinity, and then we will get an n-dimensional Koch curve shown as Fig. 50.1.

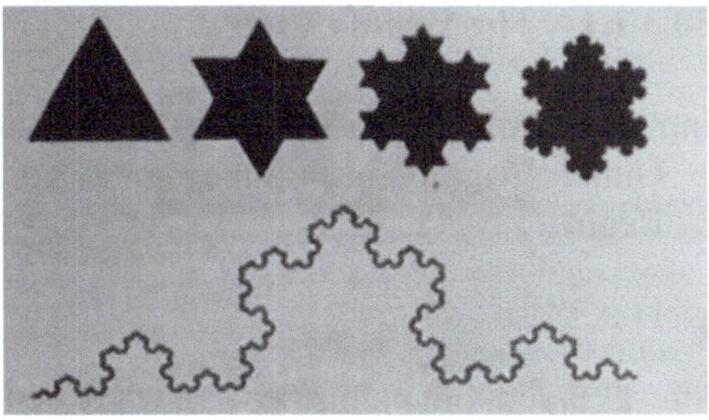

Fig. 50.1 Koch curves [4]

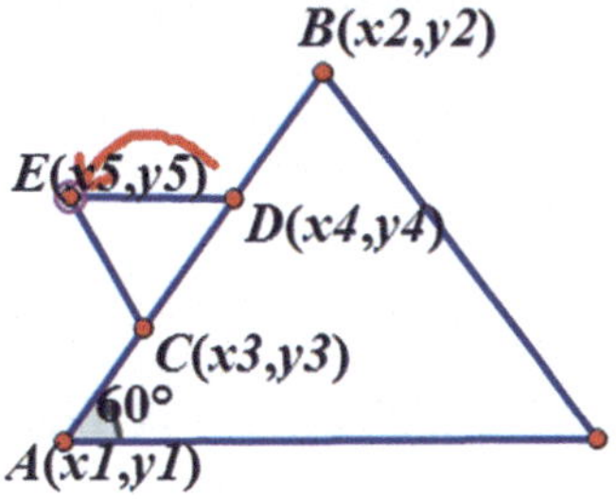

Fig. 50.2 Illustration of algorithm

For the convenience to design program, coordinates of vertices are in need. See Fig. 50.2. Given that we input coordinates of A(x1, y1), B(x2,y2),we easily figure that the points of trisection C(x3,y3),D(x4,y4):

$$x_3 = (x_2 - x_1) \div 3 + x_1$$
$$y_3 = (y_2 - y_1) \div 3 + y_1$$
$$x_4 = (x_2 - x_1) \times 2 \div 3 + x_1$$
$$y_4 = (y_2 - y_1) \times 2 \div 3 + y_1$$

Consider point E as a point which is point D rotated 60° anticlockwise around point C. Given a point a(a_x, a_y), it rotates α degrees around point o(o_x, o_y), then coordinate of the point a'(a'$_x$, a'$_y$) after rotation is calculated:

$$a'_x = (a_x - o_x) \times \cos\alpha - (a_y - o_y) \times \sin\alpha + o_x$$
$$a'_y = (a_x - o_x) \times \sin\alpha + (a_y - o_y) \times \cos\alpha + o_y$$

The formulas above derived from the sum-to-product trigonometric identity [5]. Trivially, the coordinate of point E(x_5,y_5) is calculated:

$$x_5 = (x_4 - x_3) \times \cos 60^{\circ} - (y_4 - y_3) \times \sin 60^{\circ} + x_3$$
$$y_5 = (x_4 - x_3) \times \sin 60^{\circ} + (y_4 - y_3) \times \cos 60^{\circ} + y_3$$

The author uses a simple recursive algorithm to draw the n-dimensional Koch curve because of Koch curve's self-similar nature.

50.3.1.1 The Pseudo Code

Input: two initial points(x_1,y_1),(x_2,y_2); recursive depth: **m**.
Output: A graph of m-dimensional Koch curve.
Begin

Step 1: If m = 0, draw a line at (x_1, y_1), (x_2,y_2); go to End. (0-dimensional Koch curve).
Step 2: Figure out points (x_3, y_3), (x_4, y_4) of trisection and the 'rotation point' (x_5,y_5); current depth + 1;
Step 3: If current depth = m, draw lines at the 5 vertices one by one;
Go to the End. (The exit of recursion).
Step 4: Recursively call this procedure using two adjacent points.

End

50.3.2 Program & Demo

Here is core code to recursively draw a Koch curve in JavaScript.

```
function koch(ctx, x1, y1, x2, y2, n, m){
    if(m==0){
            ctx.moveTo(x1,y1);
            ctx.lineTo(x2,y2);
            ctx.stroke();
            return false;//0-dimensional Koch curve
    }
    //coordinates of other 3 points
var x3=(x2-x1)/3+x1;
      var y3=(y2-y1)/3+y1;
      var x4=(x2-x1)*2/3+x1;
      var y4=(y2-y1)*2/3+y1;
      var x5=(x4-x3)*Math.cos(Math.PI/3)-(y4-y3)*Math.sin
      (Math.PI/3)+x3;
      var y5=(x4-x3)*Math.sin(Math.PI/3)+(y4-y3)*Math.cos
      (Math.PI/3)+y3;
n=n+1;     //increase current recursive-depth-counter
if(n==m){//the base case, drawing lines between two adjacent points
    ctx.moveTo(x1,y1);
            ctx.lineTo(x3,y3);
            ctx.lineTo(x5,y5);
            ctx.lineTo(x4,y4);
            ctx.lineTo(x2,y2);
            ctx.stroke();
            return false;
     }
```

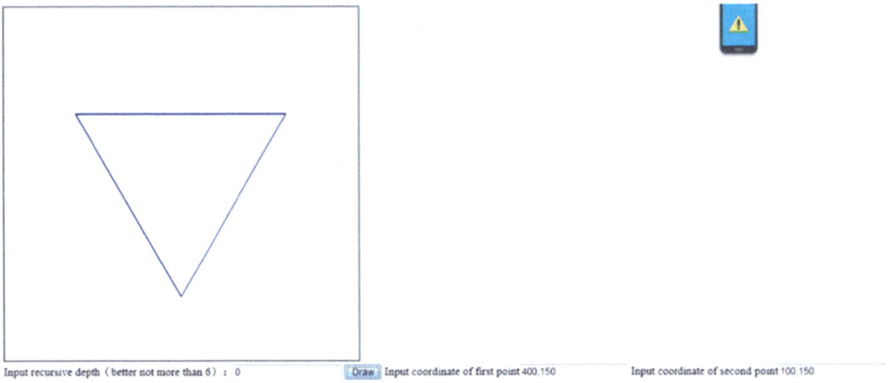

Fig. 50.3 Demo

```
koch(ctx,x1,y1,x3,y3,n,m);
koch(ctx,x3,y3,x5,y5,n,m);
koch(ctx,x5,y5,x4,y4,n,m);
koch(ctx,x4,y4,x2,y2,n,m);
  }
```

A canvas tag has been taken as main body of the web and adds input boxes for supporting users inputting data. A demo of program has been shown as Fig. 50.3.

See the following figures for Koch curves with various depths on IE9 (Fig. 50.4).

With the increment of recursive depth, we can't gradually distinguish the difference between two Koch curves by naked eye.

50.3.3 Analysis

50.3.3.1 Advantage of Big-Data Drawing

With the increment of recursive depth, it consumes more internal memory to save data on the recursive stack. Tables have been shown below (Tables 50.1, 50.2).

It has been shown that internal storage used increases rapidly due to the increment of recursive depth. In normal personal computer, the 7th depth recursion is almost the performance limit. But, it merely costs 620 s to draw the Koch curve and free memory of browser.

Using other graphic library, such a large amount of vertices, it has been used more than 800 s to draw 7-dimensional Koch curve. In the case of enough hardware devices, <canvas> element has more advantages in drawing big-data drawing owing to its high-effective in pixel processing.

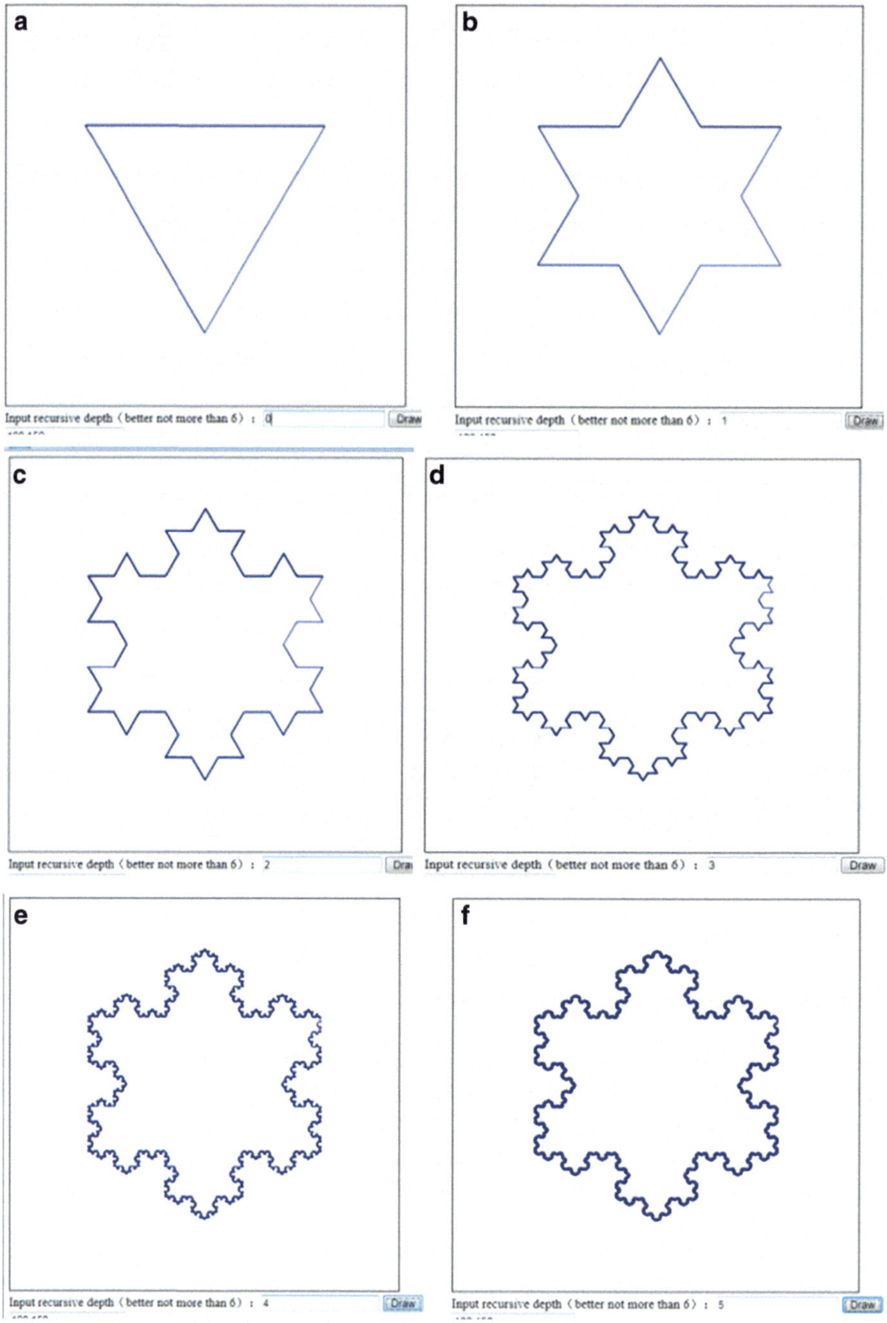

Fig. 50.4 (**a**) 0-dimensional Koch curve, (**b**) 1-dimensional Koch curve, (**c**) 2-dimensional Koch curve, (**d**) 3-dimensional Koch curve, (**e**) 4-dimensional Koch curve, (**f**) 5-dimensional Koch curve

Table 50.1 Storage and time consumed

Recursive depth	Internal memory used/ MB (before drawing)	Max internal memory used/MB (during drawing)	Multiple of memory Increment	Time Used/s
1	57.3	57.4	0.0018	0.001
2	57.0	57.6	0.01	0.003
3	57.2	61.3	0.07	0.02
4	74.1	78.7	0.06	0.1
5	77.5	101.3	0.31	4
6	82.1	470.3	4.73	25
7	91.3	1238.6	12.56	620

Table 50.2 Time consumed in various programming language

Programming language with graphic library	Average time used drawing 7-dimensional Koch curve (same recursive algorithm)
Matlab [4]	1127
C++	906
Java	889

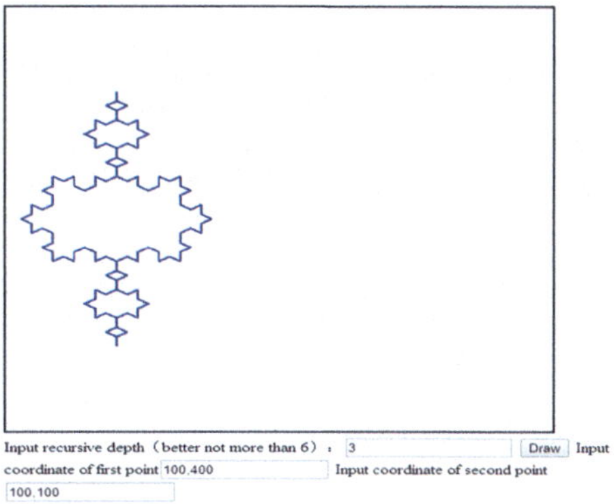

Fig. 50.5 A variant of Koch curve

50.3.3.2 Advantage of Interactive User Experience

Using <canvas> element, drawing Koch curves is combined with operating web pages. Users not only can input various coordinate of initial points to draw variant of Koch curves like Fig. 50.5, but also change the value of recursive depth. After clicking on the 'Draw' button, it draws a curve immediately on the web pages.

Html5 devotes itself in improving users' experience and <canvas> element is able to improve the visual experience. The demo of drawing Koch curves can be extended as a individual application via <canvas> API. For further polish, input box of stroking color, animation of drawing steps, axis of canvas can be added.

50.4 Conclusion

The <**canvas** > elements embed image of Koch curve in Web. Different from other graphic library, <**canvas**> element has advantages in dynamic, interaction and high capability. With <**canvas**>, users can automate the drawing commands and see the Koch curve results on the page immediately, without loading external file. It means that showing graphs like Koch curve by <**canvas** > technique brings users nice experience.

In the fractal theory side, research of algorithms and programming designs of drawing fractal graphs is becoming a popular tendency. Readers can gain inspiration of drawing fractal graph like Koch curve from the reports, and it recommends <**canvas** > API to researcher as a tool to implement fractal graphs.

Acknowledgements This work is supported by the Natural Science Foundation of Jiangsu Province (BK20151160), the Six Talent Peaks Project of Jiangsu Province (2013-JY-003), 333 High-Level Talents Training Program of Jiangsu Province(BRA2016275) and Undergraduate Training Programs for Innovation and Entrepreneurship of Jiangsu Province(XCX2017004).

References

1. Mandelbrot, B. B. (1982). *The fractal geometry of nature*. San Francisco: Freeman.
2. Milošević, N. T., & Ristanović, D. (2007). Fractal and nonfractal properties of triadic Koch curve. *Chaos Solitons & Fractals, 34*(4), 1050–1059.
3. Qiu, W., An, N., & Qi, X. (2010). Research on Algorithms of Generating Koch Fractal Graph based on MATLAB. China University of Mining and Technology, Xu Zhou.
4. Zhiying, W. (2000). *Mathematical foundations of fractal geometry*. Shanghai: Shanghai Scientific and Technological Education Publishing House.
5 https://en.wikipedia.org/wiki/List_of_trigonometric_identities.

Chapter 51
The Proposal of Centralization Deployment of Cadre Management Information System of SGCC

Jian Zhang, Tianjing Sun, Wentao Liu, and Sunnie Ren

Abstract In order to comprehensively improve the level of information-based cadre management, increase the strength of cadre management and control of SGCC, State Grid Corporation of China started to build Cadre Management Information System-Phase I, II and III projects. Pre-Cadre Management Information System used state-province two class deployment and multi-level application. In this mode, the system deployment was too scattered, and didn't match the goal of "Three Intensified Managements (in HR, Finance and materials), Five Large-Scale Movements (in Programming, Construction, Operation, Overhaul, and Production), and Two centers (Operation and monitoring center, Customer service center)", especially in the intensive uses of resources, coordination across and through different departments, and information sharing. It was why to initiate the centralization deployment work. The proposal of centralization deployment of Cadre Management Information System is the guiding principal article, which indicate the way to build the system. The article introduces the overall infrastructure, deployment scheme and application integration scheme of the centralization deployment of Cadre Management Information System Proposal. The application results supported the rationality of centralization deployment of Cadre Management Information System.

Keywords Cadre management · Centralization deployment proposal · Application integration

51.1 Introduction

By the opportunity of SG-ERP information system construction year, finding the breakthrough of deepening cadre personnel information management, and according to the standard requirements of corporate information-based construction, Director

---

J. Zhang (✉) · T. Sun · W. Liu · S. Ren
Beijing China Power Information Technology Co., Ltd., State Grid Corporation, Beijing, China
e-mail: jianzhang@sgitg.sgcc.com.cn

M. Tavana, S. Patnaik (eds.), *Recent Developments in Data Science and Business Analytics*, Springer Proceedings in Business and Economics,
https://doi.org/10.1007/978-3-319-72745-5_51

Department of Personnel, together with the Department of Communications of SGCC, organized and carried out the construction of Cadre Management Information System-Phase I, II and III projects. Facing the 12th Five-Year Plan, company's party group proposed to speed up the outstanding leadership construction work to meet the requirements of "World-Class Power Grid, International First-Class Enterprise". Information-based technology as an important carrier, will be responsible to practice and proceed advanced management thoughts, shoulder an important historical mission.

Under the background of keep proceeding information-based technology construction of SGCC, follow the spirits of 3rd Plenary Session of 18th CPC Central Committee, in accordance with the requirements to promote the deployment of national information-based technology construction of the company, the advantages of integrated forces and unified operation should be fully developed. Present, Cadre Management Information System used state-province two class deployment and multi-level application. In this mode, the system deployment was too scattered, and didn't matched the goal of "Three Intensified Managements, Five Large-Scale Movements, and Two centers", especially in intensive uses of resources, coordination across and through different departments, and information sharing etc.

According to the strategy and development goal of SGCC, and the requirements of "12th Five-Year Plan "information development planning, the implement of centralized deployment of Cadre Management Information System and the integration of business system like ERP, further promote business sharing and intensive control, effectively support the construction of "Three Intensified Managements , Five Large-Scale Movements, and Two centers", improve the utilization efficiency of IT resources and reduce the cost of informationization [1].

51.2 Overall Infrastructure

Overall infrastructure of Cadre Management Information System contains business infrastructure, technique infrastructure, data infrastructure, application infrastructure and security infrastructure, details are as below.

51.2.1 Business Infrastructure

On the foundation of corporate's human resources "12th Five-Year Plan", Cadre Management Information System includes business of cadre management, selection and appointing, evaluation and assessment, reserve cadres management, echelon building management, education and training, staff management, The Thousand Talents Plan, experts management, real-time release of information, inquiry and statistics, shot massage management, system management and such modules [2]. The business framework figure of Cadre Management Information System shows as below (Fig. 51.1).

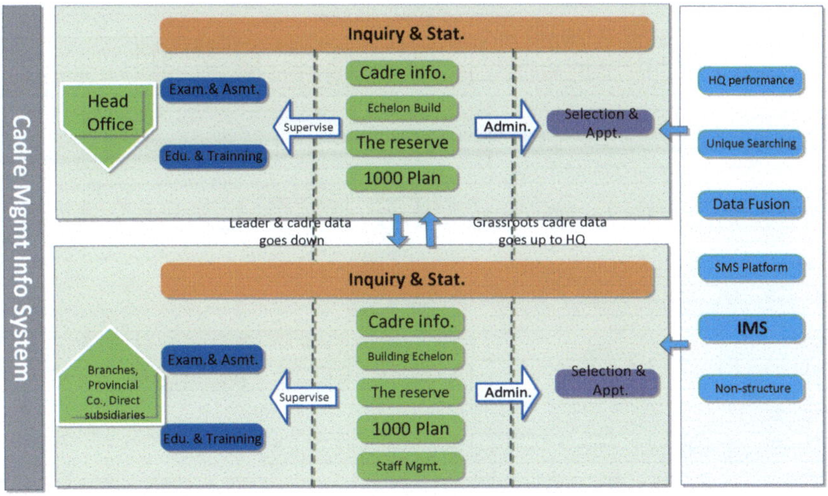

Fig. 51.1 Operation framework of cadre management information system

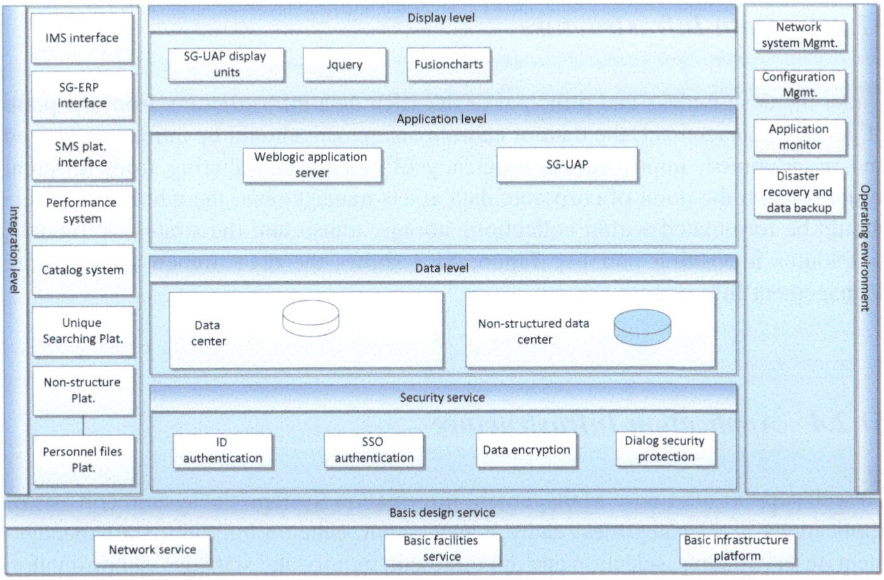

Fig. 51.2 Technique framework of cadre management information system

51.2.2 Technique Infrastructure

The main units of Cadre Management Information System of SGCC include display level, application level, integration level, data level [3]. The descriptions of each units of the different level are as below (Fig. 51.2).

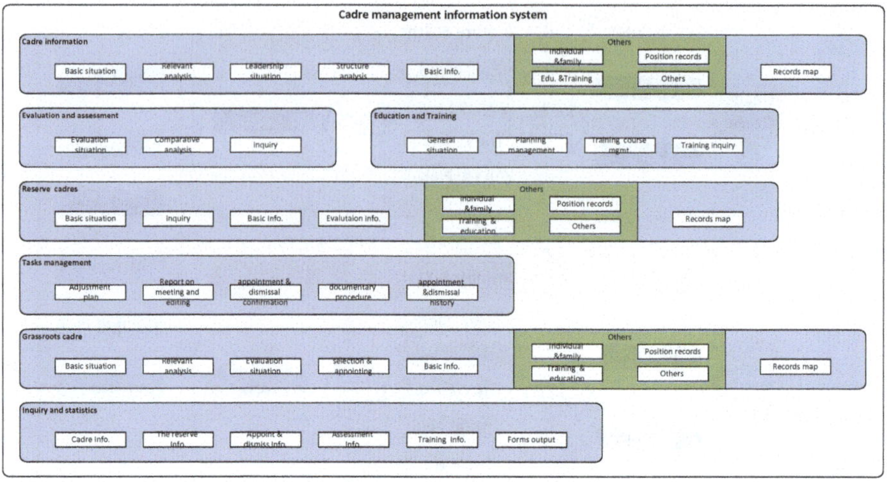

Fig. 51.3 Data framework of cadre management information system

51.2.3 Data Infrastructure

Under the design ideas and principal of SG-ERP data infrastructure, from the point of general management, the data of cadre management should be unified organized and programmed, improved the efficiency of storage and sharing from different systems; from the point of corporate data assets management, the whole data circle should be formulated within collection, storage, transit and the strategies, models, procedures in visiting part [4]. Figure 51.3 shows the data framework of Cadre Management Information System.

51.2.4 Application Infrastructure

To well support the Cadre Management Information System, the system consists of applications in 12 categories: cadre management, echelon building, staff management, reserve cadres, selection and appointment, inquiry and statistics, subordination functions, evaluation and assessment, education and training, The Thousand Talents Plan, system management etc.. On the fruits of the construction of phase I, II and III, the covering area and functions are reformed and improved. Figure 51.4 shows the application framework of Cadre Management Information System.

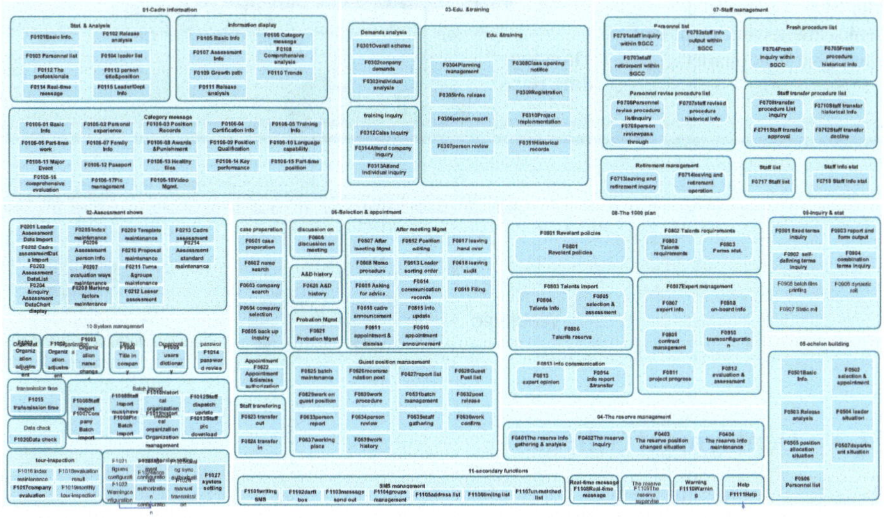

Fig. 51.4 Application framework of cadre management information system

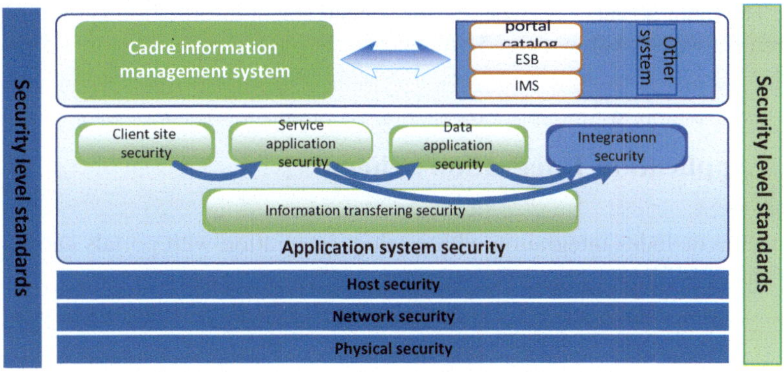

Fig. 51.5 Safe framework of cadre management information system

51.2.5 Security Infrastructure

Cadre Management Information System requires a higher level of data security, that's why the security infrastructure was designed from 5 sectors: physical security, network security, system security, application security and data security. Figure 51.5 shows the phase one security infrastructure of Cadre Management Information.

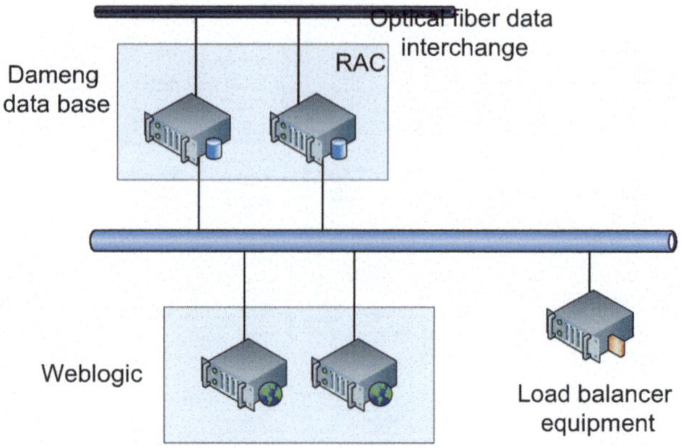

Fig. 51.6 Centralized deploy framework picture of State Grid headquarters

51.3 Deployment Scheme

The centralization and high-availability system deployment: servers located in production area, application business included two nodes which served as centralization and high-availability, load balancer were realized by F5 or middleware cluster. Data base server was in centralization and high-availability, involved two nodes. The main servers involved in cadre management information system production area showed as below (Fig. 51.6).

51.4 Application Integration Scheme

The scheme includes integration with catalog, integration with portals and integration with ESB [5].

51.4.1 Integration with Catalog

When users logged in portals, filled the cadre management information system form by catalog. When the user firstly passed through the confirmation of identity confirmation management server, the catalog sent the account message which asked by the user through several ways as below:

1. If the user's account in application system is same with the account in confirmation catalog, then automatically fills the value of properties like account name and password, and submits to application system.
2. If the user's account in application system is not same with the account in confirmation catalog, then there are two treatments:

   - The first login needs to input information like account name and password, these information will be encrypted and stored in the confirmation system.

Later, when user visit the application system again with the same URL, confirmation system will automatically fill the form with the information used before and submits;
- The user's account information in application system is saved as user's extension properties value in catalog, inquires account information in application system corresponding to user's extension properties value in confirmation catalog, and submits to the system.

51.4.2 Integration with Portals

Cadre management information system sent the message of to do task to the handling platform by calling the service from web service to-do-list, and then through data interface, the to-do-list of corporate portals comprehensively displayed the messages which sent from cadre management information system.

51.4.3 Integration with ESB

Cadre management information system integrated with application integration- ESB, and then integrated with headquarters' cadre management information system, headquarters' portals and other headquarters' business through headquarters' ESB. During the integration, the service consumers used the service which registered on ESB by service providers, through ESB realize the functions like united management, message transformation and intelligence router etc.

51.5 Conclusion

Currently, under the instruction of this proposal, Cadre Management Information System has been built out and got gradually achievements since the end of 2015. System's models and units have been developed and will go online after integration test, the third party test and confirmation test.

References

1. The 12th Five-Year Information Technology Plan of SGCC, 6–8
2. The Selections of Cadre Personnel Institutions of SGCC, 17–19
3. Information Technology SG-ERP Construction Proposal Approved by the Party Group, 31–32
4. The Methods of Overall Infrastructure of Information Technology Design of SGCC, 20–22.
5. The Standards of Application Software Integration Design of SGCC, Page 45–46.

Chapter 52
Research and Designing of Grounding Fault Diagnosis for Small Current System Based on DsPIC33 and MCP3903

Yin Hui, Shi Changkai, Guan Shilei, Lv Liping, Liu Manyu, and Wu Guoping

Abstract For the 10 kV boundary switch controller reliability fault isolation boundary switch controller designed in this paper on the hardware of CPU dsPIC33EP512MU810 with high performance and special electric power quality measurement chip MCP3903, software gives a practical TDFT non synchronous sampling measurement method. The boundary switch controller judges the ground fault according to the first 1/4 cycle, zero sequence voltage and zero sequence current direction of the small current grounding.The master station is communicated with each boundary switch controller through the 4G network, and the boundary switch controller is managed and maintained so as to facilitate data management. The experimental results show that the design is accurate and reliable in separating user fault.

Keywords Boundary switch controller · TDFT asynchronous sampling measurement method · MCP3903 · Power quality · 4G

52.1 Introduction

The domestic distribution network of the 10 kV in the neutral grounding mode, the advantages of this type of power grid is a single-phase grounding fault occurs, the fault current is small, single-phase grounding fault easily disappear, thereby

Y. Hui · S. Changkai · G. Shilei
China Electric Power Research Institute, Haidian District, Beijing, China

L. Liping · L. Manyu
State Grid Beijing Electric Power Company, Electric Power Research Institute, Fengtai District, Beijing, China

W. Guoping (✉)
Beijing Bolijie Electric Co., Ltd., Tongzhou District, Beijing, China
e-mail: 279445767@qq.com

M. Tavana, S. Patnaik (eds.), *Recent Developments in Data Science and Business Analytics*, Springer Proceedings in Business and Economics,
https://doi.org/10.1007/978-3-319-72745-5_52

increasing the possibility of running, but the disadvantage is due to the small fault current, resulting in 10 kV fault point the system is very difficult to find. When faults occur, unbalanced three-phase ground voltages may cause more serious faults in other areas. Therefore, when the single-phase permanent earth fault occurs in the actual power grid operation, it is necessary to complete the location of the fault point as soon as possible [1].

This paper developed a cut-off switch controller 16 bit microcontroller dsPIC33EP512MU810 based on high performance and measurement based on MCP3903 chip as the core, using TDFT software design method for measurement of non synchronous sampling, in determining the small current grounding fault is simple and reliable.

52.2 Hardware Design

52.2.1 Hardware Structure

Boundary switch controller comprises a three-phase voltage and current input signal, AD sampling circuit, dsPIC33EP512MU810 CPU, ICD2 simulation debugging interface, UART interface, LCD display part, RS232, RS485, opening into and out of control and power supply part.

CPU main control chip selects Microchip dsPIC33EP512MU810's high performance 16 bit microcontroller and meter specific metering chip MCP3903 as the core of the three-phase four wire multi-function meter module. The terminal is provided with a 3 way switch input terminal, which is used for detecting the circuit breaker's integral and no stored energy signals. The terminal is also provided with two relay outputs for controlling capacitor switching and providing a RS485 and RS232 serial port for data transmission of the 4G module.

The hardware structure of the demarcation switch controller is shown in Fig. 52.1. A, C, PT, CT after the zero phase current transformer after the signal processing circuit in AD MCP3903 conversion module, the signal sampling and conversion by SPI bus power quality data sent to the main CPU dsPIC33EP512 MU810.

52.2.2 CPU Monitoring Circuit

Figure 52.2 is the CPU monitoring circuit, the use of MAX706, which has power on automatic reset, manual reset, watchdog and voltage detection function. When the voltage is below 1.25 V, the threshold detects the output /PFO and the output is low to CPU. The watchdog WDI is connected to the CPU, and when the signal remains high or low over 1.6 s, the timer overflow makes the /WDO end output low, thereby making the /WR lower than the 0.6 V and the /RESET end outputting a reset signal.

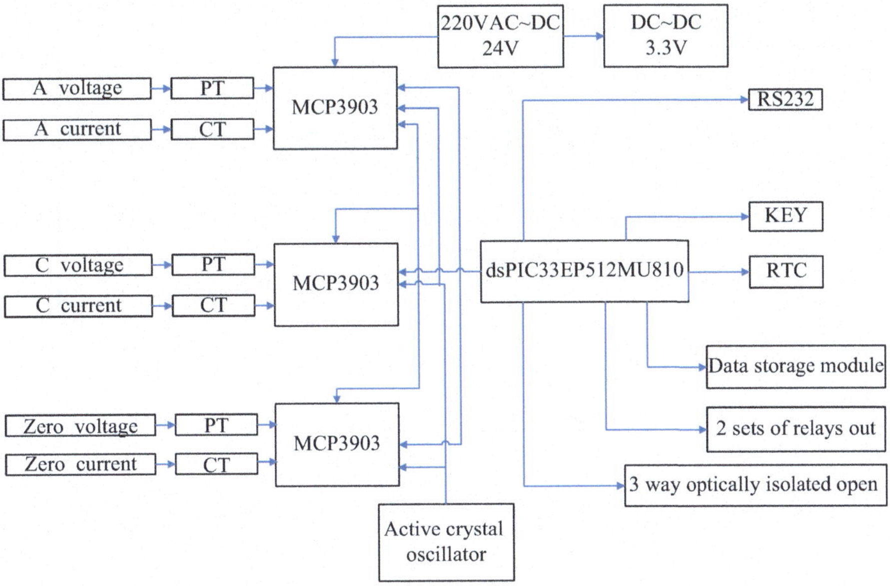

Fig. 52.1 Hardware structure diagram of the boundary switch controller

When any change in the level of the WDI input allows the timer to be reset, the watchdog is not allowed to overflow.

52.2.3 *Temperature Compensation*

Figure 52.3 is a digital temperature sensor circuit. In the measuring instrument, the temperature examination is essential. The circuit adopts AD7416 digital temperature sensor produced by AD company. It includes a band gap temperature sensor and 10 as AD converter. The resolution is 0.25°, the measuring range is −55~125°, and the accuracy is degree. The temperature sensor is used to measure the ambient temperature and to compensate the temperature of the crystal.

52.2.4 *The Opens and Goes in of the Boundary Switch Controller*

Figure 52.4 is a 3 way switch input. The external circuit breaker is integrated and divided, and the energy storage bit is isolated by DI1~DI3 through optocoupler, and then is input and connected to the CPU through PIN0~PIN2, and then the corresponding calculation and operation are carried out.

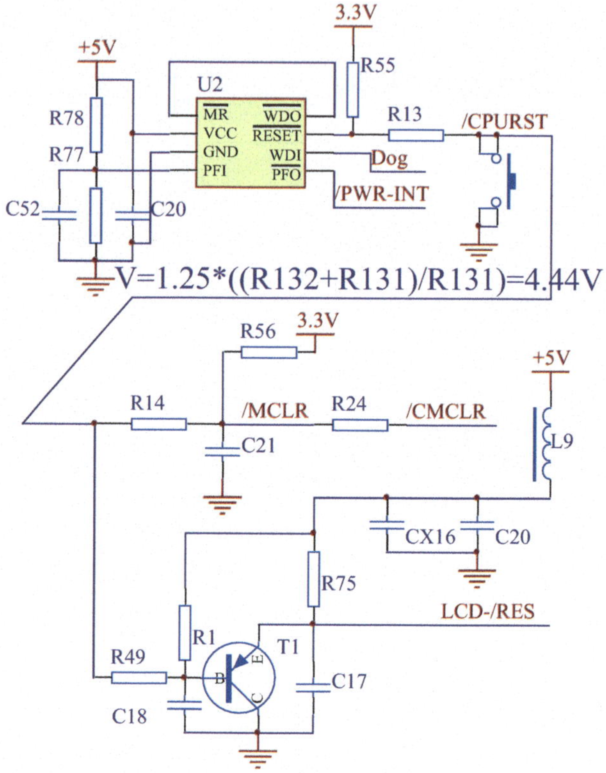

Fig. 52.2 Monitoring circuit for the CPU

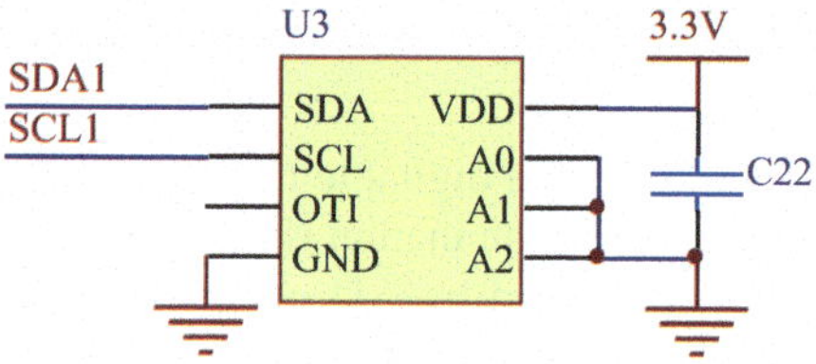

Fig. 52.3 Digital temperature sensor circuit

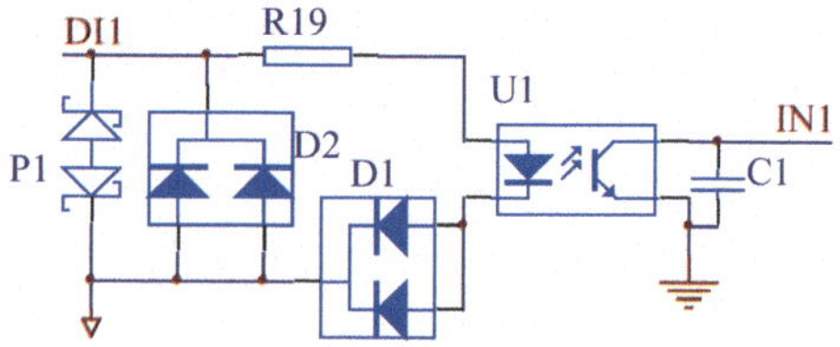

Fig. 52.4 Opened to the outside for the Boundary switch controller

Two way CPU output, O0, O1 through triode 8050, control relay output, used for circuit breaker out of the output, as shown in Fig. 52.5.

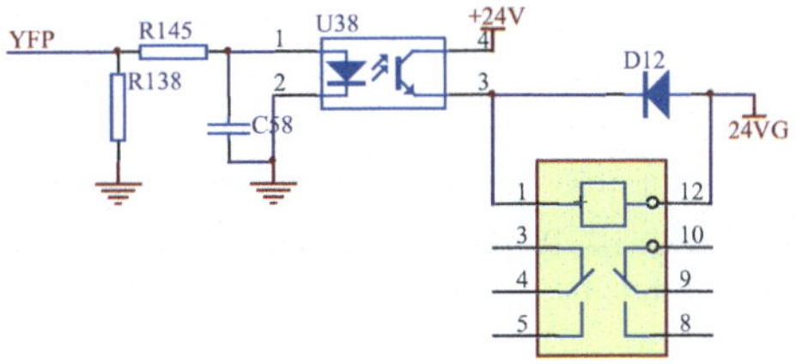

Fig. 52.5 Relay out of the Boundary switch controller

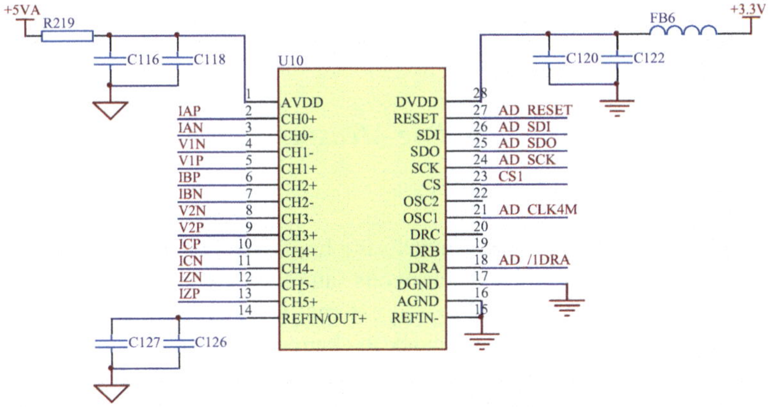

Fig. 52.6 AD sampling circuit of the Boundary switch controller

52.2.5 AD Sampling of the Boundary Switch Controller

AD sampling circuit as shown in Fig. 52.6, MCP3903 clock provided by 3.2768 MHz active crystal, the output of the sampling frequency is 12.8 K, the clock line of three-phase MCP3903 and reset lines connected together, can ensure that the three-phase sampling synchronization. MCP3903 once each sampling is completed, the SDO pin outputs a pulse indicating the end of the ADC conversion.

The MCP3903 can be used as the 16 bit ADC of the six channel, and each MCP3903 completes a phase synchronous sampling of the current and voltage. The three MCP3903 clock signals are provided by the 3.2768 MHz active crystal oscillator, and the reset signal is controlled by the dsPIC33EP512MU810 (MU810) so as to realize the three-phase synchronous sampling. MCP3903 through the SPI interface and MU810 connection, MU810 is the SPI master controller, MCP3903 current channel gain can be independently controlled by the MU810 of I/O, and can adjust the appropriate gain according to the different input current amplitude.

52.2.6 4G Communication Module

Each demarcation switch controller are installed a 4G module, boundary switch controller can be a circuit breaker switch, terminal fault data and log files through the

4G module to the 4G network, and then through the establishment of INTERNET network and the master control center, this mechanism can upload the master control center to manage and maintain the remote terminal. In addition, the master station staff can also be convenient for demarcation switch controller software updates and upgrades, eliminates the demarcation switch controller installation and maintenance personnel to upgrade software update operations in remote or environmental conditions [2].

52.3 Software Designing

52.3.1 Asynchronous Sampling Measurement Method Based on Tdft

Compared with TDFT non synchronous sampling measurement method, synchronous sampling method requires synchronous sampling interval and grid frequency, the hardware PLL circuit so need to add external to track the grid frequency, and automatically update the MCP3903 clock to change the sampling frequency. Since the output frequency of the PLL is lagging behind the frequency of the power grid, there will be synchronous error in the system, so the strict synchronous sampling is difficult to achieve. When the power grid is in nonsinusoidal condition, it will affect the zero crossing detection. In severe cases, the PLL will fail and the system will not work properly [3]. In comparison, the TDFT asynchronous sampling measurement method has the advantages of engineering practicability. This method uses equal interval sampling without synchronization with the grid frequency. TDFT non synchronous sampling measurement method can not ensure periodic sampling of the measured signal cycle and strict synchronization, in order to reduce the measurement error caused by non synchronous, the fundamental and harmonics can obtain higher accuracy in data processing by increasing the number of iterations to improve the measurement accuracy, and effectively reduce the synchronization error affect the accuracy the degree of measurement, accurate measurement and harmonic frequency. Quasi synchronous window function [4].

$$U(t) = \frac{U_{a0}}{2} + \sum_{k=1}^{\infty} \left(u_{ak} \cos\left(kwt\right) + u_{bk} \sin\left(kwt\right)\right). \tag{52.1}$$

After determining the number of samples, the number of iterations, and the numerical quadrature method, the coefficients of the quasi synchronous function can be determined, and then a quasi synchronous function array is established. The raw data is weighted by the quasi synchronous window function, which is equivalent to a synchronous processing of data [5]. The original function data and the quasi synchronous function array are multiply operated, and the new periodic signal after processing is the same as the original signal period. The frequency components are the same, and the synchronization error of the new signal is smaller than [6].

52.3.2 TDFT Asynchronous Sampling Measurement Method Harmonic Calculation

TDFT based asynchronous sampling harmonic measurement method uses TDFT interpolation to suppress the asynchronous sampling error, so as to achieve accurate measurement of harmonic parameters. The method does not need hardware phase locking, saves hardware equipment cost, and reduces hardware equipment volume. Meanwhile, the method is not affected by the sampling length, and can be used for high resolution measurement of harmonics. In addition, the TDFT interpolation method does not need to construct and store window functions, thus avoiding the finite word length effect in the storage and computation of window functions, so it is more accurate than the traditional windowed interpolation method.

In order to overcome the problem of windowed algorithm, this paper uses the TDFT Fu Liye Discrete (Transformed Fourier Transform) to suppress the spectral leakage error. The principle of the method is as follows: a new sequence X1 (n) is obtained by weighting the adjacent three terms in the discrete Fu Liye transform result X (n).

$$X1(n) = X(n) - \frac{X(n-1) + X(n+1)}{2} \tag{52.2}$$

$$X(n) = \frac{M}{L} e^{j(\theta_m + r\pi)} \tag{52.3}$$

Typing:

$$L = n - k1 - r \tag{52.4}$$

$$M = \frac{A_m N \sin(r\pi)}{2\pi} \tag{52.5}$$

Get next:

$$\begin{aligned} X1(n) &= \frac{M}{L} e^{j(\theta_m + r\pi)} - \frac{1}{2}\left(\frac{M}{L-1} e^{j(\theta_m + r\pi)} + \frac{M}{L+1} e^{j(\theta_m + r\pi)}\right) \\ &= Me^{j(\theta_m + r\pi)}\left(\frac{1}{L} - \frac{L}{L^2 - 1}\right) \\ &= -\frac{M}{L(L^2 - 1)} e^{j(\theta_m + r\pi)} \end{aligned} \tag{52.6}$$

The frequency measurement method based on TDFT for asynchronous sampling measurement is shown in the following figure. It can be seen that this method is more effective and accurate than traditional and windowed DFT methods (Figs. 52.7 and 52.8).

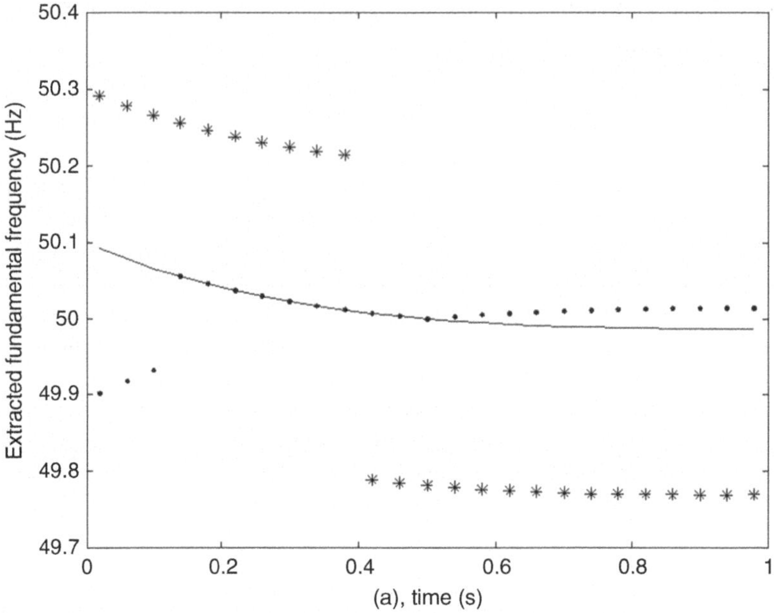

Fig. 52.7 Fundamental frequency detection results (">" traditional DFT method, "*", window DFT method, "-" TDFT method)

The boundary switch controller of small current grounding system of the flow chart, power on initialization after the calculation of voltage and current sampling in three after Zhou Bo, after the zero sequence voltage and zero sequence current to determine whether the small current grounding, such as fault wave recording storage controller, can be used in fault diagnosis and subsequent upload station.

Small current grounding fault criterion mainly depends on the limit value of zero sequence voltage and zero sequence current limits, reactive power direction of the three conditions into the small current software to determine the entrance, three conditions "and" relationship, there is a condition that does not enter the small current grounding fault [7]. According to the method of judging the transient component of the fault current in the previous 1/4 cycle, the characteristics of the fault line and the polarity of the transient zero sequence current and the zero sequence voltage of the perfect line are determined. In the event of a fault, for the resonant grounding system, the capacitor and ground zero sequence voltage source are charged to the arc suppression coil, so the arc suppression coil can not be compensated, and can be regarded as open circuit. Therefore, the former 1/4 cycle method is not affected by arc suppression coil. After the grounding fault line zero sequence current flow to the bus as the value of the system of non fault line on non fault components of capacitive current and the capacitive reactive power direction and non fault lines on the contrary, the actual direction of flow to the bus line [8].

Figure 52.9 as an example, the equivalent zero sequence network graph topology, a zero sequence voltage in fault at U0 (vector), each component in a circuit of

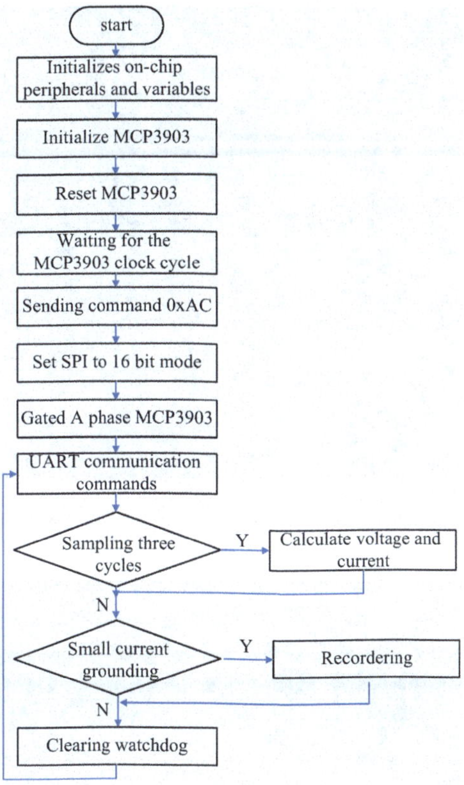

Fig. 52.8 Flow chart software

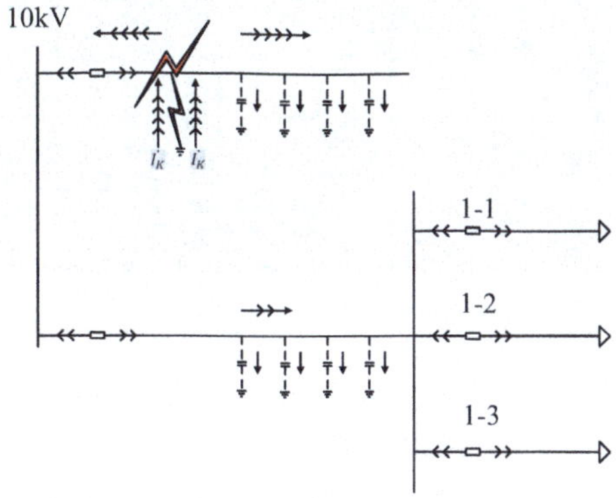

Fig. 52.9 Topological diagram of small current grounding

capacitor to constitute the zero sequence current loop, the zero sequence impedance impedance and capacitance line compared to the small, so you can ignore the influence of zero sequence impedance. So each element in neutral point grounded system in which the capacitance current of zero sequence current of the whole network.

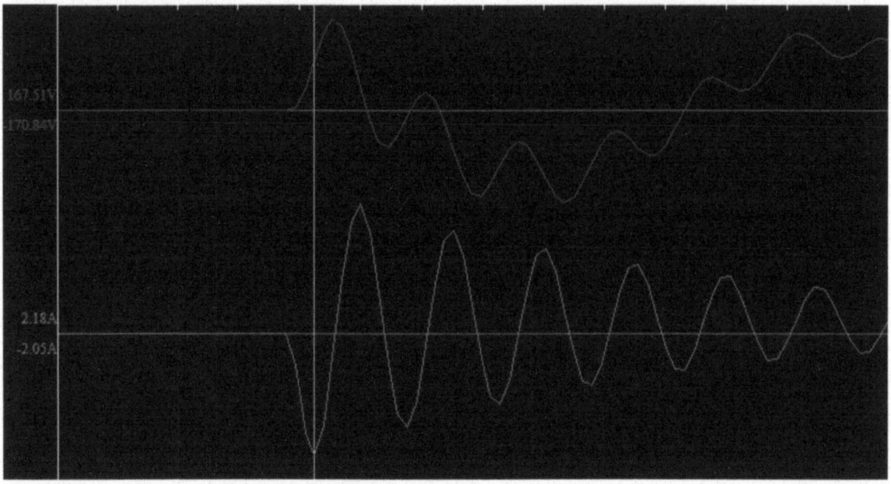

Fig. 52.10 The Intra regional failure waveform for the arc suppression coil

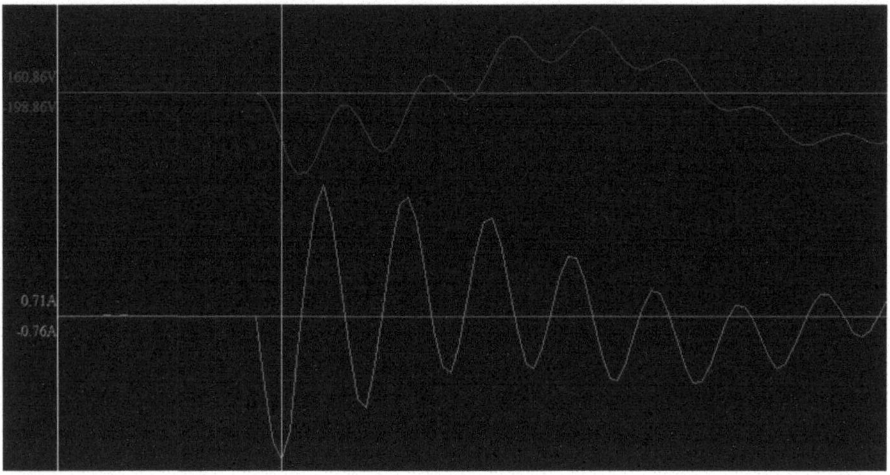

Fig. 52.11 The out of band waveform for the arc suppression coil

52.4 Conclusion

In conclusion, the characteristics are embodied in the following four aspects [9]:

1. When the single-phase grounding fault occurs, the whole system has zero sequence voltage;
2. The zero sequence current on the normal line is the capacitance current of the line itself, and the capacitive reactive power direction is opposite to the fault line, which shows that the bus flows to the line;

3. The zero sequence current on the fault line is the sum of zero sequence current of all normal lines, and the direction of capacitive reactive power is opposite to the normal line, and the line flows to the bus.
4. When the single-phase grounding fault happens, the zero sequence circuit basically is purely capacitive, all normal lines have 90° ahead of the zero sequence current of zero sequence voltage, and zero sequence current fault line is non fault line zero sequence current of all phase lag and zero sequence voltage of 90° [10] (Fig. 52.10).

The boundary switch controller of small current grounding fault in the region when the recorded waveform graph, can be seen from the figure of zero sequence current in the first 1/4 cycle of a peak value of 40.96A (CT zero sequence current ratio is 20:1), reactive power to meet the current conditions within the region, the controller action outlet protection (Fig. 52.11).

The boundary switch controller of small current grounding fault zone in the waveform diagram, it can be seen from the figure of zero sequence current in the first 1/4 cycle of a peak value of 6.4A (CT zero sequence current ratio is 20:1), does not meet the conditions of small current grounding area in reactive power direction, the controller does not move the export atresia.

Acknowledgments Thanks Yang Rengang, Feng Xiaoming, Tang Yunfeng, Zhang Zongsheng and so on to give the help and the support.

References

1. Hui, C., Shiying, Y., & Jun, Y. (2005). Quasi- simultaneous sampling method in the distribution of the application of integrated monitor. *REALAY, 33*(17), 45–48.
2. Xiaobo, M., Wenli, Z., & Junjie, H. (2005). A.C. Samping technology and result of realizing using DSP. *Measurement & Control, 21*(2), 54–55.
3. Jiashen, L., Shijie, C., & Yuxing, D. (2008). The study of quasi-simultaneous algorithm based on interpolation in the application of harmonic detection. *Electrical Measurement & Instrument, 45*(510), 1–4.
4. Liheng, W., Yongfeng, R., Shengkun, L., et al. (2009). The design of multiple channel measurement system based on FPGA. *Electrical Measurement & Instrument, 46*(518), 20–24.
5. HaiLin, Z., & Lin, L. (2009). Design of power load management and power monitor system based on ECC. *Electrical Measurement & Instrument, 46*(519), 53–57.
6. Jun, C., Wei, Z., & Renxin, C. (2000). Power measurement and harmonic frequencies of the new method. *Tsinghua University Scince & Technology, 40*(1), 25–27.
7. Yong, L., & Jian, Z. (2004). Quasi synchronous sampling method of harmonic analysis based on adaptation of parameter. *21*(18), 18–19.
8. Guangyu, D., Xiao-qiao, C., & Zhongtian, W. (2001). Application of phase lock multiple frequency synchronous sampling and quasi-synchronous sampling to measurement of harmonics. *Engineering Journal of Wuhan University, 34*(5), 39–44.
9. Shiji, W., & Xicai, S. (2003). Quasi-synchronous sampling and its application in the digital measurement of common electronical parameter. *Applied Science and Technology, 30*(12), 10–12.

10. Zhanshi, S., & Yuqin, F. (2009). Single-phase reactive rapid detection method based on the instantaneous Reactive Theory. *Electrical Measurement & Instrument, 46*(518), 1–4.

Brief introduction of the author:

Yin Hui (1989) male, graduate student, assistant engineer, engaged in distribution automation, distribution automation system, equipment inspection technology research.

Shi Changkai (1983) male, graduate student, senior engineer, engaged in distribution network planning, rural power network intelligence, intelligent power consumption and interaction.

Guan Shilei (1984) male, graduate student, engineer, engaged in distribution automation, distribution automation system, equipment inspection technology research.

Lv Liping (1983) male, master, engineer, mainly engaged in power distributionAutomation, relay protection and so on.

Liu Manyu (1986) female, master, assistant engineer, mainly engaged in Distribution automation work.

Wu Guoping (1979) male, master's degree, research interests include power system automation, power quality, etc.

Chapter 53
Mobile Phone Addiction in Youngsters: Concept and Measurement

Menglong Li and Yujia Ren

Abstract Smart phone has brought great convenience to our life. It has become an indispensable part of our life, and also has negative impact, such as mobile phone addiction. Mobile phone addiction will greatly impact youngsters in study and life. Due to excessive indulgence in or misuse of mobile phone, more and more youngsters indulge themselves in the virtual world created via mobile phone, and even get addicted, which has a serious negative effect on youngsters' mentality, physiology, social relations, study and life. This study aims to summarize the concepts in relation to mobile phone addiction and measuring tools, with a view to providing a reference for learning about, understanding, researching, preventing and treating mobile phone addiction, and promoting physical and psychological health of youngsters.

Keywords Youngsters · Mobile phone addiction · Concept · Measurement

53.1 Introduction

With the continuous improvement of society, economy, science and technology, phones with increasingly perfect functions are being produced, mobile phones become more and more popular, and have gradually become an indispensable tool in our life [1]. The popularization of mobile phone has greatly facilitated our life, and has a far-reaching and extensive influence on our lifestyle. Meanwhile, however, overuse of mobile phone leads to mobile phone addiction, which has an adverse impact on people's body, psychology, and social behavior [2].

As one of the research topics in the field of health psychology, addiction refers to that an individual is badly eager for some activity or abuse of some substance uncontrollably, regardless of the adverse impact of such behavior. Addiction is classified into substance abuse and behavioral addiction [3]. Mobile phone addiction

M. Li (✉) · Y. Ren
Physical Education Institute, Hunan First Normal University, Changsha, China
e-mail: lml0713@yeah.net

M. Tavana, S. Patnaik (eds.), *Recent Developments in Data Science and Business Analytics*, Springer Proceedings in Business and Economics,
https://doi.org/10.1007/978-3-319-72745-5_53

is behavioral addiction caused by overuse of modern technology, similar to internet addiction and game addiction. Specifically, it means a state that a mobile phone user becomes obsessed with mobile phone due to overuse of mobile phone for some purposes, in which the addict suffers from impairment of psychology and social function. In serious case, the addict even will suffer from pins and needles, palpitation, dizziness, sweating, gastrointestinal dysfunctions, and so on. Mobile phone addiction, as a new social problem following internet addiction, has gradually become one of the focuses of attention in the society [4].

The physical and psychological health of youngsters objectively reflects the comprehensive national strength of a country, and is the mark of prosperity of a nation. We have made great efforts in improving youngsters' physical and psychological health, but there still are a lot of problems. Mobile phone addiction is a major problem impacting youngsters' physical and psychological health. Some researches show that more than 25% youngsters have got addicted to mobile phone [5]. Researches show that about 40% youngsters in Spain have got addicted to mobile phone, who spend more than 4 h in making phone call or texting every day. More than 33% Italian youngsters own at least two mobile phones and are deeply addicted to mobile phone; 36% British youngsters insist that they can't do without mobile phone [6]. Undoubtedly, mobile phone addiction has a great impact on youngsters' study and life. Due to excessive indulgence in or misuse of mobile phone, more and more youngsters become obsessed with the virtual world created via mobile phone, and then become addicted, which has an adverse impact on youngsters' psychology, physiology, social relations, study and life. Youngsters' addiction to mobile phone has become a serious psychological and social problem. Hence, it is of great theoretical and practical significance to study youngsters' addiction to mobile phone.

53.2 Concept of Mobile Phone Addiction

Defining the concept of mobile phone addiction is the basis for studying mobile phone addiction. There are many researches on this topic. However, the concept of mobile phone addiction has not clearly defined yet. Some scholars defined "mobile phone addiction" to be "mobile phone addiction", "problem use of mobile phone", or "pathological phone use", etc. [7]. Ezoe et al. believed that addiction and dependency are two similar concepts defining mobile phone addiction from two aspects: excessive use and eagerness for discontinuous use of mobile phone [8]. Walsh et al. pointed out that the manifestation of mobile phone addiction is that an individual is badly for or has a strong desire for mobile phone use, which is uncontrollable [9]. It is also considered that mobile phone addiction refers to obsession with various activities with mobile phone as media, or strong and continuous eagerness for and dependency on mobile phone use, which greatly impairs individual's social and psychological functions. Foreign researches get used to describing mobile phone addiction to be problem use of mobile phone [10]. Some researchers also define mobile phone addiction to be behavioral addiction caused by interaction between human and mobile phone [4]. The behavior of improper use of mobile phone on occasion where it is explicitly prohibited to use mobile phone or there is perceptible potential danger due to mobile phone use is called "problem use of mobile phone" by some people [10]. According to the definitions of mobile phone addiction above, mobile phone addiction essentially means behavioral addiction arising in the interaction process between human and mobile phone, and is a manifestation of improper mobile phone use, which is at mental level.

Chinese scholars tend to focus on the state or consequences of mobile phone addiction for defining mobile phone addiction. Research defined mobile phone addiction to be a state of chronic or periodic obsession caused by continuous use of mobile phone, which leads to a psychological state and behavior of strong and continuous eagerness and addiction, and is classified into mobile phone entertainment addiction, mobile phone relationship addiction and mobile phone information gathering addiction [11]. Han insisted that there should be three criteria for judging mobile phone addiction: (1) misuse of mobile phone; (2) undue influence of mobile phone on living, work and study; (3) a series of adverse physiological and psychological reactions in the case of mobile phone being out of service or absence of mobile phone, which relatively describes the concept of mobile phone addiction completely [12]. Wang, based on summary of concepts of mobile phone addiction defined by foreign and Chinese scholars, came to the conclusion that mobile phone addiction means a state of obsession in which an individual suffers from significant physiological, psychological and social function impairment due to uncontrollable use of mobile phone, which is a kind of mental and psychological disease newly emerged with the employment and rapid popularization of mobile phone [13]. It is essentially similar to internet addiction, but the former is more extensive and in disguised forms. This definition is widely recognized. According to the definitions of mobile phone addiction set out above, mobile phone addiction has a lot of adverse impacts on people, physiologically, psychologically and socially (including life, study and family), and impairs people's healthy life, physical and psychological health, and social life.

Although the criteria for defining mobile phone addiction are still controversial in the academic circle, it is uniformly regarded that mobile phone addiction falls into the scope of behavioral addiction, because it doesn't involve matter intake, and is a new type obsessive-compulsive dependent behavior. Based on the previous researches, we can think that mobile phone addiction is a state of obsession in which mobile phone users suffer from physiological, psychological and social function impairment due to uncontrollable excessive mobile phone use for some purposes, which is similar to internet addiction and game addiction [14].

53.3 Measuring Tool of Mobile Phone Addiction

Measuring tools of good reliability and validity also are a basis for studying mobile phone addiction. As the study on mobile phone addiction deepens, researchers have developed some tools for measuring mobile phone addiction of good reliability and validity, which lays a good foundation for studying mobile phone addiction of middle school students.

Relatively speaking, foreign researchers have developed related tools before Chinese researchers do. The first tool for measuring mobile phone addiction is mobile phone dependence questionnaire (MPDQ), which was developed by Toda et al. [15]. It consists of 20 items, and the Likert 4-level scoring method is adopted, ranging from 0 (hardly any) to 3 (always). The aggregate score ranges from 0 to 60. High score indicates high degree of mobile phone addiction. The questionnaire is widely applied, and turns out to be of good reliability. Bianchi and Phillips [16], in accordance with literatures on addiction and in combination with social factors related to mobile phone, developed the MPPUS according to the measuring methods of behavioral and technological addiction, which consists of 27 items from five perspectives, i.e. tolerance, avoidance of other problems, withdrawal, eagerness and negative consequence. Many researchers adopted this tool as basic measuring tool for deeper research. For it, the Likert 10-level scoring method is adopted, ranging from 0 (completely out of line) to 10 (completely in line). High score indicates intense attention to problem use of mobile phone. The coefficient of internal consistency of this tool is 0.9, indicating that the tool is of high reliability. It was referred to by researchers in Japan, the United States and the UK, and has been proved to be of good internal consistency and construct validity [17]. Billieux et al. developed the PUMPQ, which is for comprehensively measuring the actual and potential problems of mobile phone use [18]. It consists of 30 items from four perspectives, i.e. ultralimit use, dangerous use, financial problems, and withdrawal symptoms, and the Likert 4-level scoring method is adopted, ranging from 1 (strongly agree) to 4 (strongly disagree). It was widely applied in Turkey, Switzerland, and South Korea, and has been proved to be of good reliability [19]. Leung et al. [20] developed the mobile phone addiction tendency scale (MPATS), which consists of 17 items from four perspectives, i.e. uncontrollability, withdrawal, avoidance, and inefficiency. The Likert 5-level scoring method is adopted, ranging from 1 (never) to 5 (always). The coefficient of internal inconsistency α of the tool is 0.87, and it is of good reliability and validity.

Initially, Chinese researchers introduced scales for studying mobile phone addiction, and then also started to develop measuring tools by themselves in accordance with the actual situation. Wang [21] developed the middle school student mobile phone addiction tendency scale, which consists of 16 questions from three perspectives, i.e. withdrawal (6 questions), salience (5 questions) and compulsivity (5 questions). The coefficient of internal consistency of the general scale is 0.824, the test-retest reliability of subscales ranges from 0.642 to 0.853, and the result of confirmatory factor analysis shows that all data of the theoretical model of the scale fit well. It is thus clear that the scale is of high reliability and validity. Shi [22], by referring to clinical experience, developed the mobile phone dependence diagnosis scale based on Young's Internet Addiction Test (LAT). It consists of nine items. That at least five items are satisfied indicates mobile phone dependence [22]. Tao developed the youngsters mobile phone dependence self-assessment questionnaire, which consists of 13 items from three perspectives, i.e. withdrawal symptoms, eagerness and physical and psychological effect [23]. The Cronbach's coefficient α of the general questionnaire is 0.87, and that of each perspective ranges from 0.58 to 0.83. The result of confirmatory factor analysis shows that the questionnaire is of high reliability and validity. Xu et al., based on the diagnosis of substance dependence and behavioral addiction in the fourth edition of the Diagnostic and Statistical Manual of Mental Disorders, developed the Mobile Phone Dependence Inventory (MPDI) from four perspectives, i.e. tolerance, withdrawal, social function and physiological reaction [24]. To sum up, there are many measuring tools of mobile phone addiction of high reliability and validity developed by both foreign and Chinese researchers, but such tools have not been popularized in practical application. Thus, there is no universally accepted tool yet.

53.4 Conclusion

The popularization of smart phone has brought great convenience to us, and also leads to mobile phone addiction, which attracts great attention and concern. People have realized the adverse impact of mobile phone addiction on the physical and psychological health of youngsters, which also attracts great attention from the society. There are a lot of researches on the concept of measuring tools of mobile phone addiction, substantial progress has been achieved, and such researchers are being deepened. But there are still some problems: (1) the concept of mobile phone addiction has not been uniformly defined, which needs to be further researched; (2) concerning the measuring tool, on one hand, there lacks measuring tools specific to particular groups; on the other hand, there lacks effective measuring tools with comprehensive functions and easy to operate. In a word, the existing researches provide an important theoretical and practical foundation for this research, and there are still many problems to be researched and solved.

Acknowledgments This study was financially supported by Hunan Provincial Natural Science Foundation of China (2017JJ2054).

References

1. Li, M. L., & Lu, L. Y. (2017). The influence of mobile phone addiction on left-behind middle school students' sleep quality: The mediator role of loneliness. *Revista Argentina de Clinica Psicológica, 26*(1), 71–81.
2. Bohman, B., Forsberg, L., Ghaderi, A., et al. (2013). An evaluation of training in motivational interviewing for nurses in child health services. *Behavioural and Cognitive Psychotherapy, 41* (3), 329–343.
3. Li, M. L., Nie, J. S., & Ren, Y. J. (2015). Effects of exercise dependence on psychological health of Chinese college student. *Psychiatria Danubina, 27*(4), 413–419.
4. Augner, C., & Hacker, G. W. (2012). Associations between problematic mobile phone use and psychological parameters in young adults. *International Journal of Public Health, 57*(2), 437–441.
5. Ge, X. H., & Zhu, Z. H. (2014). An empirical study on the relationship between adolescents' social support and mobile phone addiction. Chinese journal of health. *Statistics, 31*(5), 830–832.
6. Birdwell, A. (2010). *Addicted to phones? Cell phone use becoming a major problem for some, expert says*. Retrieved December 3, 2011, from http://news.ufl.edu/2007/01/18/cell-addiction/
7. Thomée, S., Harenstam, A., & Hagberg, M. (2011). Mobile phone use and stress, sleep disturbances, and symptoms of depression among young adults-a prospective cohort study. *BMC Public Health, 11*(1), 66.
8. Ezoe, S., Toda, M., Yoshimura, K., et al. (2009). Relationships of personality and lifestyle with mobile phone dependence among female nursing students. *Social Behavior and Personality: An International Journal, 37*(2), 231–238.
9. Walsh, S. P., White, K. M., & Young, R. M. D. (2009). The phone connection: A qualitative exploration of how belongingness and social identification relate to mobile phone use amongst Australian youth. *Journal of Community and Applied Social Psychology, 19*(3), 225–240.
10. Luo, Y. H., Huang, Y. P., & You, M. H. (2011). Mobile Phone's influence on college students and its countermeasures. *Journal of Chongqing University of Posts and Telecommunications (Social Sciences Edition), 23*(2), 22–26.
11. Adriana, B., & Phillips, J. G. (2005). Psychological predictors of problem mobile phone use. *Cyber Psychology and Behavior, 8*(1), 39–51.
12. Hang, D. L., & Qi, Z. F. (2005). Psychological analysis of college students' "mobile-phone addiction". *Contemporary Young Research, 12*, 34–38.
13. Wang, C., Wang, S. Y., & Li, W. H. (2013). Study on the mobile phone dependence syndrome and its distribution among 2213 college students in Guangzhou. *Chinese Journal of Epidemiology, 34*(10), 949–952.
14. Li, M. L., Jiang, X., & Ren, Y. J. (2017). Mediator effects of positive emotions on social support and depression. *Psychiatria Danubina, 29*(2), 201–207.
15. Toda, M., Monden, K., Kubo, K., & Morimoto, K. (2004). Cellular phone dependence tendency of female university students. *Japanese Journal of Hygiene, 59*, 383–386.
16. Bianchi, A., Phillips, J. G., & Psych, P. G. D. (2005). Psychological predictors of problem mobile phone use. *Cyberpsychology & Behavior, 8*(1), 39–51.
17. Smetaniuk, P. A. (2014). Preliminary investigation into the irevalence and irediction of iroblematic cell phone use. *Journal of Behavioral Addictions, 3*(1), 41–53.
18. Billieux, J., Lander, M. V. D., & Rochat, L. (2008). The role of impulsivity in actual and problematic use of the mobile phone. *Applied Cognitive Psychology, 22*(9), 195–210.
19. Güzeller, C. O., & Coşguner, T. (2011). Development of a problematic mobile phone use scale foe Turkish adolescents. *Cyberpsychology Behavior and Social Networking, 15*(4), 205–211.
20. Leung, L. (2008). Linking psychological attributes to addiction and improper use of the mobile phone among adolescents in Hong Kong. *Journal of Child Media, 2*(2), 93–113.

21. Wang, X. H. (2011). *Research on the situation of mobile phone dependency and the relationship among MPD, social support and social adaptation for middle school student*. Fuzhou: Fujian Normal University.
22. Shi, J. G. (2009). Mobile phone dependency syndrome. *Journal of Clinical Psychiatry, 19*(2), 138–139.
23. Tao, S. M., Fu, J. L., Wang, H., et al. (2013). Development of self-rating questionnaire for adolescent problematic mobile phone use and the psychometric evaluation in undergraduates. *Chinese Journal of School Health, 34*(1), 26–29.
24. Xu, H., Wu, X. N., Lan, Y. T., & Chen, Y. H. (2008). Development of mobile phone dependence inventory for college students. *Chinese Journal of Clinical Psychology, 16*(1), 26–27.

Chapter 54
Measurement and Empirical Research of High Technology Industry Development for Mainland China Region

Ming Luo and Rui Luo

Abstract High technology industry is the strategic and guiding industry of economic and social development; it is a core symbol of measure economic development level and competitiveness of a country or a region.

This paper uses statistic data from 2000 to 2014, results show that High Technology Industry Competitiveness by industry and regional development are uneven from an empirical analysis.

Keywords Mainland China Region · High Technology Industry · Industry Development

54.1 Regionalization of Mainland China

54.1.1 The Meaning of the "Region" Division

Human economic and social activities are carried out in a certain time and space, "region" refers to a certain geographical space of human activities, some aspects of the region within the homogeneity, regional and regional differences between the certain.

Regional economic High Technology Industry innovation refers to the conditions under certain conditions, the regional High Technology Industry through the adjustment of various innovative elements of the independent movement and the associated movement, so that innovation elements affect the independent development of other elements, to achieve innovation between the elements Mutual cooperation,

M. Luo (✉)
Department of Management, Guangdong University of Science and Technology, Dongguan, Guangdong, China
e-mail: luoming8654@163.com

R. Luo
Wealth Center (Shanghai) Zhongtai Trust Co, LTD, Shanghai, China

M. Tavana, S. Patnaik (eds.), *Recent Developments in Data Science and Business Analytics*, Springer Proceedings in Business and Economics,
https://doi.org/10.1007/978-3-319-72745-5_54

mutual development of the trend. This innovation determines the development trend of High Technology Industry in the whole region, so that the regional High Technology Industry from the old structural state to develop into a new structural state, to achieve the development of innovative elements in accordance with the law, regional High Technology Industry mutual benefit, Comprehensive and coordinated development. The coordinated development of the regional High Technology Industry reflects the synergy of the innovative elements in the industry, and reflects the synergies of the various elements after the interaction of the various innovative elements. It is the dialectical unity of the various links among the elements of the regional High Technology Industry.

54.1.2 History of Mainland China Regionalization

China is broad. Due to historical and realistic development and other reasons, the level of economic development between the Chinese mainland region there is a big difference. According to the different development level, the Chinese mainland is divided into different economic regions, is conducive to the study of regional High Technology Industry innovation.

In the 1950s and 1960s, the division of the economic region was different. In the 1960s, with the deepening of the study of economic development, the people of China were divided into two parts: the north, the northeast, the east, the middle, the middle, the southwest and the northwest. Proposed a number of regional division method. Some of the mainland is divided into first-line, second and third areas; also divided into East, Central and West three zones; six comprehensive economic zone; seven economic zones; nine economic zones; nine "metropolitan economic circle "And so on, different periods, different division method covers the geographical range is different.

54.1.3 Definition of "Region" in This Paper

Based on the existing research on regional economic development and economic growth in China, this paper will build an infrastructure to support its economic and social development according to the characteristics of economic situation, natural resources and population distribution in various economic zones. Economic zone within the integration of the effective allocation of factors of production, and then lay the foundation for the national economic integration. Circum-Bohai Sea Region (Beijing, Tianjin, Hebei, Shandong), Yangtze River Delta Region (Shanghai, Jiangsu, Zhejiang, Anhui), Southern Coastal Region (Guangdong, Fujian, Guangxi, Hainan) and Middle Region (Henan, Jiangxi, Hunan, Hubei) four economic zones with a total land area of 202.8 square kilometers, population 895.37 million, the largest population density, 441 per square kilometer; Secondly, Northeast Region

(Heilongjiang, Jilin, Liaoning) and Southwest Region (Chongqing, Sichuan, Yunnan, Guizhou) two economic areas of the total land area of 193.57 million square kilometers, population 303.29 million people, that findings and the population density of 156 people per square kilometers; the sparsely populated Northern Region (Inner Mongolia, Shaanxi, Shanxi, Gansu, Ningxia), Xinjiang and Qinghai-Tibet Plateau region (Tibet, Qinghai), three largest economic region, with a total area of 539.74 square kilometers, population 163.8 million people, population density, at least 30 people per square kilometer.

54.2 Condition of High Technology Industry in Mainland China

In the process of regional High Technology Industry development, collaborative development is an important feature of the development of High Technology Industry. This collaborative development trend, so that different regions of high-tech industries to varying degrees of development. The implementation of open policy areas, High Technology Industry to achieve knowledge sharing, business development faster, and vice versa is the opposite.

54.2.1 Southern Coastal Region Compare with the Middle and Western Regions Have a Large Difference in Scale

After the reform and opening up, especially since the beginning of the twenty-first century, China has made great achievements in the development of regional High Technology Industry, and successfully explored a road of development with Chinese characteristics. Since 2000, Revenue from Principal Business (hereafter referred to as the Revenue) of China's high technology industry has increased by nearly 12 times between 2000 and 2014, with an average annual growth rate of 19.89%. In 2014, the revenue reached CNY12736.767 billion, creating the development of High Technology Industry in the world the miracle. However, while the economy is growing at a high speed, the gap between the high-tech industries in mainland China is expanding, and the income from the main business of the High Technology Industry in Circum-Bohai Sea Region is the highest, reaching CNY1,320.2 billion; the other is the Yangtze River Delta Region, which is CNY 4049.6 billion, Southern Coastal Region reached CNY 354.83 billion the central region reached CNY 1368.7 billion, the southwest region reached CNY 997.9 billion, the northeast region reached CNY 465.2 billion, the northern region reached 299.6 billion CNY, the Qinghai-Tibet plateau region reached 7.3 billion CNY, Xinjiang reached 2.69 billion CNY (Table 54.1).

Table 54.1 Comparison of Revenue from Principal Business of High Technology Industry in each region (2000–2014) Units: 100 million CNY

Region\Time	2000	2005	2011	2012	2013	2014
Nationwide	**10050.1**	**33916.2**	**87527.2**	**102284.0**	**116048.9**	**127367.7**
Circum-Bohai Sea Region	10,651	34,983	90,817	106,161	120,453	132,020
Yangtze River Delta Region	2862	12,066	31,122	35,352	37,869	40,496
Southern Coastal Region	3104	11,975	26,849	29,234	32,664	35,483
Middle Region	513	1130	6491	9023	11,584	13,687
Southwest Region	458	902	4792	6428	8448	9799
Northeast Region	600.9	1067.2	3289	3877	4405	4652
Northern Region	305	640	1739	2277	2599	2996
Qinghai-Tibet Plateau Region	4	11	27	46	63	73
Xinjiang	4.7	10.5	31.1	16.9	20.7	26.9

54.2.2 The Central and Western Regions Grew Rapidly and the Southern Coastal Regions Grew Steadily

From 2000 to 2014, the average annual growth rate of High Technology Industry in each region increased. Especially in the central and western regions in the central region to speed up the development of major policies in the central and western regions under the strong incentives, high technology industry grew rapidly, of which Middle Region of high technology industry, the revenue from principal business has grown 25.68 times, an average annual growth of 26.44% in 15 years; Southwest Region has grown 20.4 times, an average annual growth of 24.46%; Qinghai-Tibet region has grown 17.25 times, an average annual increase of 23.05%. The Yangtze River Delta region has grown 13.15 times, an average annual growth of 20.84%; the Circum-Bohai Sea Region has grown 11.40 times, an average annual increase of 19.70%; Southern Coastal Region inhas grown 10.43 times, an average annual growth of 19.01%. North region, northeast region, Xinjiang respectively has grown 8.82 times, 6.74 times, 4.72 times; annual average growth of 17.73%, 15.74%, 13.26% in the recent 15 years. But the regional gap widened, Circum-Bohai Sea Region, the Yangtze River Delta region, the Southern Coastal Regions continue to lead. Northeast region has been beyond the central region and southwest region; Xinjiang has also been beyond the Qinghai-Tibet Plateau region, while Xinjiang and Qinghai-Tibet Plateau region and developed areas of High Technology Industry gap widened further.

54.3 Comparison of High Technology Industry Development in Different Region

54.3.1 Medical and Pharmaceutical Products Manufacturing

State Council "second five" national development plan, the Medical and Pharmaceutical Products Manufacturing was classified as a strategic emerging industries. Refers to the raw material by physical changes or chemical changes after becoming a new pharmaceutical products containing commonly referred to in the manufacture of Chinese and Western medicine, veterinary drugs also contain pharmaceutical raw materials and health materials. Into the twenty-first century, the expansion of industrial scale by leaps and bounds. During the period of 2000–2014, China's medical and pharmaceutical products manufacturing The revenue grew 13.35 times, with an average annual growth rate of 20.96%. In 2014, the revenue from principal business of pharmaceutical manufacturing reached CNY 233.533 billion. In 2014, the revenue ranking, Circum-Bohai Sea Regional, Yangtze River Delta region, the central region were ranked one, two, three, the revenue reached CNY 581.1 billion, CNY 538.4 billion and CNY 440.9 billion. More than CNY 100 billion in Shandong, Jiangsu, Henan, Jilin, Guangdong, Sichuan, Zhejiang and Jiangxi, the revenue of pharmaceutical manufacturing reached CNY 3715.83,304.55 billion, CNY 166.37 billion, CNY 148.04 billion, CNY 129.46 billion, CNY 110.44 billion, CNY 109.25 billion, CNY 102.77 billion in 2014 (Table 54.2).

54.3.2 Manufacture of Aircrafts and Spacecrafts and Related Equipment

In China, manufacture of aircrafts and spacecrafts and related equipment industry to produce aircraft, helicopters, airborne systems, aircraft engines and aviation weapons is complete, is the integration of China's most advanced technology manufacturing industry, belonging to the manufacturing industry in high-end equipment manufacturing. The Manufacture of Aircrafts and Spacecrafts and Related Equipment has also been included in the national development plan in the "twelfth five-year" development planning (2011–2015 plan). In recent years, the state and local governments have introduced a series of relevant policies to support and encourage the Manufacture of Aircrafts and Spacecrafts and Related Equipment. They have made policy guidance and institutional arrangements from the aspects of overall deployment, tax incentives, personnel training, funding guarantee and infrastructure guarantee respectively, to establish the manufacture of aircrafts and spacecrafts and related equipment with Chinese characteristics.

During the period from 2000 to 2014, the revenue from China's manufacture of aircrafts and spacecrafts and related equipment has grown 10.53 times, with an average annual growth rate of 19.08%. The revenue of aerospace, spacecraft and

Table 54.2 Revenue from Principal Business of Medical and Pharmaceutical Products Manufacturing in each region (2014)

The Revenue\Region	Nationwide	Circum-Bohai Sea Region	Yangtze River Delta Region	Middle Region	Northeast Region	Southwest Region	Southern Coastal Region	Northern Region	Qinghai-Tibet Plateau Region	Xinjiang
100 million CNY	**23,350**	5811	5384	4409	2665	2026	1969	838	58	24.3

Table 54.3 Revenue from Principal Business of Manufacture of Aircrafts and Spacecrafts and Related Equipment in each region (2014)

The Revenue\ Region	Nationwide	Circum-Bohai Sea Region	Northern Region	Yangtze River Delta Region	Northeast Region	Southwest Region	Southern Coastal Region	Middle Region
100 million CNY	**3028**	822	610	438	404	381	196	174

equipment manufacturing industry reached CNY302,266 million in 2014. The revenue ranking in the country, Circum-Bohai Sea Region, Northern Region, Yangtze River Delta region, Northeast Region, Southwest Region were ranked first, second, third, fourth, fifth, the revenue reached CNY 82.2 billion, CNY 61 billion, CNY 43.8 billion, CNY 40.45 billion and CNY 38.1 billion. More than CNY 10 billion in Shaanxi, Tianjin, Jiangsu, Sichuan, Liaoning, Beijing, Heilongjiang, Shanghai and Guizhou, manufacture of aircrafts and spacecrafts and related equipment The revenue reached CNY 60.11 billion, CNY 58.31 billion, 297.3 Billion, CNY 26.03 billion, CNY 25.56 billion, CNY 18.88 billion, CNY 14.89 billion, 12.34 billion, CNY 12.01 billion in 2014 respectively (Table 54.3).

54.3.3 Manufacture of Electronic Equipment and Communication Equipment

Manufacture of electronic equipment and communication equipment as a part of the information industry, mainly for communication equipment manufacturer (including cable transmission and wireless transmission, switching equipment, wired communication and wireless communication terminal manufacturing three parts), radar and equipment manufacturing, broadcasting and television equipment manufacturing, electronics manufacturing (including electronic integrated circuit manufacturing, vacuum device manufacturing and semiconductor discrete device manufacturing), electronic components manufacturing, home audio-visual equipment manufacturing etc. Manufacture of Electronic Equipment and Communication Equipment with independent development combined with imported technology in China, it has formed a complete system under the atmosphere of rapid growing internet age.

During 2000–2014, Manufacture of Electronic Equipment and Communication Equipment The revenue has grown 10.51 times, the average annual growth rate of 19.07%. electronic and communication equipment manufacturing business revenue has reached CNY 6758.42 billion in 2014., Southern Coastal Region, the Yangtze river Delta Region leading at Manufacture of Electronic Equipment and Communication Equipment of The revenue ranked, more than one trillion CNY, CNY 25085 billion and CNY 20619 billion respectively; Circum-Bohai Sea Region's The revenue reached CNY 963.4 billion, Middle Region reached CNY 749.2 billion,

Southwest Region reached CNY 255.1 billion, Northeast Region reached CNY 103.34 billion in other region is more than one hundred billion CNY; The revenue of more than one trillion CNY in Guangdong and Jiangsu provinces, reached CNY 22284 and 1428.5 billion respectively.; the rest of The revenue of more than hundreds of billions CNY in Shandong, Henan, Zhejiang, Tianjin, Shanghai, Fujian, Beijing, Sichuan, Hunan, Hubei, Jiangxi, Anhui provinces or city (Table 54.4).

54.3.4 Manufacture of Computer and Office Equipments

The scale and level of development of manufacture of computer and office equipments marked the level of national economic development. It includes the manufacture of office equipment, the manufacture of electronic computers and the manufacture of computer peripherals. It is a national strategic industry.

2000–2014 during the period, China's manufacture of computer and office equipments. The revenue has grown 13.70 times, with an average annual growth rate of 21.16%. The fastest growing regions were Southwest, Middle and Yangtze River Delta Regions, with an average annual increase of 46.18%, 27.44% and 24.68% respectively. In manufacture of computer and office equipments of The revenue reached CNY 2349.91 billion in 2014, the top four of Yangtze River Delta Region, Southern Coastal Region, Southwest Region and Circum-Bohai Sea Region, the revenue reached CNY 887.9 billion, CNY 726.1 billion, CNY 451 billion and CNY 220.8 billion respectively (Table 54.5).

More than one hundred billion CNY in Guangdong, Jiangsu, Shanghai, Sichuan, Chongqing and Shandong provinces or city in 2014, the manufacture of computer and office equipments of The revenue reached CNY 586.86 billion, CNY 4 69.37 billion, CNY 341.4 billion, CNY 235.66 billion, CNY 213.47 billion and CNY 117.69 billion respectively.

54.3.5 Manufacture of Medical Equipments and Measuring Instrument

Manufacture of medical equipments and measuring instrument is a capital intensive, knowledge intensive, multi discipline, cross technology industry, mainly including medicine, electronics, machinery and after-sales services and other fields of business. Under the impetus of the new medical reform policy, the medical device industry has developed rapidly, and China has become a huge consumer market for medical devices.

2000–2014 during the period, China's Manufacture of The revenue of medical equipments and measuring instrument has grown 16.75 times, with an average annual growth rate of 22.81%. It's the Manufacture of medical equipments and

Table 54.4 Revenue from Principal Business of Manufacture of Electronic Equipment and Communication Equipment in each region (2014)

The Revenue\Region	Nationwide	Southern Coastal Region	Yangtze River Delta Region	Circum-Bohai Sea Region	Middle Region	Southwest Region	Northeast Region	Northern Region	Qinghai-Tibet Plateau Region	Xinjiang
100 million CNY	67,584	25,085	20,619	9634	7492	2551	1031	561	14	1.5

Table 54.5 Revenue from Principal Business of Manufacture of Computer and Office Equipments in each region (2014)

The Revenue\ Region	Nationwide	Yangtze River Delta Region	Southern Coastal Region	Southwest Region	Circum-Bohai Sea Region	Middle Region	Northeast Region	Northern Region
100 million CNY	**23,499**	8879	7261	4510	2208	507	127.9	3

measuring instrument reached CNY 990.65 billion in 2014. The Yangtze River Delta Region thriving, the revenue reached CNY 517.7 billion, followed by the Circum-Bohai Sea Region and the Middle Region, the revenue reached CNY 167.9 billion and CNY 110.5 billion; the Southern coastal region is close to one hundred billion CNY; the Northeast region and the southwest region reached CNY 42.4 billion and CNY 33 billion (Table 54.6).

In provincial of city medical equipment and instrument manufacturing business revenue highest in Jiangsu amounted to CNY 379.45 billion; followed by Shandong amounted to CNY 104.77 billion; CNY more than 70 billion in Guangdong and Zhejiang province; CNY 300–500 billion in Henan, Shanghai, Beijing, Liaoning.

54.4 Countermeasures and Suggestions to Promote the Development of Regional High Technology Industry

54.4.1 Reasonable Planning, Achieve Regional Balanced Development of High Technology Industry

From the above analysis can be seen, the Circum-Bohai Sea Region the Yangtze River Delta Region ranked first in the two industries, they have two industries ranked first in the country. Circum-Bohai Sea Region were the pharmaceutical manufacturing and aviation, spacecraft and equipment manufacturing, the Yangtze River Delta Region were the computer and office equipment manufacturing and medical equipment and instrumentation manufacturing. The Southern Coastal Region of electronic and communications equipment manufacturing industry ranked first in the country. Ranked second, the Yangtze River Delta has two industries, the Circum-Bohai Sea Region has an industry, the southern coast also has an industry. It can be known that the development of high-tech industry is better concentrated in the region around the Circum-Bohai Sea, Yangtze River Delta and Southern Coastal Region. And Xinjiang has three industries ranked ninth; the Qinghai-Tibet Plateau has three industries ranked eighth; northern three industries ranked seventh, the regional economic development is very uneven. Therefore, the government in the planning of regional high-tech industry innovation, according to China's regional resources,

Table 54.6 Revenue from Principal Business of Manufacture of Medical Equipments and Measuring Instrument in each region (2014)

The Revenue\Region	Nationwide	Yangtze River Delta Region	Circum-Bohai Sea Region	Middle Region	Southern Coastal Region	Northeast Region	Southwest Region	Northern Region	Qinghai-Tibet Plateau region	Xinjiang
100 million CNY	**9906.5**	5177	1679	1105	971	424	330	191	1.1	1.1

industrial base and scientific and technological conditions, to avoid weaknesses, give full play to their comparative advantage. The government plays an important role in macroeconomic management and socio-economic development. According to the local advantages and characteristics, it is consciously adjusting the overall planning for development.

54.4.2 Optimization of Coordination Mechanism, the Promotion of Independent Innovation

Regional high-tech industry in the innovative elements of linkage, this linkage can be achieved through industry collaboration mechanism. Industrial cooperation mechanism, on the one hand, the performance of regional industrial enterprises innovation and resource conditions, such as regional innovation in the field of market, product innovation and technological innovation; on the other hand the performance of enterprise and the system of collaborative innovation, embodied in the enterprise In the relationship between the organization, technical relations and cultural relations. Enterprises can build an industry collaborative innovation system in a certain area, strengthen the cooperative relationship between enterprises and government, scientific research institutions, customers and suppliers, as well as competitive enterprises and other participants, through the optimization of this collaborative relationship, cultivate independent innovation capability. Enterprises and the government, the government can increase the guidance of high-tech industries, high-tech enterprises to create the conditions for tax or other aspects of preferential policies to promote enterprises to carry out independent innovation. The most ideal collaborative innovation is the industry within the innovative elements of a high degree of integration, innovation and resources in the system to achieve free flow.

54.4.3 Increase Investment in R & D, Attract the Talents of Science and Technology

At present, China's financial institutions in regional technological innovation and high-tech industries to promote the role of development is not obvious, this is the current financial system is not perfect, and cannot serve the High Technology Industry. We can learn from the foreign high-tech enterprise investment and financing system, such as the United States and Britain to use venture capital to promote the development of high-tech enterprises, through the implementation of government finance, government procurement or government loans, risk investment to create a favorable environment, and according to the situation of our country, the investment and financing system and perfect the construction of high tech industry them. First of all, the government is based on the market to guide the construction of a reasonable investment environment, establishing risk investment mechanism; secondly, through the listing of financial institutions, various types of companies, the use of innovative financial products and tools to absorb the social and private funds to support High Technology Industry; thirdly, according to the development of high-tech industries of different combination according to the need of industrial structure, the proportion of R & D funds, gradually formed a business investment as the main body, government guidance funds, financial institutions and listed companies and other channels of investment and financing system.

54.4.4 Increase R & D Investment to Attract Scientific and Technological Personnel

The success of technological innovation requires the support of R & D investment and the construction of innovative environment. Governments in various regions need to increase their investment in science and technology.

And to strengthen the combination of scientific research institutions and universities, high-tech enterprises, the establishment of innovation system of science and technology, talent is the most important factor in the construction of innovation system and the development of High Technology Industry, from the empirical results also show that the talents of science and technology investment in high technology industry development play a significant positive effect, so we should formulate relevant policies to attract talents, take out the way of culture. First of all, to play the role, for the high technology enterprise with the needs of the people, professional training programs to set or adjust, intensify training of technical personnel; secondly, to enable a competitive high-tech enterprise talent introduction mechanism of talent, in addition to solve the housing, household registration, and the reward mechanism set up, should also set up the working system the difference and flexible in each region of the high technology enterprise salary system, establish the quality of technical personnel and management mechanism, grasp the changes of the quantity and quality of the talents; thirdly, support enterprises to establish research and development institutions and increase investment in research and development, in the conditions of the enterprise, the establishment of the National Engineering Center and industrial base. To attract talents to flow to the enterprise, cultivating a number of R & D, design, manufacture and provide system integration services for large enterprise To support and help enterprises to improve R & D personnel, and to encourage enterprises and research institutes to attract R & D personnel to participate in scientific research activities through financial subsidies and other forms. Fourth, from the world of attracting talent in science and technology system, attract more foreign workers to work with scientific technology, provide necessary office space and work environment, retain talent for the development of High Technology Industry in china.

References

1. National Bureau of Statistics, et al. (2016). *China statistics yearbook on high technology industry*. (2015) (1). Beijing: China Statistics Press.
2. Liu, P.-L. (2011). Empirical study on influence of critical success factors on ERP knowledge management on management performance in high-tech industries in Taiwan. *Expert Systems with Applications, 38*(8), 10696–10704.
3. Narayanan, K., & Bhat, S. (2010). Technology sourcing and outward FDI: A study of IT industry in India. *Technovation, 31*(4), 177–184.
4. Vaidya, K., Bennett, D., & Liu, X. (2007). Is China's manufacturing sector becoming more high-tech?: Evidence on shifts in comparative advantage, 1987–2005. *Journal of Manufacturing Technology Management, 18*(8), 912–932.
5. Lai, H.-C., & Shyu, J. Z. (2005). A comparison of innovation capacity at science parks across the Taiwan Strait: The case of Zhangjiang High-Tech Park and Hsinchu Science-based Industrial Park. *Technovation, 25*(7), 805–813.
6. Yin-jie, X. U., & Zheng, J. (2011). *Innovative resources within industrial clusters: How they are agglomerated and diffused*. Proceedings of 2011 I.E. the 18th international conference on Industrial Engineering and Engineering Management (Vol. 3).
7. Guo, T., & Niu, X. (2011). *Research on the formation mechanism of the innovative enterprises growth path*. Mechanical, industrial, and manufacturing engineering–Proceedings of 2011 international conference on Mechanical, Industrial, and Manufacturing Engineering (MIME 2011).

Chapter 55
Sina Weibo User Influence Research

Zhu Yangpeng and Li Peng

Abstract This paper takes Weibo users as the research object, and extracts two dimensions from the point of view of user relationship: Micro-bo concern number, micro-blog fans number, and on this basis to generate fans rate indicator. Through the Python Network crawler to acquire and analysis data, this paper has obtained the exponential function model and cumulative distribution model of the fans rate distribution, and verified the correlation between fans rate and Weibo influence. It has higher practical application value.

Keywords Micro-blog · User Influence · Fans rate · MCI

55.1 Introduction

As a kind of information sharing and exchanging platform, Sina Weibo has become a new information sharing and social fashion. By December 2016, Weibo registered more than 500 million users and the monthly active users reached 340 million. With the micro-blog in the industry, agriculture, politics, business and other fields continuing to heat up, a huge amount of user data has been quickly gathered. In order to discover the hidden value behind the data, scholars began to explore the social law behind the volume of micro-data, among which the most important and fundamental research direction was the research of user influence.

Z. Yangpeng · L. Peng (✉)
School of Economics and Management, Xi'an Shiyou University, Xi'an, China
e-mail: mzlp_xa@126.com

M. Tavana, S. Patnaik (eds.), *Recent Developments in Data Science and Business Analytics*, Springer Proceedings in Business and Economics,
https://doi.org/10.1007/978-3-319-72745-5_55

55.2 Related Research

Some research findings may not be directly applicable to Sina Weibo users in China compared to foreign Twitter's influence studies, mainly in content and form [1, 2]: The content of domestic microblogs can contain pictures, videos and links, but content posted on Twitter contains only text and links; Domestic microblogs content is mainly entertainment and leisure category, however Twitter discussed more political topics and news. Due to the number of registered users and the use of different habits, the domestic microblogs volume is very large. So the study is of particular importance to domestic scholars on Sina Weibo.

The paper [1] mainly studies the problem of automatic classification of ordinary users and machine users in Weibo. This paper takes the machine user as the research object, extracts the feature from 4 dimensions, such as the behavior pattern, the microblog content, the user relation and the publishing platform, and uses the decision tree algorithm in the machine learning to make classification experiment, which obtains the good classification effect.

The paper [3] takes the zombie fans as the research object, thinks that the form of the zombie fans has shifted from the low-level form to the advanced form, and by constructing the eigenvector of the fans number, the attention number, the microblogs number and the forwarding number, this paper proposes a user behavior recognition algorithm based on the vector machine.

The paper [4] gathers more than 649 thousands records of Weibo users and takes ten factors including user gender, certification attributes, region distribution, user ID and nickname, fans number and attention number of users, the way to sending Weibo, the time to sending Weibo, the contend of Weibo, communication and quote behaviors, into consideration to systematically show the behavior characteristics of Sina Weibo users. The results have been verified by the current situation.

The paper [5] uses SPSS software to test the distribution status and correlation of four characteristic variables, such as fans number, concern number, micro-blog number, collection number. It is found that the significant probability of fans number and attention number under Kolmogorov-smirnov is less than 0.05, which is not normal distribution. Further analysis found that the number of fans and attention shows moderate positive correlation.

The paper [6] uses the principal component analysis method to collect and analyze the information of the enterprise officials, and finds that the number of fans is negative correlated with the number of concerns, which shows that the enterprises and celebrities often show more attention number and less number of fans on Weibo.

Literature [7] study the correlation between user influence and the number of fans, obtaining some viewpoints through the scatter map: Overall, there is little correlation between user influence and the number of fans, or there is only very weak positive correlation; segmented, the influence of the smaller users has no significant correlation with the number of fans, and when the influence is greater, the user influence has obvious positive correlation with the number of fans.

The literature [8] empirically analyzes the behavioral characteristics of Sina Weibo users from several angles, and observes through large sample data that the number of fans is not correlated with the number of attentions, the number of fans is related to the influence of users and the information they contribute, and the amount of attention depends on the behavior of users.

The literature [9] analyzes the correlation variables of blog influence and establishes the regression equation of influence. The correlation coefficient of attention number and fans number is 0.723, the concomitant probability is 0, and the positive correlation is highly significant by spearman correlation coefficient analysis.

The literature [10] chooses the number of fans, the number of concerns, the number of tweets being forwarded, the total number of micro-blogs as the index to evaluate the influence of Weibo users, which is more reasonable to embody the actual influence of users through the AHP method.

The paper [11] constructs the evaluation index of user influence by studying user behavior characteristics, and proposes an improved model based on PageRank algorithm. The analysis shows that the improved algorithm can evaluate the user's influence more accurately and objectively.

In the research of user influence algorithms and influence variables in the above literatures, both the influence of Twitter users and the study of user influence on Sina Weibo in China, number of fans and number of attention are the important indexes that cannot be neglected. But for the distribution of fans number and attention number, the literature [5] thinks its distribution is not normal distribution, and it does not delve into the correct distributing state. As for the correlation between the amount of fans and the attention, the different scholars' viewpoints differ greatly, even the opposite, as the literature [5] considers that the amount of fans is positively correlated with the amount of attention, the literature [6] considers negatively correlated, and the literature [8] considers positive correlation.

In view of this problem, this paper focuses on the two indicators of fans number and attention, combines the fans quantity with the attention quantity to construct the fans rate index drawing on the literature [1] to eliminate the correlation. The research on the distribution of fans rate and the correlation between fans rate and user influence were studied. The probability distribution function and the cumulative distribution function of the fans rate index are obtained by researching and analyzing the data acquisition by Python network Crawler. The exponential function models of fans rate and user influence is obtained, which can make up the autocorrelation of fans number and attention number in model analysis, and reduce the redundancy attribute of machine learning algorithm for user classification, which is significant to improve the efficiency and recognition rate of the algorithm.

55.3 Feature Extraction

The larger the amount of fans, the greater the speed and coverage of micro-blogging and impact; the greater the concern number, the greater the user's ability to spread information, the greater the likelihood of being affected. Therefore, if the amount of fans is far greater than the amount of attention, the influence on behalf of the user is greater. If the number of attention is more lager than the volume of fans, the user is mainly the microblogs propagator, the ability to be affected is greater. In this paper, we use the concept of ratio to synthesize a index--fans ratio, and the following formula is used to construct the fans rate index:

$$\mathrm{FansF} = \frac{Fans\_number}{Attention\_number + \mathrm{Fans\_number}} \tag{55.1}$$

Among them, Fans_number for the number of fans, Attention_number for attention number, FansF for fans rate the ranged from 0 to 1. The bigger value of FansF, the greater the potential impact. The smaller value, otherwise. This index can eliminate the autocorrelation of fans number and concern number in empirical research, and the value has been normalized and comparable.

In order to further measure the correlation between fans rate index and Weibo influence, this paper uses MCI (the content's Charm index of microblog)" [12] to analyze, the formula is:

$$\mathrm{MCI} = \frac{\mathrm{Fans\_number}}{\mathrm{Weibo\_number}}. \tag{55.2}$$

Among them, Fans_number is the number of fans, Weibo_number for users to publish the total number of micro-blog, MCI for the micro-blogs charm index. The greater the MCI value, the greater the ability of the microblogs content to attract readers, the greater the user influence; the smaller the MCI value, the less the lack of the content of micro-blog, not to attract the attention from readers, Weibo users less influence.

55.4 Experimental Analysis

55.4.1 Data Collection

This paper uses the Python platform to compile the network crawler to collect data including the user ID and fans number and concern number in the fans list. This data collection collects altogether 3000 records of micro-blog user data. But some user information has the collection failure situation, in the experiment, the user information was removed, and some users follow each other resulting in multiple collections, such as user and user fans to pay attention to each other. In order to ensure the uniqueness of the user ID, this experiment will delete the failed user information and duplicate user information and eventually get 2821 valid data.

55.5 Data Analysis

55.5.1 Data Preprocessing and Descriptive Statistical Analysis

This paper uses EVIEWS9 software to analyze data, export data from database into text file, and then import EViews software to analyze the distribution. The fans rate index is located between 0 and 1. In order to ensure the accuracy of data grouping, data of this paper is divided into 40 groups by interval 0.025, and does the frequency distribution histogram, as shown in Fig. 55.1.

It can be seen from the graph that the frequency distribution histogram of fans rate is a typical exponential distribution. The fans rate is low, the frequency is higher, the person with low fans rate is relatively few; the fans rate is higher, the frequency is relatively low. This is also in line with the actual situation of micro-blog users, the low fans rate is generally ordinary users who occupied more number among all Weibo users; Entertainment stars, Big V and other people have a higher ability to absorb fans, fans rate is relatively high, but the number of these users is small.

In order to make the research on the distribution index of fans rate more universal, this paper deals with the frequency of fans rate and the rate of fans ratio.

$$\text{Fans Rate frequency} = \text{Fans Rate Number}/2821 \tag{55.3}$$

The average fans rate for each group of data is used as a representative value for the group:

$$\text{FansF\_avg} = \frac{1}{n}\sum\nolimits_{i=1}^{n} FansF_i. \tag{55.4}$$

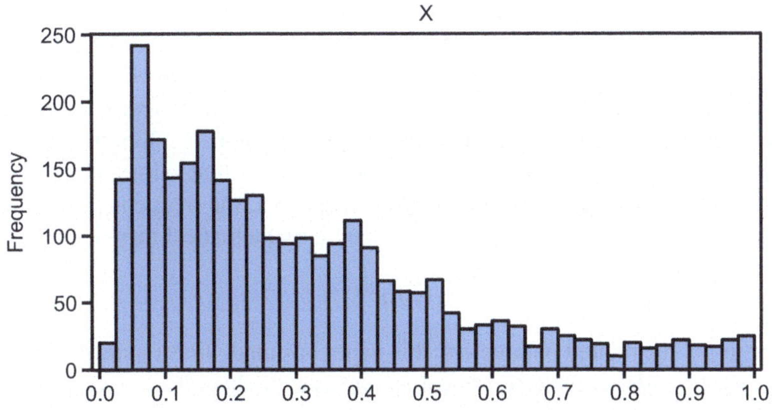

Fig. 55.1 Frequency distribution histogram of fans rate

Table 55.1 Analysis table of fans rate

Average fans rate FansF_avg	Number	Frequency	Cumulative frequency	Average value of MCI MCI_avg	Range
0.0164	20	0.0071	0.0071	0.3686	0~0.025
\\\\\\	\\\\\\	\\\\\\	\\\\\\	\\\\\\	\\\\\\
0.9894	25	0.0089	1.0000	0.0090	0.975-0.1

The FansF_avg is the average fans rate, n is the actual frequency of each group, and $FansF_i$ is the first fans rate value of each grouping.

For MCI exponents, this article first calculates the MCI exponent for each user using Formula 55.2, and then uses the average MCI value for each group as the representative value for that group:

$$\text{MCI\_avg} = \frac{1}{n}\sum\nolimits_{k=1}^{n} MCI_k \tag{55.5}$$

MCI_avg is the average MCI value, n is the actual frequency of each group, and MCI_k is MCI value indexed by k of each grouping.

After the above index calculation and conversion, a fans rate analysis table has been generated, as shown in Table 55.1. But in Fig. 55.1, you can see that the first set of values is generally small, so the two sets of values are merged into one group in the subsequent model fans process, which is still computed using the formula (55.3, 55.4, and 55.5).

55.6 Model Analysis

1. Fans Rate Distribution Model

Through descriptive statistical analysis, it is found that the histogram of frequency distribution has exponential distribution, and the Table 55.1 is generated after data preprocessing. Then the FANSF_AVG and frequency are inputted EVIEWS9 software, and the average fans rate-frequency scatter map is depicted. It can be found that the exponential distribution of scatter graphs in graphs is obvious, but the first set of values is very abnormal, so this paper combines the first values with the second value, taking FansF_avg as the independent variable x, the frequency as the dependent variable y, and constructing the exponential mathematical model.

$$\mathrm{Y} = \mathrm{A}e^{BX+u}. \tag{55.6}$$

In order to get the related parameters, this paper has made linear statistical analysis, variance test, model correction, model restore and finally get the following model (Fig. 55.2).

Fig. 55.2 Fans Rate and MCI index Scatter diagram

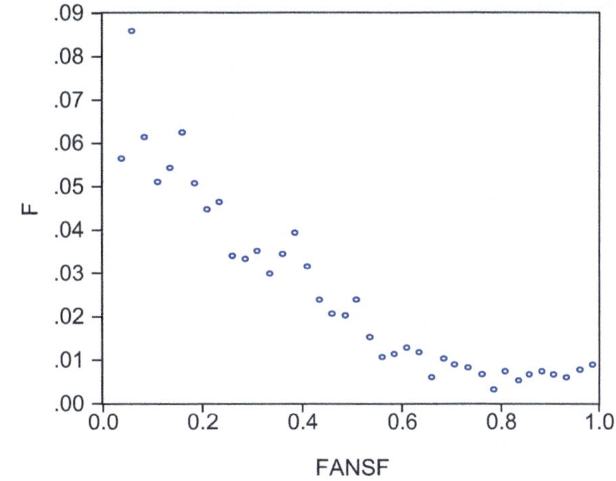

Fig. 55.3 Fans rate cumulative distribution map

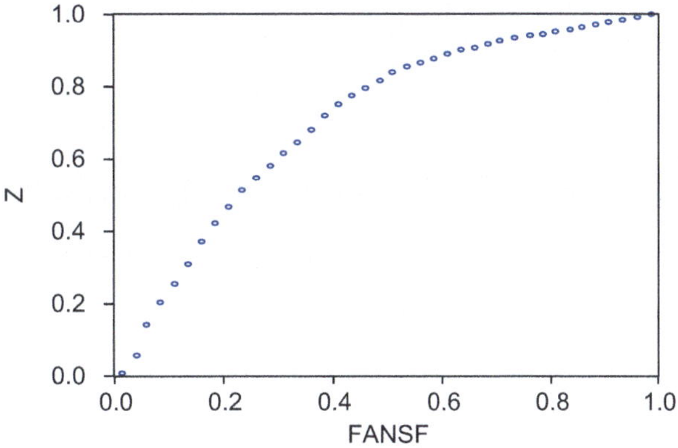

$$\widehat{Y} = 0.0734e^{-2.604543X}. \text{或} \ \widehat{Y} = e^{-2.612136-2.604543X}. \tag{55.7}$$

2. Fans Rate Cumulative Distribution Model

Through the study of the fans rate in the Model Table 55.1, the average fans rate and cumulative frequency were studied, and the EVIEWS9 statistical analysis software was used to do the scatter plot as shown in Fig. 55.3, and it is easy to find that the cumulative distribution of fans rate conforms to the logarithmic function characteristics. So this paper takes the fans rate FansF as the independent variable, and accumulates the frequency as the dependent variable to construct the mathematical model.

$$\text{Y} = \text{B} \ln FansF\_avg + C + u. \tag{55.8}$$

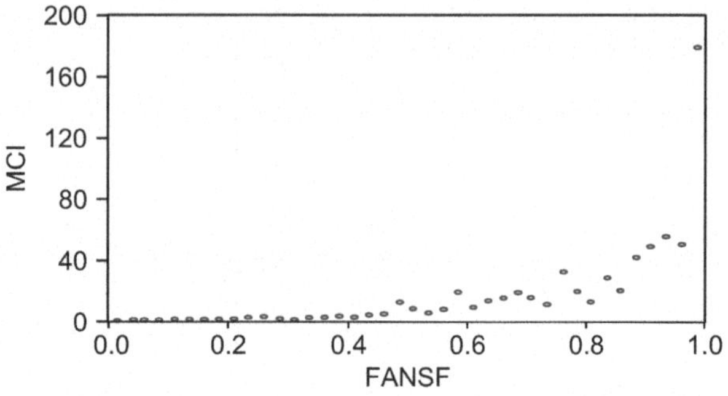

Fig. 55.4 Fans rate and MCI index scatter diagram

The analytical method is similar and the following model is finally obtained:

$$\widehat{Y} = 1.003056 \quad + \quad 0.284001 \ln Fanf\_avg. \tag{55.9}$$

3. Correlation Model between Fans Rate and MCI index

According to the fans rate and MCI value Scatter map Fig. 55.4, this paper takes the fans rate as the independent variable and the MCI value as the dependent variable to construct the exponential model, because the fans rate distribution model is the same as the exponential model, so the model construction is similar to the model estimation, so the model correction model is given directly as follows:

$$\mathrm{Y} = 0.68796e^{4.567152\,X}. \tag{55.10}$$

55.7 Conclusion

Based on the network crawler technology, this paper extracts Sina Weibo users data, analyzes the basic data of the fans number, concern number and total micro-blog number, and combines the attributes index to generate the fans rate index and the micro-Bo Charm index MCI. In this paper, the distribution characteristic of fans rate index and its correlation with MCI index are studied in detail, and the relationship between fans rate and Weibo user influence is studied by MCI Index, and two distribution models and a correlation logarithm model are obtained.

Based on the results of the analysis, the fans rate index generated by attention number and fans number is exponential function distribution. The accurate function model parameter is estimated which reveals the higher the fans rate with the lower the frequency or the lower the user's number in the whole network. For example, according to the fans rate distribution model, The user group with a fans rate of 0.8 accounted for about 0.0091 percent of the whole Sina Weibo network users, while the fans rate was 0.2% accompanying with 0.0436 percent. The cumulative distribution model of the fans rate was expressed by the logarithm function model. When the cumulative probability was 0.8, according to the cumulative probability distribution, the fans rate is about 0.4892, that is, in the entire network, fans rate is less than 0.5 for 80% of the users. In other words, the user's attention number is close to the number of fans approximately, this may be caused by "follow each other". Therefore, there is reasonable to believe that The user ratio who are able to attract fans unilaterally with their charms, the fans rate of who is more than 0.5, is about 20% of the all users, this is in line with the sociology of "two-eight theorem". Finally, based on the correlation model analysis of the fans rate and MCI index, the exponential model of the fans rate and MCI index was obtained. With the increase of the fans rate, the MCI exponential value became larger, and the function was slower increase in the early stage, and the later was accelerated rapidly. When the fans rate was 0.8, the MCI index was 26.5677, which shows that the number of fans was about 26.57 times than the total amount of Weibo. These models have great guiding significance in evaluating the size of the affected population in terms of academic research, enterprise evaluation of market value and government policy prediction.

This paper focuses on the research of fans rate, which is the early stage of research of user influence and user type recognition based on Bayesian inference. As an important attribute of the Weibo user, the fans distribution model and the cumulative distribution model can be directly applied to Bayesian inference formula, and the deep one-step study on Bayesian inference is the next work of the author.

Acknowledgment Thanks for support from Science and technology department of Shaanxi province granted by 2016GY-106, social science foundation of Shaanxi province with number 15JZ047 and key laboratory research plan of Shaanxi province department numbered by 2015R026.

References

1. Yan, L., & Ying, Y. Y. (2015). Liu. Research on identification of micro-blogging machine user based on stochastic forest classification. *Journal of Peking University (Natural Science Edition), 02*, 289–300.
2. Yu, L., Asur, S., & Huberman, B. A. (2011). *What trends in Chinese social media*l*i Proceeding of the 5th ACM Workshopon Social Network Mining and Analysis* (pp. 978–988). San Diego.
3. Xiying, Z., Xin, C., & Xianyu, T. (2014). A method of identifying zombie fans based on microblogging user behavior. *Journal of Natural Science, Heilongjiang University*, (02), 250–254.
4. Yang, C. (2011). An empirical analysis of user behavior characteristics of micro blogs. *Library and information Service, 12*, 21–25.
5. Lin, Z., & Xie, Z. (2016). Research on the type and influence of microblogging user based on clustering. *Information Sciences, 08*, 57–61.
6. Xiaoling, H., & Yijie, C. (2013). Research on the influence index of Enterprise micro-blog. *Journal of Intelligence*, (07), 64–68.
7. Jing, Y. (2013). An empirical analysis of the characteristics of user influence in micro-blogging communication. *Journal of Intelligence*, (08), 57–61+38.
8. Yang, C. (2011). An empirical analysis of user behavior characteristics of micro blogs. *Library and Information Service, 12*, 21–25.

9. Wang, X. (2010). An empirical analysis of user behavior characteristics and relationship characteristics of micro-blog–taking "Sina Weibo" as an example. *Library and Information Service, 14*, 66–70.
10. Yaoting, L. (2008). *Research on the structure of social network [D]*. Hangzhou: Zhejiang University.
11. Social, R. L. (2007). influence. In Blackwell Encyclopedia of Sociology. *Demography, 16*(4), 208–210.
12. Bin, L., Jingyuan, Z., Qiang, L., et al. (2015). A review of micro-blogging analysis. *Journal of Hebei University of Science and Technology, 01*, 100–110.
13. Yang, X., & He, Y. (2012). Characteristics Analysis of Tencent and Weibo users. *Journal of Intelligence, 3*, 84–87.

Index

M. Tavana, S. Patnaik (eds.), *Recent Developments in Data Science and Business Analytics*, Springer Proceedings in Business and Economics,
https://doi.org/10.1007/978-3-319-72745-5

The manufacturer's authorised representative in the EU is Springer
Nature Customer Service Centre GmbH, Europaplatz 3, 69115 Heidelberg,
Germany. If you have any concerns regarding our products, please
contact ProductSafety@springernature.com

Printed and bound by CPI Group (UK) Ltd, Croydon, CR0 4YY
05/10/2026
02228141-0015